Earth System Responses to Global Change

Earth System Responses to Global Change

Contrasts between North and South America

Edited by

HAROLD A. MOONEY
Department of Biological Sciences
Stanford University
Stanford, California

EDUARDO R. FUENTES
Departamento de Ecología
Pontificia Universidad Católica de Chile
Santiago, Chile

BARBARA I. KRONBERG
Department of Earth Sciences
Lakehead University
Thunder Bay, Ontario
Canada

ACADEMIC PRESS, INC.
A Division of Harcourt Brace & Company
San Diego   New York   London
Toronto   Sydney   Tokyo

Copyright © 1993 by ACADEMIC PRESS, INC.

All Rights Reserved.
No part of this publication may be reproduced or transmitted in any form or by any means, electronic or mechanical, including photocopy, recording, or any information storage and retrieval system, without permission in writing from the publisher.

Academic Press, Inc.
1250 Sixth Avenue, San Diego, California 92101-4311

United Kingdom Edition published by
Academic Press Limited
24–28 Oval Road, London NW1 7DX

Library of Congress Cataloging-in-Publication Data

Earth system responses to global change   :   contrasts between North and
   South America   /   edited by Harold A. Mooney, Eduardo R. Fuentes,
   Barbara I. Kronberg.
         p.   cm.
     Includes bibliographical references and index.
      ISBN 0-12-505300-2
      1. Climatic changes--America.   2. Science--America.   3. Ecology-
   -America.   I. Mooney, Harold A.   II. Fuentes, Eduardo R.
   III. Kronberg, Barbara I.
   QC981.8.C5E16   1993
      557--dc20   92-43350
                                                                             CIP

PRINTED IN THE UNITED STATES OF AMERICA
93 94 95 96 97 98   B B   9 8 7 6 5 4 3 2 1

Contents

Part II: Climate Controls

Part IV: Biogeochemistry

Part V: Intertidal

Part VI: Plants

Part VII: Animals

Contributors

Numbers in parentheses indicate the pages on which the authors' contributions begin.

Patricio Aceituno (61), Departamento de Geophysica, Universidad de Chile, Casilla 2777, Santiago, Chile

Paul B. Alaback (299, 347), USDA Forest Service, Pacific Northwest Research Station, Juneau, Alaska 99802

Belisario Andrade J. (101), Instituto de Geografia, Pontificia Universidad Catòlica de Chile, Vicuño Mackenna 4860, Santiago, Chile

Juan J. Armesto (195, 239), Departamento de Biología, Universidad de Chile, Casilla 653, Santiago, Chile

Mary T. Kalin Arroyo (239), Departamento de Biología, Universidad de Chile, Casilla 653, Santiago, Chile

Richard T. Barber (17), Duke University Marine Laboratory, Durham, North Carolina 27706

Patricio A. Bernal (1), Instituto de Fomento Pesquero, Chile, Casilla 1287, Santiago, Chile

James H. Brown (267, 295), Department of Biology, University of New Mexico, Albuquerque, New Mexico 87131

Juan Carlos Castilla (147, 167, 189), Departamento de Ecología, Pontificia Universidad Catòlica de Chile, Casilla 114-D, Santiago, Chile

Luis C. Contreras[1] (285, 295), Departamento de Biología, Universidad de La Serena, Casilla 599, La Serena, Chile

Eduardo R. Fuentes (329, 349), Departamento de Ecología, Pontificia Universidad Catòlica de Chile, Casilla 114-D, Santiago, Chile

Humberto Fuenzalida (61), Departamento de Geophysica, Universidad de Chile, Casilla 2777, Santiago, Chile

Julio Gutiérrez (239), Departamento de Biología, Universidad de La Serena, Casilla 599, La Serena, Chile

Sterling C. Keeley (209), Department of Botany, University of Hawaii at Manoa, Honolulu, Hawaii 96822

Barbara I. Kronberg (121, 143), Department of Earth Sciences, Lakehead University, Thunder Bay, Canada P7B 5E1

R. G. Lawford (73, 115), Hydrometerorological Processes Division, Atmospheric Environment Service, National Hydrology Research Centre, Saskatoon, Saskatchewan, Canada S7N 3H5

Jane Lubchenco (147, 167, 189), Department of Zoology, Oregon State University, Corvallis, Oregon 97331

Michael H. McClellan (299), USDA Forest Service, Pacific Northwest Research Station, Juneau, Alaska 99802

Harold A. Mooney (209, 265), Department of Biological Sciences, Stanford University, Stanford, California 94305

Mauricio R. Muñoz (329), Departamento de Ecología, Pontificia Universidad Catòlica de Chile, Casilla 114-D, Santiago, Chile

Sergio A. Navarrete (147, 167, 189), Department of Zoology, Oregon State University, Corvallis, Oregon 97331

Humberto Peña T. (101), Direcciòn General de Aguas, Ministerio de Obras Públicas, Morandé 59, Santiago, Chile

Charles H. Peterson (17), University of North Carolina at Chapel Hill, Institute of Marine Sciences, Morehead City, North Carolina 28557

Benjamín Rosenblüth (61), Departamento de Geophysica, Universidad de Chile, Casilla 2777, Santiago, Chile

Eugenio Sanhueza (131, 143), Instituto Venezolano de Investigaciones Científicas (IVIC), Caracas 1020-A, Venezuela

Gregory A. Skilleter (17), University of North Carolina at Chapel Hill, Institute of Marine Sciences, Morehead City, North Carolina 28557

[1]*Present address:* Comisión Nacional del Medio Ambiente, Casilla 520-V, Santiago 21, Chile.

Francisco Squeo (239), Departamento de Biología, Universidad de La Serena, Casilla 599, La Serena, Chile

Brian N. Tissot (147), Department of Biology, University of Hawaii at Hilo, Hilo, Hawaii 96720

Kevin E. Trenberth (35, 71), National Center for Atmospheric Research, Boulder, Colorado 80307

Carolina Villagrán (195) Laboratorio de Sistemática y Ecologia Vegetal, Departamento de Biología, Universidad de Chile, Casilla 653, Santiago, Chile

Preface

In many ways the Northern and Southern Hemispheres are mirror images of one another in geologic, climatic, and biological patterns. This is particularly true for the western parts of North and South America. In both regions there is a progression from low to high latitudes of arid, semiarid, Mediterranean, temperate, and subpolar to polar climates and a corresponding match of functionally equivalent biotic systems. The biotic systems, although comparable type for type in terms of function, are, to a large degree, independently derived through evolutionary time. The latitudinal climatic gradients in both hemispheres are controlled by equivalent ocean–atmospheric circulation patterns. In both regions, to a large degree, equivalent orientations of mountain ranges produce comparable climatic and hydrologic patterns west to east.

Despite these similarities, there are differences that may prove to be an important probe for understanding and predicting the response of the earth system to global change. The differential landmasses of the two hemispheres have comparable differences in the total amount of vegetative cover present, as well as the proportions of any given biome. This is evidently reflected in, for example, differences in biotically driven seasonal amplitudes of atmospheric CO_2, with large amplitudes in the Northern Hemisphere and considerably reduced ones in the Southern Hemisphere. This hemispheric difference in land-to-ocean mass also may mean that the rates of change in climatic patterns in response to global change may be dissimilar due to greater oceanic buffering in the south, as discussed in this volume. This would also mean a possibly different rate of change of biotic patterns.

This book examines the differences and similarities in the earth system components, the ocean, the atmosphere, and the land, between the western portions of the northern and southern Western Hemisphere, past, present, and projected. This volume had its beginnings at a workshop held in La Serena, Chile, hosted by the Universidad de La Serena, and sponsored by the Chilean Academy of Sciences and the AAAS through a grant from the MacArthur Foundation.

The general structure of the book is an examination of each earth system component, in both the north and the south, with a summary comparison. It was not possible, however, to do this in all cases; for example, the oceans are treated in a global manner for both the physical and biological components and in a couple of other sections the desired symmetry was not possible for one reason or another. The book addresses first the physical components of the earth system, examining the physical and biological patterns and responses of the open ocean, and then the climate system, land forms, and hydrology. The physical and biotic systems are interrelated in a section on biogeochemistry. The past and present patterns of the biota of the land margins and of the land surface, and their response to climatic change, are evaluated. The biotic systems are discussed in greater detail than the physical components of the earth system. Finally, the book concludes with an assessment of the direct impacts of humans on the natural systems of North and South America, giving full consideration to the land-use drivers of global change.

The book opens with a consideration of ocean function and the response of the physical and biological responses of oceans to climate change. Unlike subsequent sections of the book, the ocean chapters on the physical and biological responses to change treat the north and south units together, rather than individually, reflecting the shorter time responses and continuity of ocean phenomena.

In the initial chapter Bernal reviews the development of coupled ocean–atmosphere general circulation models and shows their increased sophistication with time. The addition of realistic land-to-water ratios between hemispheres into the models has had profound effects on the predictions of ocean and air temperatures with a doubling of atmospheric CO_2. Importantly, for virtually all of the considerations given in the rest of this book the results of modeled climate scenarios of the future predict a large asymmetry in ocean temperatures with considerably greater warming in the Northern versus the Southern Hemisphere.

Peterson, Barber, and Skilleter show the interaction of large-scale Pacific Ocean basin phenomena with small-scale wind and upwelling patterns. They illustrate how the relatively well-understood El Niño phenomenon provides a clear window for understanding the potential impacts of climate change on biological diversity, distribution, and productivity. Very rapid oceanic biological responses are predicted in response to climate change because of the mobility and sensitivity of most marine organisms. Dissimilar biological responses between north and south are predicted due not only to the differential oceanic temperature responses but also to differential coastal wind patterns established because of differences in coastal topography.

In the treatment of the atmosphere, Trenberth broadly outlines the kinds of climatic changes predicted for the future, including predictions of increased frequencies of severe climatic events. He shows the importance of sea surface temperatures in controlling atmospheric temperature and circulation, and he outlines the physical basis for the predicted differential warming of the Southern and Northern Hemispheres, namely ocean heat capacity and differential land/water ratios. He shows how this in turn will differentially affect the polar to lower-latitude thermal gradients in the north and south, and consequently atmospheric circulation patterns. Finally, he, as others in this volume, utilizes the El Niño events to illustrate the kinds of interactions and changes that the future might bring. In discussing the Southern Hemisphere specifically, Aceituno, Fuenzalida, and Rosenblüth focus on whether we are already able to detect a change in temperatures in Chile and they find differential time trends with latitude. Considerable attention is now being given to how to distinguish the signal of climate change from the noise of interannual and decadal natural variability. As Trenberth notes, we should expect to see considerable lag in reaching equilibrium temperature values with a change in climate forcing and that the lags will be greatest over the southern oceans.

Lawford gives a comprehensive overview of the interactions of land surface features and the hydrologic processes in North America and shows how the latter will be affected by potential climate change as well as land use change. There is considerable uncertainty regarding the future nature of the hydrologic cycle in any particular region because of limitations of global circulation models in predicting precipitation distribution. Further, as Lawford notes, we still lack some fundamental information on the sensitivity of various hydrologic components, as well as their interactions, to a changing climate. Andrade and Peña describe the changing physiographic and hydrologic patterns along the impressive latitudinal and elevational gradients of Chile. For this analysis Chile is taken as the unit of analysis since, for many processes, particularly hydrologic, it can be considered a closed system due to the extent and elevation of the Andes. This contrasts greatly with the complex hydrologic patterns described by Lawford for North America. Andrade and Peña, utilizing worst-case scenarios of climate change with CO_2 doubling, describe the substantial changes that would be expected in hydrological patterns particularly due to changes in snowpack duration and increased precipitation. Lawford, summarizing the differences expected between Chile and North America, notes, as have others in this volume, the land/sea ratio difference and its impact on climatic, and hence hydrologic, patterns. He also calls attention to the fact that the continent of South America terminates at 54°S, whereas the landmass and mountains of North America extend north of 70°N.

The hydrology–geomorphology section illustrates the richness of comparative possibilities between North and South America. There are great similarities in the contemporary patterns, but yet there are also significant differences. This similarity/dissimilarity is equally apparent for the projections of the rates of response of climate to a change in climatic forcing elements.

Whereas the ocean and atmospheric sections deal with predictions of future responses of these earth system components, principally to changes in radiative transfer functions, the biogeochemistry section gives equal focus to the other major driving force of change, land use. Furthermore, the biogeochemistry chapters give considerable attention to the feedback to the atmosphere of changes in terrestrial systems. In the discussions by Kronberg on the north and by Sanhueza on the south, it becomes clear that different ecosystems will dominate these processes on the respective continents—tropical forests in the Southern Hemisphere and northern forests and tundra in the Northern Hemisphere. These differences will be reflected in dissimilar changes in the emissions of gases from these systems.

The section on intertidal ecosystems introduces the complexities of predicting biotic response to global change. Lubchenco, Navarrete, Tissot, and Castilla clearly outline the basic problems of predicting responses of organisms to a changing climate. Utilizing the responses of intertidal organisms to the 1982–1983 El Niño, they show how simple predictions based on the responses of individual species to temperature change were not realized because of the overriding influence of either secondary physical environmental changes that occurred, such as changes in wave action or, more commonly, changes in various biotic interactions including the loss of predators or keystone species in controlling distribution and abundances.

For the Southern Hemisphere, Castilla, Navarrete, and Lubchenco outline the main physical and biological trends in the intertidal zone along the west coast of South America. They note that some of the large-scale predictions for global change involve profound changes in the major biotic controlling factors of the intertidal zone, such as currents, intensity of upwelling, and frequency of upwelling. They utilize the El Niño responses for the south to demonstrate further the large biotic impacts of temperature increase and associated phenomena.

Navarrete, Lubchenco, and Castilla, in comparing the potential responses of the northern and southern intertidal systems, make a plea for the establishment of a network of coastal observation stations for long-term studies of the dynamics of particularly critical species. They demonstrate that the dynamics and mechanisms of responses of intertidal organisms to the changing temperatures of El Niño were most clearly understood when there was a long-term record of the dynamics of organisms under more normal climatic conditions.

The past, present, and predicted configurations of the vegetation of the Northern and Southern Hemispheres are addressed in three chapters. For the south, Villagrán and Armesto examine the past history of the Chilean vegetation, focusing on the Glacial-to-Holocene transition. They show that there was an asynchrony in deglaciation and the thermal maximum in each hemisphere with faster melting of ice in the Southern Hemisphere—the thermal equator was shifted to the south during the late Glacial. It should be noted that this historical trend is not the trend predicted for future global warming in other chapters in this volume. However, the driving forces for change are different, of course.

Arroyo, Armesto, Squeo, and Gutiérrez examine the present and predicted floristic and vegetation patterns of Chile. They show comparabilities in the general vegetation patterns of the Northern and Southern Hemispheres but with important differences. They note the impact of geographic isolation on the terrestrial vegetation of Chile. The massive Andean range to the west and the hyperarid Atacama Desert have provided substantial migrational barriers and a high degree of biotic endemism in Chile. The Chilean climate is apparently more equable than its Northern Hemisphere geographic counterpart, resulting in dissimilarities in plant growth forms and in their distributional patterns. There are proportionally fewer annual species in Chile than in California, for example, and the distributional patterns, both elevationally and latitudinally, are broader. These differential patterns have important implications for the future response to change. Also, the fact that the Andes extend continuously to much lower latitudes than the Northern Hemisphere analogs will differentially influence the migrational possibilities of various plant groups.

Keeley and Mooney reiterate the conclusions of Lubchenco et al., but for terrestrial systems, showing the complexity of physical and biotic interactions that determine the distribution of any plant and how difficult it is to predict the precise distributional outcome of the impact of the change of one or more controlling factors on the cascade of events that would follow. The problem is even more acute in terrestrial systems, where movement is more constrained and life spans are greater. Keeley and Mooney show that the vegetation of the Northern Hemisphere is a rich assemblage of diverse growth forms, from annuals to species with life spans in the millennia, that will, like their southern counterparts, have very dissimilar responses to global warming. They also emphasize that the environments of the future will be unlike any of the past, resulting in novel biotic configurations. The landscapes considered in the north are larger in extent and have a greater continuity than those of the south because of the smaller barriers to distributional movement due to mountains that are lower and more dissected.

In assessing the potential effects of global change on the mammals of western North America, Brown calls upon his vast experience in the study of the biogeography of the animals of the southwestern United States. He notes the island nature of many of the dissected mountain ranges of this region and the potential loss of species restricted to high-elevation sites because they do not possess the dispersal means to bridge the developing intervening climatic barriers. In contrast, Contreras shows that the mammal species of most of Chile have continuous north–south range distributions, because of the configuration of the Andes, except in the far north, where there are discontinuities in habitats due to the patchy distribution of sites with available water. The Andes do, however, represent an east-west barrier to the migration of mammals because of their high elevation and the intervening snow zone. Global warming may break this barrier and result in more homogeneous distributions west to east as now occur north to south. Brown and Contreras predict greater sensitivity of the animals of the Southern Hemisphere to global change because of their smaller distributional ranges, the greater potential of habitat fragmentation by human activities, and the fewer potential new habitats available at higher latitudes to serve as migrational sinks.

Alaback and McClellan provide a comprehensive assessment of environmental patterns past and present and of the managed ecosystems along the entire west coast of North America, emphasizing forest systems. They show how many of these systems are driven by disturbance regimes, including fire, wind, landslides and avalanches as well as the natural moisture patterns, including snowpack duration. They show how many of these disturbance patterns, in addition to the climatic patterns, are expected to change with concomitant effects on forestry practices. They note that the west coast of North America has undergone an unusually rapid and massive land transformation, with "human encroachment, competition for water resources, habitat fragmentation, air and water pollution, and increasingly challenging pest and weed problems." Furthermore, the greatest population expansion is occurring in areas with existing severe environmental problems, such as water limitations and air pollution, which will be exacerbated with climate change. Current land-use practices are not sustainable even under current climatic conditions. Alaback and McClellan call for a new approach to managing these landscapes and for the development of adaptive strategies to handle the problems that a changing climate will bring.

Fuentes and Muñoz provide a description of the patterns of land use in Chile at present and the potential land-use responses of the population to a changing climate. In global change research there has been increasing attention on how humans are driving global change but much less on how they will respond to these changes. The Fuentes and Muñoz chapter is a valuable contribution to this emerging discussion. They note that Chile is extremely mountainous and the options for response are limited. They further note that the devastation of the arid zone of northern Chile by human activities may well be repeated as the population spreads to the now colder southern regions.

In a comparison of the responses of Northern and Southern Hemispheres to global change, Alaback notes that there are substantial differences in the functional characteristics of the extant managed systems, and of the social systems driving them, and that they will diverge even further in the coming decades.

The concluding remarks of the book reiterate the enormous richness and unusual utility of making comparisons between the western Northern and Southern Hemispheres in the quest to monitor, understand, and predict the consequences of global change to the earth system.

Part I: The Ocean

C H A P T E R   1

Global Climate Change in the Oceans: A Review

PATRICIO A. BERNAL

I. Introduction

Here I describe the possible effects on the oceans of global climatic change, with particular emphasis on the Southern Hemisphere. Because CO_2 and other greenhouse gases show differential absorption of radiant energy at different wavelengths, a shift in the global radiation budget between incoming and outgoing (mostly infrared) radiation is expected to cause a net increase of global temperature. This change in the energy budget will have a profound effect on climate and consequently on the earth system as a whole. Direct instrumental measurements show a clear monotonic trend of increasing CO_2, from 280 to 300 parts per million in 1880 to 335 to 340 in 1980 (Hansen *et al.*, 1981), a trend that has already exceeded the between-centuries range of variation (Webster, 1985). Insights into how these changes will manifest themselves and evolve in time are obtained through numerical modelling of climate, a field of research that has seen dramatic achievements in the past 10 years. General circulation models (GCMs), which are coupled ocean–atmosphere numerical models, include oceanic compartments. Early versions considered a static ocean as an infinite source or sink of heat and humidity. Newer versions of these models considered stratified, two-layer oceans. However, only recently have GCMs begun to include explicit formulations of the dynamics of the ocean interior. This chapter deals exclusively with models that incorporate, to different degrees, ocean interior dynamics.

EARTH SYSTEM RESPONSES TO GLOBAL CHANGE
Contrasts between North and South America

1

II. The Ocean Response to Increased CO_2

Modeling the oceans is difficult because there are major gaps in our knowledge of some oceanic processes, and their parameterization in the absence of new information is uncertain. The most crucial of these uncertainties are the various types of mixing in the oceans, particularly cross-isopycnal mixing (Stewart and Bretherton, 1985). Despite these difficulties, significant progress has been achieved using models ranging from rather simplified ones involving a highly idealized geography on a single hemisphere (Bryan et al., 1982; Spelman and Manabe, 1984; Bryan and Spelman, 1985; Manabe and Bryan, 1985), to a model with quasi-realistic geography (Bryan et al., 1988), to models with fully realistic geography (Stouffer et al., 1989; Manabe et al., 1990).

The usual numerical experiment with these models involves running the model until it reaches an asymptotic equilibrium under normal conditions. That is, a geographically stable configuration of properties is obtained for an atmosphere with a concentration of CO_2 similar to the one observed today. The second step is to run the same model using an atmosphere in which the CO_2 concentration has been suddenly increased (or decreased as in Manabe and Bryan, 1985) by a given factor (e.g., a factor of 2 or 4). Then these two results are compared to evaluate the transient response of the coupled system to a sudden increase of CO_2 at different times from onset of the altered conditions.

Simulations in these models represent the asynchronous combined integration of three compartments: the atmosphere, the upper ocean (<1000 m), and the deep ocean (>1000 m). Each of these compartments reacts with a characteristic response time; for example, because of the large heat capacity of the ocean, its response is quite slow compared to the atmosphere. Table I shows characteristic times to reach asymptotic equilibrium for each compartment in the simulations.

Although extremely valuable, the experiments just described still suffer from the unrealistic condition of a sudden increase of CO_2, compared to the monotonic increase actually observed. This limitation was removed by Stouffer et al. (1989) in an experiment in which they increased the concentration of atmospheric CO_2 at a rate of 1% per year (compounded) in order to simulate the trend observed in nature.

TABLE I
Characteristic Times (in years) to Reach Equilibrium in the Ocean–Atmosphere Model

Compartment	Normal CO_2	$4 \times CO_2$
Atmosphere	7.8	11.8
Upper ocean (<1000 m)	850	1,290
Deep ocean (>1000 m)	23,000	35,000

Source: Spelman and Manabe (1984).

As would be expected, the four stages of modeling the ocean–atmosphere response that I have just described (idealized geography, sudden increase; quasi-realistic geography, sudden increase; realistic geography, sudden increase; realistic geography, gradual increase) produce different results. These results are summarized in this chapter.

III. Idealized Geography Models Responding to a Sudden Increase of CO_2

The "geography" of these models consists of paired wedge-shaped continents and oceans that start at the pole and span 60° of longitude at the equator. This unrealistic geography was due to the limited computer capacity at the time and by the complexity of the calculations of the coupled dynamic equations involved. In fact, the computations were conducted on a single pair of wedges and numerical results were extrapolated for one hemisphere by amplifying the results by a factor of 3. Furthermore, to generalize results for the two hemispheres (north and south), this family of models imposed a condition of mirror symmetry at the equator. This severely limited the confidence with which the results can be interpreted for the Southern Hemisphere, where the land/ocean ratio departs radically from 1:1, continental effects are less prominent, the Drake Passage exists, and the Antarctic continent is centered at the pole.

A comparison of normal and increased CO_2 responses at asymptotic equilibrium (Spelman and Manabe, 1984) shows that zonally averaged ocean temperatures are higher everywhere compared to the normal climate. The warming in the deep ocean of 7.5°C is about equal to the large warming of the ocean surface at 65–70° latitude, where ocean stratification is weak and the heat anomaly can penetrate from the surface into the deeper ocean (Spelman and Manabe 1984). Deep-ocean warming is significantly greater than that experienced by the surface layer expressed per unit area. The smallest temperature increase occurs in surface tropical waters, coinciding with the smallest atmospheric warming. In contrast to what happens in the atmosphere, the largest increase in temperature at the ocean surface does not occur at the pole (a distinct possibility in this model because of its geometry) but in a very shallow layer near 75° latitude. This difference is explained by the strong ocean stratification observed at this latitude, which partially insulates the surface water from the deeper ocean.

Figure 1 shows the transient response after 25 years of warming following a sudden increase of CO_2. Major results are as follows: (1) the surface layer of the model atmosphere at high latitudes is almost decoupled from the ocean; (2) near the sea–ice margin at 75° latitude, warming of the surface is limited to a shallow layer because the strong stratification caused by the sharp halocline insulates deeper layers; (3) maximal warming in the ocean occurs at subarctic (subantarctic) latitudes around 55–60°; (4) penetration of surface warming into the deep ocean is exceedingly slow and has only a small influence on the latitudinal distribution of zonally averaged temperature in the upper (<1000 m) model ocean (Spelman and Manabe, 1984).

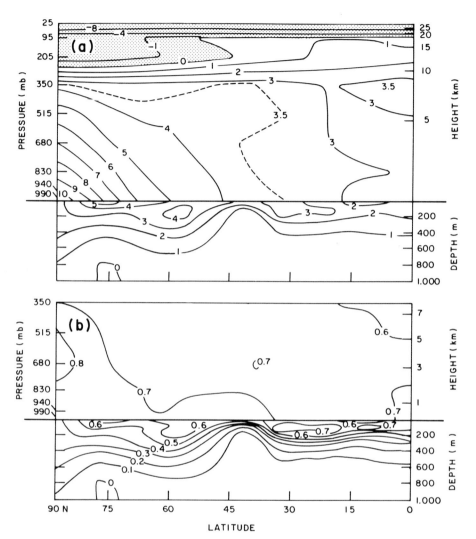

Fig. 1.   Transient response after 25 years of a fourfold increase in CO_2. Latitude height–depth distributions from the transient response study showing (a) the zonally averaged temperature at 25 years minus initial temperature (kelvin) and (b) the fractional response of zonally averaged temperature, $R = (T_i - T_0)/(T_{as} - T_0)$, where T_i is the temperature at year i of the simulation run, T_0 is the temperature at year 0 of the simulation run, and T_{as} is the temperature at asymptotic equilibrium. R measures the fraction of the total response realized at the picture time. (Redrawn from Spelman and Manabe, 1984.)

The surface circulation pattern established after 20 years in response to a fourfold increase in atmospheric CO_2 (Bryan and Spelman, 1985) shows the presence of the subtropical anticyclonic gyre and two cyclonic gyres at higher latitudes, the latter a major difference from the normal pattern of circulation; the salinity distribution at the surface shows higher salinities in the subtropical anticyclonic gyre and two pools of very low salinity water (<32 per mil) close to the west and east boundaries caused by increased runoff from the continents. This pool of fresher water implies an increased input of nutrient from runoff and should have an impact on primary production.

Figure 2 shows the thermohaline circulation pattern. During the transient response period (Fig. 2b), water formed at the surface in high latitudes is not dense enough to sink to the bottom and flows as intermediate water toward the equator at mid-depths (approximately 600 m). This leads to (1) formation of a clockwise cell in the deep ocean and (2) shallowing and partial collapse of the "equilibrium" counterclockwise cell, which has two nuclei at 25° and 60° latitude. The shallow cells induced by Ekman transport remain much the same in all three cases of Fig. 2 because zonal wind stress does not diminish after the onset of the increased CO_2 conditions. Bryan and Spelman (1985) highlight the far-reaching consequences of the partial collapse of the thermohaline circulation, which decreases the capacity of the ocean to take up CO_2.

IV. Quasi-Realistic Geography Model Responding to a Sudden Increase of CO_2

Bryan *et al.* (1988) developed an improved version of the idealized geography model in which an ocean basin is differentiated into two hemispheres preserving the actual ratio of continents and oceans. The upper panel of Fig. 3 shows the geographic pattern used in this model and the asymptotic equilibrium result for the vertically and zonally integrated mass transport under normal CO_2. The pattern of flow shown by the horizontal stream functions defines two subtropical gyres centered at 30°, clockwise in the north and counterclockwise in the south, and a strong circumpolar current between 50 and 60°S with a total transport of 120 Mt s^{-1},* a value falling within the observed range of 118 to 146 Mt s^{-1} (Whitworth, 1983).

The vertical meridional circulation shown in the bottom panel of Fig. 3 is asymmetric north and south of the equator. There are two Ekman cells associated with the trade winds that transport surface waters toward the poles and cause equatorial upwelling. However, the existence of a "Drake Passage" in this model shows fundamental differences between both hemispheres: associated with the westerlies, Ekman transport is equatorward, and surface flow is compensated by a shallow return flow in the north but by a very deep cell in the south.

---

*Units of mass transport used in this chapter are Mt s^{-1}, million tons per second (10^{12} g s^{-1}). This mass transport unit, also used as volume flow, 10^6 m^3 s^{-1}, is known in oceanography as 1 sverdrup, abbreviated Sv.

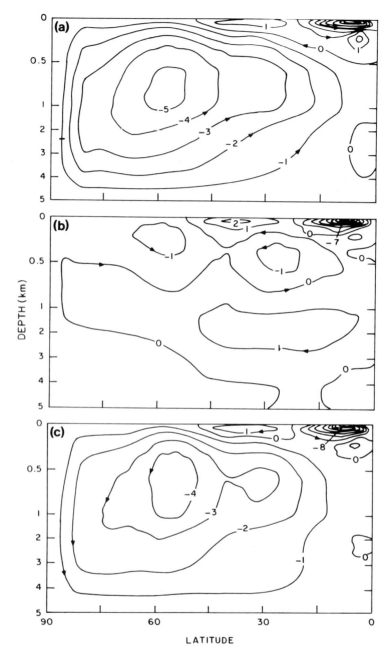

Fig. 2. Total meridional circulation. The zonally integrated equilibrium circulation, mostly thermohaline below the surface layer, in units of Mt s^{-1}. Overturning in the meridional plane shows a main counterclockwise circulation, with a concentrated downward branch near the pole and a broad upward branch close to the equator. Close to the surface, smaller circulation cells are associated with Ekman pumping or suction. Simulations for normal and increased CO_2 tend to show similar patterns, whereas the transient response after 20–30 years gives rise to radically different ones. (a) Equilibrium solution for a fourfold increase in CO_2; (b) average circulation for years 21–30 after "switching on," showing the partial collapse of the thermocline, reduced thermohaline circulation (<600 m), and a clockwise deep circulation; (c) equilibrium solution for normal CO_2. The upper 1 km of the ocean depth is expanded. (Redrawn from Bryan and Spelman, 1985.)

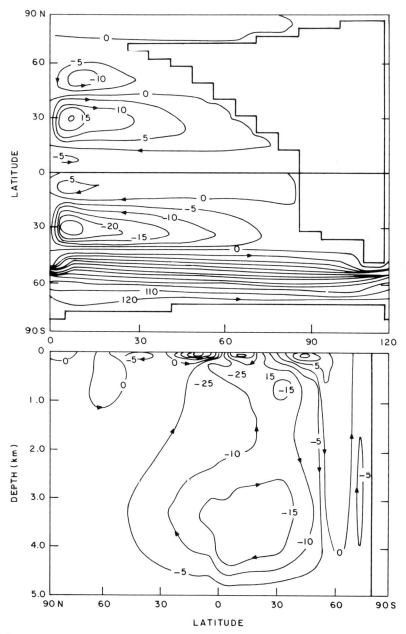

Fig. 3. Surface and meridional circulation patterns in the two-hemisphere model under normal CO_2 conditions. (Top) The surface circulation pattern is shown by the streamlines of the vertically integrated mass transport in Mt s^{-1}. Dominant features are the circumpolar current with a total transport of 120 Mt s^{-1}, two subtropical gyres centered at 30°, and the subarctic gyre. (Bottom) The meridional circulation is shown by the streamlines of zonally integrated mass transport. The main deep cell is characterized by sinking in the Southern Hemisphere and rising motion in the Northern Hemisphere. Associated with the trades and westerlies, Ekman cells are observed in the surface layer. In the Southern Hemisphere the Ekman flow under the westerlies (40–45°S) is compensated by a very deep cell (60°S). (Redrawn from Bryan *et al.*, 1988.)

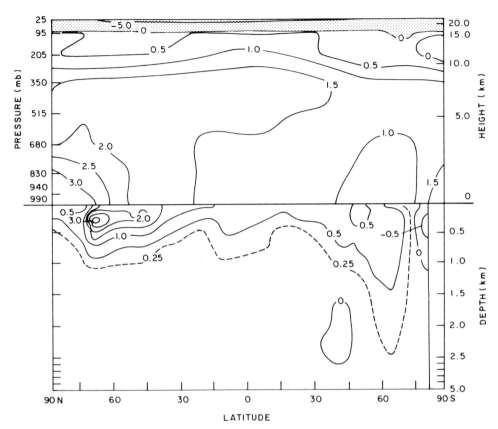

Fig. 4. Transient response after 41–50 years in the two-hemisphere model. A highly asymmetric pattern of the zonally averaged temperature minus initial temperature (K) emerges. Polar amplification, although strong in the equilibrium response (not shown), is much reduced during the transient period. Warming of the surface ocean is much less than predicted in the one-hemisphere model (Fig. 3). Notice "cooling" at the edge of the Antarctic continent. (Redrawn from Bryan *et al.*, 1988.)

Figure 4 (from Bryan *et al.*, 1988) shows the transient response of this model after 41–50 years of a sudden increase in CO_2. There is a clear asymmetry in the response of Northern and Southern Hemispheres. The maximum upper ocean heating of 3.0°C occurs around 60°N. Heating in the Southern Hemisphere is less than 1°C, and surprisingly, after four decades, cooling (–0.5°C) is observed close to the Antarctic continent. An interesting result associated with the vertical circulation is the deep penetration of the heat anomaly in the south (2.5 km) compared with north (1.2 km). This increases the thermal inertia of the ocean by removing excess heat from the surface to the deep ocean. In the words of the authors "the geography of the Southern Ocean of the model and the deep sinking and upwelling near the Drake Passage allow the Southern Hemisphere Ocean to be a very effective brake on climate change due to external forcing" (Bryan *et al.*, 1988, p. 863).

V. Realistic Geography Responding to a Sudden Increase in CO_2

Figure 5 (from Manabe *et al.*, 1990) shows the sea surface temperature pattern during the transient period after 50 to 60 years of integration. The most relevant feature of this result is the clear asymmetry of the climate response in the Northern and Southern Hemispheres. In the north the effect of polar amplification produces a positive anomaly of up to 4°C close to the Kamchatka peninsula and Greenland. In contrast, in the Southern Hemisphere close to Antarctica, a pool of negative-anomaly water is observed with a minimum of –3°C around 120°E. This sharp asymmetry can be seen in a vertical section of the atmosphere and ocean in Fig. 6. Here polar amplification causes a maximum positive anomaly greater than 4.5°C close to the northern pole at the surface, compared with a mere 1°C increase over the Antarctic continent. In the ocean, maximum heating is obtained around 60°N at the surface with +3.5°C. Maximum penetration of this warm pool into deeper water (1800 m) occurs immediately below this surface maximum. No analog of this feature is observed in the south, but a narrow, deep, cool pool of water is observed at 68°S associated with a deep downwelling–upwelling cell. Changes in the intensity of atmospheric circulation in the Southern Hemisphere intensify surface westerlies at middle latitudes, which in turn intensify an Ekman circulation cell that carries fresh surface water toward the equator. This increased Ekman drift to the north increases the stability of the water column and effectively prevents warming of the circumpolar ocean. Changes in meridional circulation in the intermediate ocean reduce the poleward heat transport toward high latitudes in the south.

A comparison of these results with those shown in Fig. 1 highlights the importance of land mass distribution and the role of real geographic features in shaping the response of the coupled ocean–atmosphere system to an external driving force. Changes with real geography are much less than those reported for models with idealized geography, despite the fact that Spelman and Manabe (1984) used a four-fold increase of CO_2. North–south interhemispheric asymmetries were predicted based on fundamental principles, and the work reviewed here has strongly corroborated this expectation. Climate change will undoubtedly display a highly heterogeneous response in space, a feature that will challenge our ability to predict the rate and intensity of expected local changes and will affect the accuracy and precision of global estimates.

VI. Realistic Geography Responding to a Gradual Increase in CO_2

Stouffer *et al.* (1989) modified the canonical numerical experiment by forcing the system with a gradually increasing perturbation consisting of a CO_2 concentration

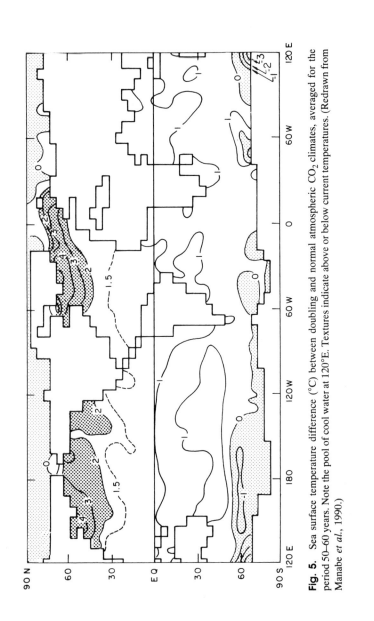

Fig. 5. Sea surface temperature difference (°C) between doubling and normal atmospheric CO_2 climates, averaged for the period 50–60 years. Note the pool of cool water at 120°E. Textures indicate above or below current temperatures. (Redrawn from Manabe *et al.*, 1990.)

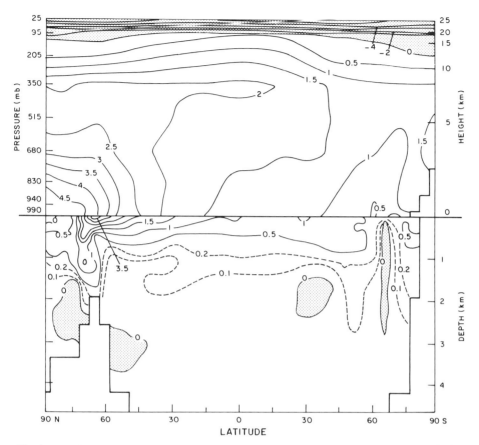

Fig. 6. Zonally averaged temperature response of the atmosphere and ocean for doubling of atmospheric CO_2. Transient response (°C) averaged over the period 50–60 years. (Redrawn from Manabe *et al.*, 1990.)

continuously increasing at a rate of 1% per year compounded. At this rate, doubling the concentration of CO_2 would take 72 years. The results were quite dramatic, displaying a marked interhemispheric asymmetry, and were quickly reported as a letter to *Nature*. After 95 years of integration, polar amplification resulted in an air temperature anomaly of +7°C close to the northern pole but of only +2°C close to the southern pole. In general, southern high latitudes showed increased thermal inertia, mainly attributable to ocean circulation and geography. Figure 7 shows the response of the system 60 years after the onset of the perturbation. Ocean warming is essentially confined to the surface layer (<800 m) except around Antarctica, where a small positive anomaly penetrates up to 4000 m. The pattern of warming of the upper ocean is similar to that in Figs. 1, 4, and 6: minimum warming is observed at the northern pole and maximum penetration is observed close to 60°N. An interesting

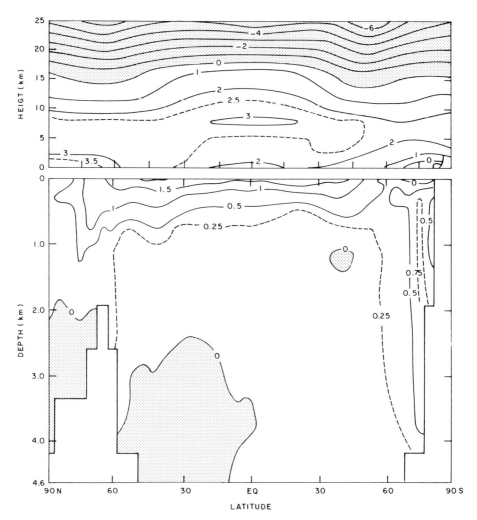

Fig. 7. Zonally averaged temperature response of the atmosphere and ocean for the gradual increase numerical experiment. The difference represents the decadal average over years 61–70. (Redrawn from Stouffer *et al.*, 1989.)

result of this experiment is the weakening of the thermohaline circulation in the North Atlantic, caused by increased runoff and excess precipitation over evaporation, which in turn lowers the salinity (density), enhancing stratification and thus slowing down thermohaline circulation. In fact, thermohaline circulation decreased about 25% by the seventh decade of the experiment in the North Atlantic. Figure 8 shows the pattern of thermohaline circulation for normal CO_2 (upper panel) and after 60–70 years of integration (lower panel). A comparison of both situations shows a reduction from 16

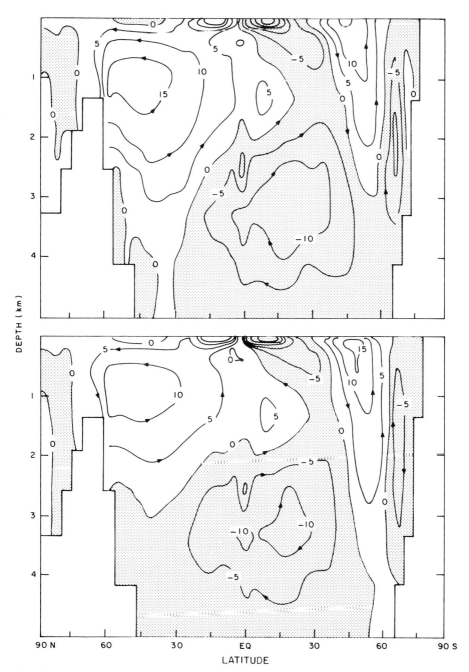

Fig. 8. Reduction of thermohaline circulation. Streamlines of zonal mean meridional oceanic circulation averaged between years 61 and 70 after onset of perturbation, for normal (top) and increased (bottom) CO_2 atmospheres. (Redrawn from Stouffer *et al.*, 1989.)

Mt s^{-1} to 11 Mt s^{-1} at 1000 m depth around 50°N. This weakening of the thermohaline circulation is in line with results reported by Bryan and Spelman (1985) and reproduced here in Fig. 2 and with other reports cited by Stouffer *et al.* (1989). This effect, if confirmed by more detailed studies, would have strong consequences for the ability of the ocean to take up CO_2.

VII. Summary

This chapter reviews the results obtained with GCMs that explicitly incorporate the dynamics of the ocean interior in order to describe the possible effects of global climate change on the oceans. Four stages of increasing degrees of realism in these models and numerical experiments are covered:

1. Idealized geography, sudden increase in CO_2
2. Quasi-realistic geography, sudden increase in CO_2
3. Realistic geography, sudden increase in CO_2
4. Realistic geography, gradual increase in CO_2

In general, warming of the upper layer of the ocean is confined to the upper 500 to 800 m, with deeper penetration of the positive anomaly (0.5°C) at higher latitudes. The models using realistic geography show strong north–south interhemispheric asymmetry because of increased thermal inertia in the south, attributable to ocean circulation (Drake Passage) and geography. A common feature of the simulations is weakening of the thermohaline circulation, a factor that, if confirmed in the future, could have strong effects on the ability of the oceans to take up CO_2, amplifying global warming.

References

Bryan, K., and Spelman, M. J. (1985). The ocean's response to a CO_2-induced warming. *J. Geophys. Res.* **90**(C6): 11679–11688.

Bryan, K., Komro, F. G., Manabe, S., and Spelman, M. J. (1982). Transient climate response to increasing atmospheric carbon dioxide. *Science* **215**, 56–58.

Bryan, K., Manabe, S., and Spelman, M. J. (1988). Interhemispheric asymmetry in the transient response of a coupled ocean–atmosphere model to a CO_2 forcing. *J. Phys. Oceanogr.* **18**, 851–867.

Hansen, J., Johnson, D., Lacis, A., Lebedeff, S., Lee, P., Rind, D., and Russel, G. (1981). Climate impact of increasing atmospheric carbon dioxide. *Science* **213**, 957–966.

Manabe, S., and Bryan, K. J. (1985). CO_2-induced change in a coupled ocean–atmosphere model and its paleoclimatic implications. *J. Geophys. Res.* **90**(C6): 11689–11707.

Manabe, S. Bryan, K., and Spelman M. J. (1990). Transient response of a global ocean–atmosphere model to a doubling of atmospheric carbon dioxide. *J. Phys. Oceanogr.* **20**(5): 722–749.

Spelman, M. J., and Manabe, S. (1984). Influence of oceanic heat transport upon the sensitivity of a model climate. *J. Geophys. Res.* **89**(C1): 571–586.

Stewart, R. W., and Bretherton, F. P. (1985). Atmosphere–ocean interaction. *In* "Global Change," (T. F. Malone and J. G. Roederer, eds.), pp. 146–156. Proceedings of a Symposium sponsored by the International Council of Scientific Unions (ICSU) during its 20th General Assembly in Ottawa, Canada on September 25, 1984. ICSU Press, Cambridge University Press, London.

Stouffer, R. J., Manabe, S., and Bryan, K. (1989). Interhemispheric asymmetry in climate response to a gradual increase of atmospheric CO_2. *Nature* **342**, 660–662.

Webster, P. J. (1985). Great events, grand experiments: man's study of the variable climate. Part II. Prospects of a warming earth. *Earth Min. Sci.* **55**, 21–24.

Whitworth, T., III. (1983). Monitoring the transport of the Antarctic circumpolar current at the Drake Passage. *J. Phys. Oceanogr.* **13**(11): 2045–2057.

C H A P T E R   2

Global Warming and Coastal Ecosystem Response: How Northern and Southern Hemispheres May Differ in the Eastern Pacific Ocean

CHARLES H. PETERSON   RICHARD T. BARBER
GREGORY A. SKILLETER

I. Introduction

A common approach of biological oceanographers has been to compare geographically different, but physically similar, areas of the oceans as a means of testing the generality of models of physical–biological coupling and of ecosystem dynamics. For example, many studies have detailed the similarities in physical forcing, biological production processes, and ecosystem structure in the eastern boundary currents of the world's oceans (e.g., Barber and Smith, 1981; Parrish *et al.*, 1981; Bakun and Parrish, 1982; Husby and Nelson, 1982; Parrish *et al.*, 1983; Rothschild, 1986 among others). Analogous studies of convergence have contributed a great deal to our understanding of how environment dictates form and function in ecology at all levels of biological organization (e.g., Brown, 1975). Yet, the recognition that the proportion of the earth's surface covered by oceans differs greatly between the Southern and Northern Hemispheres raises the question of how future global climate change may differ between hemispheres and thus affect similar ecosystems differently.

The intent of this chapter is to provide a current scenario for how anticipated global warming might be expected to alter the coastal marine ecosystems of the

17

eastern Pacific Ocean. We focus most of our attention on the Northern Hemisphere, the California Current ecosystem, but in a context of contrast with the corresponding temperate coastal system off Peru and Chile. We review the meteorology, ocean physics, and seawater chemistry necessary to understand the functional basis of primary productivity and ecosystem dynamics in these systems. The eastern boundary current ecosystems of the Pacific Ocean are an ideal model system for evaluating the likely biological responses to global warming because the high-frequency variability in physical forcing associated with the El Niño–Southern Oscillation events in the eastern Pacific provides a window through which to view components of the future. Our relatively good understanding of how changes in the meteorology, ocean physics, and ocean chemistry drive biological change during an El Niño cycle allows realistic process-based predictions of responses to longer-term global warming in the system. Our analysis suggests the possibility of similarly dramatic change in ecosystem structure in both Northern and Southern Hemispheres with distinct differences in the magnitude of response between hemispheres.

II. Overview of Physics and Biology of Pacific Eastern Boundary Current Regions

To assess how global warming could affect eastern Pacific coastal ecosystems, we first describe the larger-scale oceanic setting in which the coastal ecosystems are embedded. An advantage in discussing this subject is that oceanographic work in the past two decades has begun to describe the relative roles of large-scale as opposed to local physical forcing in determining the nature and variability of coastal ecosystems. Work on coastal upwelling ecosystems was initiated for the purpose of understanding mesoscale physical–biological coupling and the role of coupling in the regulation of primary productivity, food web structure, and fisheries yields, but the developing understanding of mesoscale forcing led to an appreciation of large-scale, remote forcing and how it affects coastal ocean ecosystems. The emerging principles relating climate variability to ecosystem function are useful for predicting how global change could alter coastal ecosystems.

Originally, the study of physical–biological coupling in coastal upwelling ecosystems concentrated on the specific physical and biological processes taking place in the mesoscale spatial and temporal domains where upwelling occurs. It became apparent that some aspects of the physical and biological variability of coastal upwelling could not be explained when the analysis was limited to atmospheric and physical processes occurring only in the upwelling region (Barber and Smith, 1981). Observations showed that coastal upwelling was an integral part of the basinwide thermal asymmetry of the ocean and, therefore, an important driver of the global wind system (Wyrtki, 1975; Cane, 1983; Rasmusson and Wallace, 1983). In essence, coastal upwelling is a component of the ocean basin heat budget; when basinwide thermal dynamics are perturbed, the physical and biological components of coastal upwelling change and the nature of the change is predictable given an understanding of the changes in winds and the ocean's density structure. In this context, the coastal upwelling work of the past two decades provides a useful conceptual model for developing scenarios of how global warming might alter eastern Pacific coastal ecosystems.

To find that the mean biological character and major variability of coastal upwelling were regulated by basin-scale ocean and atmospheric processes was somewhat surprising because the major physical process (wind-driven coastal upwelling) and important biological processes (phytoplankton blooms, zooplankton growth, and larval fish recruitment) are mesoscale processes that involve relatively short time scales of 1–10 days and small space scales of 5–200 km (Barber and Smith, 1981). The character of both physical (Smith, 1983) and biotic variability (Barber and Chávez, 1983) indicates that coastal upwelling is an integral part of a larger system, as well as an inherently mesoscale phenomenon. The nature of the basinwide coupling was revealed clearly by the changes that took place during the 1982–1983 El Niño in both the Northern and Southern Hemispheres. Two of the most far-reaching biological changes were significant decreases in primary production and changes in phytoplankton species composition.

It has been known for a century that eastern boundary regions of the ocean basins have higher biological productivity than western boundaries (Buchanan, 1886; Coker, 1908). Sverdrup (1955) illustrated this when he made the first estimate of the global pattern of primary productivity. The principle Sverdrup (1955) used in his estimate was that the first-order process regulating ocean productivity is the supply of inorganic plant nutrients from deep water to the surface layer. Later syntheses of large-scale patterns of productivity, such as that by Koblentz-Mishke *et al.* (1970), confirm Sverdrup's prediction that the eastern Pacific is considerably more productive than the western Pacific; that is, there is a strong east–west basinwide asymmetry in biological productivity.

To predict the ecosystem consequences of global change involving warming, it is useful to examine how the east–west asymmetry in basic productivity is created and maintained, because, in simplest terms, the productivity asymmetry is a direct reflection of an east–west asymmetry in heat storage. The western Pacific contains a "warm pool" of water with an annual mean temperature of about 29°C. Solar energy falling on the 29°C water is efficiently transferred back to the atmosphere by evaporation, convection, and back radiation. Satellite measurements of outgoing longwave radiation (OLR) (Weickmann, 1983; Lau and Chan, 1985) show that the major convective center of the Pacific is located over the Indonesian "maritime continent" (Ramage, 1969). The tremendous center of convection and evaporation over the warm water of the western Pacific causes air to rise, creates the Indonesian low-pressure system, and feeds into the upward branch of the east–west atmospheric circulation cell (Newell and Gould-Steward, 1981). Meanwhile, in the eastern portion of the ocean basin, equatorward winds force upwelling of subsurface waters along the coastal margin. At the surface, this cool upwelled water extracts heat from the atmosphere, causing atmospheric subsidence and intensifying the seasonal North Pacific and permanent South Pacific high-pressure systems. The subsiding air feeds into the easterly trade winds and flows to the west toward the convective center over the warm pool of the western Pacific. The easterly trade winds drive surface water westward in the Equatorial Current system. As the water moves westward, the upper layer gains heat from the sun and the wind-driven circulation transports heat to the western Pacific. The western Pacific is kept warm and the eastern Pacific cool. Thus, the west versus east temperature gradient in the tropical Pacific sets up a pressure gradient that forces large-scale trade winds, and at the same time the trade winds are responsible for forcing the circulation that maintains the west–east temperature gradient (Rasmusson and Wallace, 1983; Cane, 1983). In combination with the midlatitude westerly winds, the equatorward coastal winds and low-latitude easterly winds set up the basinwide east–west thermocline tilt that characterizes ocean basins from midlatitudes to the tropics.

The basinwide tilt in the thermal, density, and nutrient structure makes the entire eastern portion of the ocean basin inherently richer in nutrients than the central or western portions because the subsurface pool of inorganic nutrients penetrates to the sunlit layer, where radiant energy is available for the photosynthesis of new organic material. The shallower nutricline in the eastern boundary region means that any physical process that causes additional turbulent or advective vertical transport (island wake mixing, tidal mixing, wind-driven upwelling, shelf-break mixing) will supply more nutrients to the surface layer. Therefore, the eastern portion of the ocean basin is, during climatological mean conditions, always richer than the central or western portions. Figure 9.9 in Barber (1988) shows that this characteristic east–west basinwide tilt in the nutricline is present across the Pacific at midlatitudes about 30°N as well as in the low latitudes. Since the east versus west generalization about nutricline tilt applies to the temperate ocean as well as to the tropical ocean, perturbations of this basinwide structure by global warming will probably affect both the middle and low latitudes.

The role of basinwide nutricline topography in the regulation of productivity and phytoplankton species composition was clearly revealed in the 1982–1983 El Niño, in which a marked depression of the nutricline was observed off California (McGowan, 1985; Dayton and Tegner, 1989) and off Peru (Barber and Chávez, 1983). Analysis of El Niño weather records showed that off Peru coastal winds remained favorable for upwelling (Enfield, 1989) and that upwelling continued but the upwelled water was nutrient depleted (Barber *et al.*, 1985). The hypothesis that nutricline depression is the causal process for the productivity decreases observed during El Niño is consistent with the time series data of interannual variability in primary production off the Peru coast reported by Chávez *et al.* (1989). Effects of nutrient decreases on kelp and other macroalgae are reported by Dayton and Tegner (1989), and Avaria and Muñoz (1987) have described the phytoplankton changes off Chile that accompany the change in nutrient availability. During normal upwelling the phytoplankton community had low species diversity and was dominated by a few species of diatoms, which underwent characteristic blooms; dinoflagellates were present in relatively small numbers. Off northern Chile during the 1982–1983 El Niño, diatoms decreased in abundance and dinoflagellates dominated most of the region. The diatoms that remained were concentrated in small pockets of cool, nutrient-rich water very close to the coast. Sardines were also concentrated in these pockets, but decreases in their planktonic food driven by decreases in the nutrient supply led to physiological changes in the surviving fish; off Chile sardines had a lower oil yield and lower fat content (Avaria and Muñoz, 1987), just as the anchovies did off Peru (Barber and Chávez, 1986). These observations of changes during El Niño emphasize that two conditions are necessary to maintain the nutrient-rich and highly productive character of the eastern boundary region: the shallow nutricline set up by the basinwide winds and the local upwelling-favorable winds. How will these two conditions change during global warming?

III. Major Physical and Chemical Signals and Their Ecological Consequences

A. Direct Effects of Ocean Warming

Examination of changes in the distribution of marine organisms during El Niño events along the eastern Pacific coasts gives insight into possible expected changes in biogeography as a result of increases in temperature due to global warming. Unlike sedentary terrestrial biota and the lacustrine biota, which may have significant barriers to dispersal, there are essentially no barriers to long-range dispersal of marine organisms, which tend to be mobile and have planktonic life stages. Scheltema (1986, 1988) indicated that increases in temperatures of surface waters, such as those associated with El Niño events, may provide additional opportunity for dispersal of many invertebrate larvae that may be unavailable during average conditions.

Changes in the geographic distribution of marine biotas have occurred in association with the increases in the temperature of coastal waters in the mid-latitudes of the eastern Pacific Ocean during El Niño events. The abundance of dinoflagellates along the coast of Peru declined during the 1982–1983 El Niño, but more species were present, many of which are rarely found in that region (Ochoa and Gòmez, 1987). There was also a decrease in the abundance of small diatoms with a shift to larger species of diatoms typical of subtropical oceanic waters (Alamo and Bouchon, 1987). Changes in the phytoplankton community were also observed off northern Chile. The nearshore surface waters are normally dominated by a few species of small diatoms, with only low numbers of dinoflagellates or other groups of phytoplankton. As the temperature of the seawater increased, the dominance of these small species of diatoms was restricted to areas very near the coast and the greater portion of the nearshore coastal zone was dominated by an abundance of warm-water species of diatoms and dinoflagellates (Avaria and Muñoz, 1987).

Off Peru, the zooplankton was affected qualitatively and quantitatively during the 1982–1983 El Niño event. Aggregations of species more typical of warmer water were observed. The warmer water and the associated fauna modified the specific and numeric structure of the plankton populations. Before the onset of the event, copepods were the most abundant component of the zooplankton. During the event, the abundance of copepods declined and the abundance of chaetognaths, euphausiids, appendicularians, and siphonophorans increased. These species are more typically associated with warmer, subtropical surface waters. The distribution of the larval stages of three species of penaeid prawns was extended from 6°S to 12°S (Carrasco and Santander, 1987).

In addition to changes in the distribution of phytoplankton and zooplankton, during the 1982–1983 El Niño, in low latitudes in the eastern Pacific, there were altered patterns of abundance and migrations of pelagic and benthic communities. These changes affected fishes, benthic invertebrates, seabirds, and pinnipeds; many species showed dramatic population reductions (Barber and Chávez, 1983; Pearcy and Schoener, 1987 and references therein). On the other hand, *Nyctiphanes simplex*, a subtropical euphausiid, extended its range to the north and became abundant off Oregon and Washington (Brodeur, 1986). There were migrations of species of fishes north into the waters of the California Current. Many of these species had never been recorded in these waters before. There were also increases in the abundance of some common nektonic animals, including the Pacific mackerel (*Scomber japonicus*), jack mackerel (*Trachurus symmetricus*), and bonito (*Sarda chiliensis*) (Pearcy and Schoener, 1987). Some species of larval fishes showed changes in relative abundance indicative of transport from offshore waters (Brodeur *et al.*, 1985). Active swimmers such as mackerel, bonito, and seabass were presumed to have migrated northward into the warm coastal waters resulting from the depressed thermocline (Pearcy and Schoener, 1987). Species usually found farther offshore (e.g., blue shark, *Prionace glauca*; Pacific saury, *Cololabis saira*; and ocean sunfish, *Mola mola*) were found farther inshore. Migrations such as this may also have implications for seabirds, such as the aucklets, which consume species such as the Pacific saury.

These changes in the geographic distribution of marine flora and fauna, in response to increasing water temperatures caused by intense El Niño events, were rapid and resulted in major changes in the composition of the biota throughout the eastern boundary current systems of the eastern Pacific Ocean. Similar changes in the distribution of the marine biota could be expected as water temperatures increase in the coastal zones because of global warming (see also Castilla *et al.*, Chapter 13). The shifts in the geographic boundaries and in species compositions could also result in a spatial shift of the patterns of exploitation of fisheries resources for the countries affected: the distribution of a specific fishery could be shifted into territorial waters of another country, with strong political ramifications (Glantz, 1990).

B. Direct Effects of Increased CO_2 Concentration

In ocean waters the atom ratios of the inorganic forms of carbon, nitrogen, and phosphorous are about 1000 C : 16 N : 1 P; in plankton the atom ratio is 106 C : 16 N : 1 P (Redfield, 1958). Seawater has a nearly 10-fold atom ratio excess of inorganic carbon to inorganic nitrogen and phosphorous, so the present consensus among oceanographers is that the supply of inorganic carbon does not limit (or regulate) the rate of primary production by oceanic phyoplankton. Because the present concentrations in seawater provide an atom ratio excess of carbon for phyto-plankton, it appears that increased concentrations of CO_2 in the atmosphere causing higher surface layer concentrations of inorganic carbon will not increase the photosynthetic rate. Therefore we predict that there will be no direct effect of increased CO_2 on primary production in the ocean.

C. Effects of Basinwide and Local Change in Physical Forcing as a Result of Global Warming

The premise that Bakun (1990) uses to explain an observed increase in coastal wind stress between 1950 and the 1980s is that differential greenhouse warming over the land intensified the atmospheric pressure gradient between the continental low-pressure cell and high-pressure system over the cooler ocean. This gradient is a major determinant of strength of the alongshore wind; as the gradient steepens, the equatorward coastal winds strengthen. Bakun (1990) argues convincingly that dry mediteranean-like climates of west coast regions are particularly sensitive to change because increasing concentrations of greenhouse gases inhibit the nighttime long-wave radiative cooling that is an important component of the heat budgets of these regions.

In the Bakun scenario, increased alongshore winds are due to differential land–sea heating. For the basinwide winds it is necessary now to consider how global heating will modify the east–west pressure gradient set up by the oceanic temperature gradient, that is, between the eastern boundary region and the warm pool of the western Pacific. Obviously, no differential land–sea heating effect will come into play in regard to the east versus west temperature, pressure, or wind system. In the coastal process (Bakun, 1990) the thermal low-pressure cell is intensified over land, but ocean waters of the western Pacific cannot increase in temperature beyond 30°C (Newell and Gould, 1981) because the evaporation–temperature curve steepens sharply between 27 and 30°C. Above those temperatures, evaporative heat loss transfers all the incoming solar energy back to the atmosphere. Global warming cannot increase the temperature of the western Pacific, but one possible response would be an increase in the areal extent of warm water by enlarging the warm pool toward the east (Trenberth, 1990). This increase in heat storage in the western and central Pacific will reduce the distance between the cool eastern boundary regions and the warm pool, reducing the fetch over which the trade winds transfer kinetic energy to the ocean. As the spatial domain of trade winds decreases, less momentum is transferred to the ocean. Philander (1989) mentions that the difference in length of zonal fetch best accounts for why the thermocline outcrops in the eastern Pacific and not in the eastern Atlantic. A decrease in fetch will reduce the cross-basin thermocline tilt, thereby depressing the thermocline and nutricline and reducing the concentration of inorganic nutrients in the water entrained in the wind-driven upwelling circulation. Surprisingly, the intensification of upwelling by increased land–sea temperature differentials and the decrease of trade wind fetch can occur at the same time but may have different time scales during global heating. When the decrease in basin tilt develops, stronger coastal winds will intensify the upwelling circulation (and turbulence) but the supply of nutrients will not be enhanced. In essence, there will be more upwelling, but water that is upwelled will be lower in nutrients.

Accepting for the sake of argument the Bakun scenario that equatorward coastal winds will intensify during global warming, let us also assume that global warming will expand the Indonesian Low and reduce the distance between the Indonesian Low and the North Pacific High and South Pacific High, depressing the thermocline and nutricline along the eastern margins. The processes involved in both these assumptions have been observed during El Niño events (Trenberth, 1990) and in the 30-year wind record cited by Bakun (1990). Coastal winds have also been observed to intensify during El Niño due to increased land–sea temperature differentials (Enfield, 1989) when the thermocline and nutricline were depressed because of weakening in the trade winds in the western Pacific (Cane, 1983). The physical processes we are assuming are, therefore, realistic processes in that they have been observed, but of course we are uncertain about the time course and intensity of the changes that can be forced in the coastal ocean by greenhouse warming.

Assuming increased alongshore thermal winds and decreased large-scale zonal wind stress, we can describe with some confidence the modified physical environment of the California Current system. The intensified wind-driven upwelling will cause greater offshore surface transport, and increased wind stress will increase turbulence in the surface mixed layer. Both of the changes have important biological effects. These nearshore changes will be more pronounced in the Northern Hemisphere off California, where the Central Valley will generate a more intense thermal low-pressure system (see Fig. 2 in Bakun, 1990), but the large-scale thermocline and nutricline changes probably have a greater impact in the Southern Hemisphere because the zonal extent of the South Pacific basin is considerably greater than that of the North Pacific. Despite stronger upwelling, depression of the nutricline in both hemispheres will cause the upwelled waters to have reduced nutrient concentrations. Therefore primary production will not be stimulated or, at least, not stimulated to the degree that would be expected given the intensified upwelling, leading to a likely overall reduction in primary and then also secondary production in these coastal ecosystems.

D. Ecosystem Responses to the Effect of Basinwide and Local Changes in Physical Forcing

1. Intensification of Offshore Transport and Its Implications for Pelagic Fish Stocks. Upwelling centers are characterized by strong offshore transport, intense wind-generated turbulent mixing of the euphotic zone, and reduced stability of the water column (Barber and Smith, 1981). The net offshore transport in the surface Ekman layer, caused by the alongshore wind stress, is equal to the amount of water that upwells along the coast (Hickey, 1979; Huyer, 1983). Where strong upwelling persists, water properties associated with upwelling can be carried beyond the 10 to 20 km width of the upwelling zone by advection and eddy diffusion (De Szoeke and Richman, 1981). The greatest offshore transport varies from month to month at different parts of the California Current system, and the effects of coastal upwelling can extend over 100 km out to sea (Huyer, 1983). An increase in the alongshore wind stress would, therefore, lead to an increase in the amount of offshore transport of coastal water masses and their entrained biotic communities, away from the coastal regions. Of course, an accompanying effect is an increase in the alongshore transport of coastal water (Chelton et al., 1982).

Offshore transport is more likely to be unfavorable to larval survival than alongshore transport because the habitat changes more rapidly in the offshore direction and the larvae are carried away from the favorable coastal habitats where they are normally found as adults. Parrish et al. (1981) indicated that strong selection for minimizing offshore loss of reproductive products must have exerted control on the development of seasonal and geographic spawning characteristics for many species of pelagic fishes, including sardines (Sardinops sagax), hake (Merluccius productus), jack mackerel (T. symmetricus), anchovy (Engraulis mordax), and Pacific mackerel (S. japonicus). There appears to be general avoidance of areas and/or periods of time with intense offshore flow conditions (Parrish et al., 1981; Bakun and Parrish, 1982). Examples of spawning of sardines and anchovies under conditions of strong turbulence coincided with times when surface Ekman transport was onshore (Parrish et al., 1983).

These observations indicate the importance of the timing of upwelling events. Turbulence may cause larval starvation on a time scale of a few days, and when the water is mixed it does not become unmixed by a reversal of wind direction (Parrish et al., 1983). Transport mechanisms act on a longer time scale and may indirectly cause mortality of late larvae and juvenile fishes by displacing them from favorable coastal environments. Under normal conditions, a period of offshore transport could be counteracted by a later period of onshore transport (Bakun and Parrish, 1982; Parrish et al., 1983). An increase in alongshore wind strength associated with global warming would, however, lead to general increases in offshore transport and a reduction in periods of relaxation, when surface Ekman transport is onshore, returning larvae to favorable coastal habitats (Huyer et al., 1979). This would reduce the suitability of most areas and times within the California Current system for spawning of many pelagic fish stocks.

Species of fish with reproductive strategies that reduce the effects of offshore transport, such as pleuronectids (sole) and sablefish (Anoplopoma fimbria), which have deep-water spawning, or herring and lingcod, which have demersal spawning (Pearcy et al., 1977; Parrish et al., 1981), may become more important to commercial fisheries in the California Current system.

2. Predator–Prey Interactions. The indirect effects of global warming on predator–prey dynamics in eastern boundary current ecosystems are probably best evaluated by reviewing the progress in understanding how physical dynamics influence trophic interactions in the plankton. The classical models of Lotka (1925) and Volterra (1926) assume that the rate of consumption of prey is directly proportional to both prey and predator abundance. These simple models produce oscillatory behavior of predator and prey populations. The simple assumptions of direct proportionality underlie many of the models of trophic interactions still used today by plankton modelers.

This classical set of feeding assumptions has been modified in significant ways as modelers have recognized that predator–prey contact rates also depend on the spatial distribution of the predators and their prey. Lasker *et al.* (1970) and Lasker (1975, 1978) proposed that fish larvae actually feed in patches of relatively dense phytoplankton, which increases their feeding rate greatly over what would be predicted on the basis of average densities of prey in the water column. Lasker also hypothesised that intense storms introduce sufficient turbulent mixing into the surface mixed layer to disperse the patches of prey and thereby disrupt larval fish feeding, occasionally even enough to cause recruitment failures (Lasker, 1975; Parrish and MacCall, 1978). Because the Bakun scenario for the physical response to global warming predicts much greater coastal wind velocities and enhanced turbulence, one might expect such disaggregation of plankton patches that the rate of feeding of larval fishes would suffer a decline.

This prediction of the destruction of feeding patches by heightened turbulence is the direct opposite of the prediction based on considering how the relative movements of predator and prey influence their rates of feeding. Gerritsen and Strickler (1977) first focused the attention of plankton ecologists on the effect of predator and prey mobility on encounter rates between the two. They modeled a mobile predator feeding on the plankton as a propagating cylinder; a circle defined the field of detection of prey, and the swimming velocity defined the long axis of the cylindrical swath cut through the water by the feeding predator. Again, in this model the rate of feeding is proportional to the density of the prey within the cylinder, but it now incorporates the velocity of swimming by the predator and the radius of prey detection while swimming. Rothschild and Osborn (1988) further evaluated the importance of how prey move relative to their predators by showing that, in theory, enhanced turbulent mixing substantially increases the encounter rates of planktonic predators and their planktonic prey. This model stood up to a test by Sundby and Fossum (1990), who showed that much of the unexplained variance in gut fullness of larval cod off Norway could be explained by the Rothschild–Osborn hypothesis. To our knowledge, no theory of planktonic predator–prey feeding and dynamics has yet been developed to integrate the opposing effects of changing prey patchiness with the effects of turbulence and shear on encounter rates.

A firm prediction of how enhancement of turbulence in coastal ecosystems of the eastern Pacific would affect predator–prey dynamics must await new developments in theory and new tests of the theory. New theory must not only integrate information on the effect of spatial patchiness and of turbulent mixing on feeding rate but also incorporate the effects of handling time and of multiple trophic levels. Ecological models of foraging behavior and optimal foraging have dealt in detail with the consequences of including realistic handling times in the models (e.g., Pyke, 1984; Mangel and Clark, 1988), yet these approaches have not been adapted in the models of plankton predator–prey interactions. The addition of realism will introduce non-linearities into the relationships (Levin, 1990); for example, encounter with three prey may not result in three consumptions if those encounters are not spread at intervals that allow completion of handling (Rothschild and Osborn, 1990).

Perhaps even more critical is the need to model predator–prey dynamics in the context of the multiple trophic levels of importance in the planktonic ecosystem. For example, the enhanced turbulent mixing in the surface layers of eastern boundary ecosystems might disrupt copepod populations by altering both bottom-up and top-down interactions. Intense turbulence could reduce the efficiency of copepod feeding on phytoplankton foods by disrupting the feeding currents that the zooplankton use to ingest prey (e.g., Strickler, 1985). At the same time, the turbulence could increase the rates of encounter between copepods and their own predators, further affecting copepod population abundance and production. This scenario assumes that the effects of enhanced turbulence on the rates of encounter between copepods and larval fish overwhelm the disruption of the zooplankton patches. In addition, it assumes that turbulence has negative effects on copepod feeding because of their small size but positive effects on larval fish feeding because they are large and mobile enough not to suffer feeding inhibition from reasonable levels of turbulence. Such assumptions require formal testing. Our justification for proposing such a hypothetical set of interactions is that it illustrates how new models must incorporate several trophic levels, in addition to physical dynamics, to allow confident prediction of the impact of enhanced turbulence in eastern boundary current ecosystems.

3. Alternative Food Chains for Eastern Boundary Current Ecosystems. Eco-system models not only must be elaborated to include multiple trophic levels but also must address the alternative pathways by which energy can flow through the food webs of the eastern boundary current system. Many of the basic components of the Bakun scenario lead to the prediction that recruitment of anchovies and sardines will suffer greatly from increased offshore transport induced by global warming, and therefore the energy produced by calanoid copepods and other planktonic herbivores in the system will be available for other pathways of energy flow. In particular, we suggest that euphausiids and their consumers will be the major beneficiaries of the predicted decline of anchovies and sardines. In addition, some increase in energy flow to the benthos and hence to demersal predators may be anticipated.

Copepods are the main converters of primary production of phytoplankton into secondary biomass suitable as food for organisms at higher trophic levels, such as pelagic fishes (Marshall, 1970). A sufficient decrease in consumption by small pelagic fishes such as anchovies and sardines would mean that a major portion of the secondary production in these systems is no longer consumed by these fishes. In creased offshore transport would transport these zooplankters offshore toward the frontal (transitional) zone between the upwelled waters and the warmer, offshore oceanic waters.

Euphausiids are found in abundance at or just beyond the shelf break in all major coastal upwelling regions (Thiriot, 1978). Several species of euphausiids, including *Euphausia pacifica*, *Nematoscelis difficulus*, and *Thysanoessa gregaria*, are abundant in the frontal zones of upwelling areas of the California Current and are often the dominant members of the macrozooplankton found in the southern California eddies (Brinton, 1976). Larval euphausiids feed primarily on phytoplankton, but adults are omnivorous and consume a large proportion of zooplankton, especially copepods (Lasker, 1966; Mauchline and Fisher, 1969). Euphausiids are long-lived, tend to eat whatever food is available and abundant, and can adapt their mode of feeding to utilize a variety of different types and sizes of prey (Mauchline and Fisher, 1969). In a situation in which there is reduced predation on zooplankton by anchovies and sardines, the euphausiids and other predatory holoplankton may become the major consumers of available secondary production. During periods of increased water temperatures in upwelling zones, due to El Niño events, the red pelagic crab, *Pleuronocodes planipes*, has become increasingly abundant in the waters of the California Current (Longhurst, 1967). The crabs are able to graze on the larger species of diatoms that occur in the upwelling zones but are also able to catch and feed on copepods using their chelipeds (Longhurst *et al.*, 1967; Walsh *et al.*, 1977). These changes would mean an increase in the biomass of larger crustacean prey available for fishes able to utilize them.

Pinkas *et al.* (1971) examined the diets of bluefin tuna (*Thunnus thynnus*) and albacore tuna (*Thunnus alalunga*). Euphausiids and the pelagic red crab, *P. planipes*, constituted an important part of the diets of both species, although anchovies were the most abundant food item in their guts. McHugh (1952) also noted the importance of red crabs (10%) and euphausiids (7%) in the diet of albacore tuna off Baja California. Albacore tuna migrate eastward across the Pacific, forming abundant large aggregations 1000–1500 km from the coast of California and Oregon. As the nearshore waters warm, they migrate toward the coastal regions and, by mid-July, are found aggregated in the vicinity of oceanic fronts associated with regions of upwelling (Laurs *et al.*, 1977; Laurs and Lynn, 1977). A major factor in their migration appears to be the abundance and availability of food. They spend little time in waters with temperatures less than 15°C. The aggregation of albacore in clear waters, on the oceanic side of fronts, may reflect their inability to catch their large, mobile prey in the turbid inshore waters. In the offshore waters, which are clearer, they are able to utilize the relatively large concentrations of food organisms present, such as euphausiids (Laurs *et al.*, 1984). Jack mackerel (*T. symmetricus*) are also common in these offshore waters, where their euphausiid prey is most abundant (Barber and Smith, 1981; Barber and Chávez, 1983). The occurrence of euphausiids and other large crustaceans, such as the pelagic red crab, in the diet of tuna varied from region to region and seasonally, probably as a function of the relative availability of other prey (Pinkas *et al.*, 1971). A decrease in the availability of anchovy prey, coupled with a likely increase in biomass of euphausiids and red crabs in the plankton in frontal zones, would suit the species of fish that could utilize these large crustacean prey.

The shift from a relatively short food chain of phytoplankton → anchovies/sardines, or phytoplankton → small copepods → anchovies/sardines, to a longer one characterized by phytoplankton → copepods → euphausiids and red crabs → large migratory fishes (albacore tuna, bluefin tuna, jack mackerel) would mean a decrease in the efficiency of energy (carbon) transfer from primary production to sources suitable for human consumption. The occurrence of these migratory fishes on the oceanic side of upwelling frontal zones also means that local commercial fisheries would have an additional cost associated with increased travel to these zones. Mackerel and tuna are, however, more valuable on a per kilogram basis than anchovies and sardines, so it is difficult to predict the social and economic consequences of these changes in human utilization of the system.

An alternative pathway for the secondary production not utilized by the declining anchovy and sardine fisheries may involve a major ecosystem shift to benthic consumption and demersal fish utilization of this production. Copepods produce large fecal pellets surrounded by a chitinous lining that makes them resistant to decomposition in the water column, and these fecal pellets may be a major contributor to the flux of particulate organic carbon sinking to the ocean depths (Eppley and Peterson, 1979; Pomeroy, 1979). Eppley and Peterson (1979) estimated that 20–30% of the organic carbon from phytoplankton ingested by copepods could be passed in the animals' feces, which are transported to the benthos. An increase in the production and biomass of euphausiids would also contribute to transfer of organic material to the benthos. Euphausiids molt regularly and continue to grow throughout their life (Mauchline and Fisher, 1969) and each cast skin represents a sizable loss (~10%) of the organic material of the animal (Lasker, 1966). These molts sink to the bottom, providing carbon to the benthos. Suess (1980) suggested that organic detritus, such as fecal pellets and crustacean molts, passing from the sea surface through the water column, could control nutrient regeneration and fuel the benthos. An increase in the particulate organic carbon supplied to the benthos would be processed through an extensive food web composed of bacteria, meiofauna, and macrofauna (Pomeroy, 1979).

If the biotic outfall to the deep-sea benthos increases and if such bathyal populations are food limited, as suggested by Sanders and Hessler (1969), then an increase in available organic matter could increase the carrying capacity of the benthos. This may benefit demersal fishes such as the pleuronectids (sole) and sablefish (A. fimbria), which are important commercial fish stocks in the California Current region (Parrish et al., 1981). Sablefish can respond rapidly to the presence of increased available food on the sea floor in the form of nekton, as can hagfish (Eptatretus dearii) (Smith, 1985, 1986).

Another possible sink for copepod secondary production is aggregations of salps. Using an efficient filtering system, they concentrate phytoplankton and zooplankton into large fecal pellets, which sink to the bottom, where they can be utilized by the benthos. When the animals die their bodies also sink to the bottom, providing a sizable introduction of particulate organic carbon and nitrogen to the deep sea (Wiebe et al., 1979). Harbison and Gilmer (1976) showed in laboratory studies that salps can have a grazing impact equal to or greater than that of copepods. They also noted that most plankton tows used to sample copepods and other zooplankton could not accurately sample salps, hampering efforts to estimate their relative abundance in coastal waters or their potential importance as grazers on phytoplankton. Increases in ctenophore predators have also been recorded after increases in copepod populations (Hirota, 1974). The effects of aggregations of ctenophores, siphonophorans, and appendicularians on the abundance of zooplankton would depend on how these macrozooplankton respond to increasing turbulence in the upper water column. The efficiency of their filtering mechanism may be reduced by the increasing turbulence (as discussed earlier for copepods), although feeding occurs simultaneously with swimming in salps (Madin, 1974), so they may be less likely to suffer these effects.

IV. Summary

The coastal ecosystems associated with eastern boundary currents in the Pacific Ocean are almost ideal systems for which to predict the effects of global warming. High-frequency variations in basinwide, mesoscale, and local physics and their well-studied biological consequences during El Niño–Southern Oscillation events make it possible to glimpse the future state of the system under conditions of global warming. The rapid response of essentially entire water column biotas to the short-term sea surface changes that accompany a strong El Niño event demonstrate that these marine biotas will probably exhibit relatively concordant shifts in biogeography with changing temperature. The barriers to dispersal that may prevent some terrestrial plants and lacustrine species from dispersing at rates to match the geographic march of temperatures are not likely to affect a marine water column biota for which planktonic dispersal at some life stage is virtually universal. Increasing concentrations of atmospheric carbon dioxide are not predicted to enhance primary production over the near term in the eastern boundary current ecosystems. In fact, the effects of global warming are likely to reduce the cross-basin gradient in atmospheric pressure in the Pacific Ocean and to depress the nutricline in the eastern Pacific sufficiently that upwelled waters will no longer be drawn from the nutrient-rich depths. This is expected to reduce primary, and then also secondary, production in these coastal ecosystems.

Upwelling-favorable coastal winds will probably be enhanced by the increased temperature differential between the now even warmer land and the sea surface. This Bakun effect would dramatically reduce the reproductive success of sardine and anchovies, the dominant consumers of zooplankton in these systems and the basis of extensive and economically important fisheries, because of enhanced offshore transport of eggs, larvae, and juveniles away from suitable coastal habitats. The consequent reduction in exploitation of planktonic foods by these fishes that now dominate the consumer trophic levels would lead to large changes in the ecosystem, as energy is shunted into a euphausiid, red crab–bonito, and mackerel food chain and perhaps also transferred in greater amounts to the benthos and to demersal fishes. Such a major shift in energy flow pathways would result in human harvest of fishes at a higher trophic level, with consequent reductions in biomass available for exploitation as well as alteration in fishing geography, gear, methods, and allocation of catches.

The increase in surface wind stress predicted by Bakun (1990) would also increase the turbulence in the upper mixed layer of the water column. Contradictory implications of enhanced turbulence are presented in the literature. Increased turbulence may dissipate patches of planktonic foods, thus reducing the feeding rate of predatory fishes (the Lasker hypothesis). Alternatively, increased turbulence may raise the rates of encounter between predator and prey and thus increase feeding rates (the Rothschild–Osborn hypothesis). This contradictory set of predictions must be resolved by developing applicable models of how ocean physics influence ecosystem dynamics, incorporating information on spatial distributions, mobility, prey handling times, and multiple trophic levels. It seems likely that substantial increases in both turbulence and offshore transport will produce dramatic ecosystem changes in eastern boundary current systems of both the Northern and Southern Hemispheres, with tremendous social consequences because of the disruption of highly productive fisheries. The changes in the upwelling-favorable coastal winds that drive these predicted shifts in coastal ecosystems of the eastern Pacific may be expected to be more intense in the Northern Hemisphere, where the central valley of California is so extensive. Heating in this central valley will permit development of a more intense thermal low-pressure cell in California than can be set up in the mountainous coastal land border off Chile.

Acknowledgments

We would like to acknowledge the support of the National Science Foundation Biological Oceanography Program (grants OCE-862010 to C.H.P. and OCE-8901929 to R.T.B.).

References

Alamo, A., and Bouchon, M. (1987). Changes in the food and feeding of the sardine (*Sardinops sagax sagax*) during the years 1980–1984 off the Peruvian coast. *J. Geophys. Res.* **92**, 14411–14415.

Avària, S., and Muñoz, P. (1987). Effects of the 1982–1983 El Niño on the marine phytoplankton off northern Chile. *J. Geophys. Res.* **92**, 14369–14382.

Bakun, A. 1990. Global climate change and intensification of coastal ocean upwelling. *Science*, Vol. 247, pp. 198–201.

Bakun, A., and Parrish, R. H. (1982). Turbulence, transport and pelagic fish in the California and Peru current systems. *Calif. Coop. Oceanic Fish Invest. Rep.* **23**, 99–112.

Barber, R. T. (1988). Open basin ecosystems. *In* "Concepts of Ecosystem Ecology. A Comparative View" (L. R. Pomeroy and j. L. Alberts, eds.), pp. 171–193. Ecological Studies 67. Springer-Verlag, New York.

Barber, R. T., and Chávez, F. R. (1983). Biological consequences of El Niño. *Science* **222**, 1203–1210.

Barber, R. T., and Chávez, F. R. (1986). Ocean variability in relation to living resources during the 1982–83 El Niño. *Nature* **319**, 279–285.

Barber, R. T., and Smith, R. L. (1981). Coastal upwelling ecosystems. *In* "Analysis of Marine Ecosystems" (A. R. Longhurst, ed.), pp. 31–68. Academic Press, London.

Barber, R. T., Chávez, F. R., and Kogelschatz, J. E. (1985). Biological effects of El Niño. *In* "Ciencia, Tecnologiay Agresion Ambiental: El Fenómeno El Niño," pp. 399–425. Consejo Nacional de Ciencia y Technologia, CONCYTEC, Lima, Peru.

Brinton, E. (1976). Population biology of *Euphausia pacifica* off southern California. *Fishery Bull. U.S.* **74**, 733–762.

Brodeur, R. D. (1986). Northward displacement of the euphausiid *Nyctiphanes simplex* Hansen to Oregon and Washington waters following the El Niño event of 1982–83. *J. Crust. Biol.* **6**, 686–692.

Brodeur, R. D., Gadomski, D. M., Pearcy, W. G., Batchelder, H. P., and Miller, C. B. (1985). Abundance and distribution of ichthyoplankton in the upwelling zone off Oregon during anomalous El Niño conditions. *Estuarine Coastal Shelf Sci.* **21**, 365–378.

Brown, J. H. (1975). Geographical ecology of desert rodents. In "Ecology and Evolution of Communities" (M. L. Cody and J. H. Diamond, eds.), pp. 315–341. Harvard University Press, Cambridge, Massachusetts.

Buchanan, J. (1886). On similarities in the physical geography of the great oceans. *Pro. R. Geog. Soc.* **8**, 753–770.

Cane, M. A. (1983). Oceanographic events during El Niño. *Science* **222**, 1189–1195.

Carrasco, S., and Santander, H. (1987). The El Niño event and its influence on the zooplankton off Peru. *J. Geophys. Res.* **92**, 14405–14410.

Chávez, F. P., Barber, R. T., and Sanderson, M. P. (1989). The potential primary production of the Peruvian upwelling ecosystem: 1953–1984. *In* "The Peruvian Upwelling Ecosystem: Dynamics and Interactions" (D. Pauly, P. Muck, J. Mendo, and I. Tsukayama eds.), pp. 50–63. ICLARM Conference Proceedings 18.

Chelton, D. B., Bernal, P. A., and McGowan, J. A. (1982). Large-scale interannual physical and biological interaction in the California Current. *J. Mar. Res.* **40**, 1095–1125.

Coker, R. E. (1908). The fisheries and guano industry of Peru. *Bull. Bur. Fish.* **28**, 333–365.

Dayton, P. K., and Tegner, M. J. (1989). Bottoms beneath troubled waters: benthic impacts of the 1982–1984 El Niño in the temperate zone. *In* "Global Ecological Consequences of the 1982–1983 El Niño–Southern Oscillation" (P. Glynn, ed.), pp. 433–472. Elsevier, Amsterdam.

De Szoeke, R. A., and Richman, J. G. (1981). The role of wind-generated mixing in coastal upwelling. *J. Phys. Oceanogr.* **11**, 1534–1547.

Enfield, D. B. (1989). El Niño, past and present. *Rev. Geophys.* **27**, 159–187.

Eppley, R. W., and Peterson, B. J. (1979). Particulate organic matter flux and planktonic new production in the deep ocean. *Nature* **282**, 677–680.

Gerritsen, J., and Strickler, J. R. (1977). Encounter probabilities and community structure in zooplankton: a mathematical model. *J. Fish. Res. Bd. Can.* **34**, 73–82.

Glantz, M. H. (1990). Does history have a future? Forecasting climate change effects on fisheries by analogy. *Fisheries* **15**, 39–44.

Harbison, G. R., and Gilmer, R. W. (1976). The feeding rates of the pelagic tunicate *Pegea confederata* and two other salps. *Limnol. Oceanogr.* **21**, 517–528.

Hickey, B. M. (1979). The California Current system—hypotheses and facts. *Prog. Oceanogr.* **8**, 191–279.

Hirota, J. (1974). Quantitative natural history of *Pleurobrachia bachei* in La Jolla Bight. *Fish. Bull.* **72**, 295–352.

Husby, D. M., and Nelson, C. S. (1982). Turbulence and vertical stability in the California Current. *Calif. Coop. Oceanic Fish. Invest. Rep.* **23**, 113–129.

Huyer, A. (1983). Coastal upwelling in the California Current system. *Prog. Oceanogr.* **12**, 259–284.

Huyer, A., Sobey, E. J. C., and Smith, R. L. (1979). The spring transition in currents over the Oregon continental shelf. *J. Geophys. Res.* **84**, 6995–7011.

Koblentz-Mishke, O. I., V. V., Volkovinsky, and J. G. Kabanova (1970). Plankton primary production of the world ocean. *In* "Scientific Exploration of the South Pacific" (W. Wooster, ed.), pp. 183–193. National Academy of Sciences, Washington, D.C.

Lasker, R. (1966). Feeding, growth, respiration, and carbon utilisation of euphausiid crustaceans. *J. Fish. Res. Bd. Can.* **23**, 1291–1317.

Lasker, R. (1975). Field criteria for survival of anchovy larvae: the relation between inshore chlorophyll maximum layers and successful first feeding. *Fishery Bull.* **73**, 453–462.

Lasker, R. (1978). The relation between oceanographic conditions and larval anchovy food in the California Current: Identification of factors leading to recruitment failure. *Rapp. P.-V. Reun. Cons. Int. Explor. Mer.* **173**, 212–277.

Lasker, R., Feder, H. M., Theilacker, G. H., and May, R. C. (1970). Feeding, growth, and survival of *Engraulis mordax* larvae reared in the laboratory. *Mar. Biol.* **5**, 345–353.

Lau, K.-M., and Chan, P. H. (1985). Aspects of the 40–50 day oscillation during the northern winter as inferred from outgoing longwave radiation. *Mon. Weather Rev.* **113**, 1889–1909.

Laurs, R. M., and Lynn, R. J. (1977). Seasonal migration of north Pacific albacore, *Thunnus alalunga*, into North American coastal waters: distribution, relative abundance, and association with transition zone waters. *Fishery Bull. U.S.* **75**, 795–822.

Laurs, R. M., H. S. H., Yuen, and J. H. Johnson (1977). Small-scale movements of albacore, *Thunnus alalunga*, in relation to ocean features as indicated by ultrasonic tracking and oceanographic sampling. *Fishery Bull. U.S.* **75**, 347–355.

Laurs, R. M., P. C. Fielder, and D. R. Montgomery (1984). Albacore tuna catch distributions relative to environmental features observed from satellites. *Deep-Sea Res.* **31**, 1085–1099.

Levin, S. A. (1990). Physical and biological scales and the modeling of predator–prey interactions in large marine ecosystems. *In* "Large Marine Ecosystems: Patterns, Processes, and Yields" (K. Sherman, ed.), pp. 179–187. American Association for the Advancement of Science, Washington, D.C.

Longhurst, A. R. (1967). The pelagic phase of *Pleuroncodes planipes* Stimpson (Crustacea, Coalatheidae) in the California Current. *Calif. Coop. Oceanic Fish. Invest. Rep.* **11**, 142–154.

Longhurst, A. R., Lorenzen, C. J., and Thomas, W. H. (1967). The role of pelagic crabs in the grazing of phytoplankton off Baja California. *Ecology* **48**, 190–200.

Lotka, A. J. 1925. "Elements of Physical Biology." Dover Publications, New York.

Madin, L. P. (1974). Field observations on the foraging behaviour of salps (Tunicata: Thalicea). *Mar. Biol.* **25**, 143–147.

Mangel, M., and Clark, C. W. (1988). "Dynamic Modeling in Behavioral Ecology." Princeton University Press, Princeton, New Jersey.

Marshall, N. (1970). Food transfer through the lower trophic levels of the benthic environment. *In* "Marine Food Chains" (J. H. Steele, ed.), pp. 52–66. University of California Press, Berkeley.

Marshall, S. M. (1973). Respiration and feeding in copepods. *Adv. Mar. Biol.* **11**, 57–120.

Mauchline, J., and Fisher, L. R. (1969). The biology of euphausiids. *Adv. Mar. Biol.* **7**, 1–454.

McGowan, J. A. (1985). El Niño 1983 on the southern California bight. *In* "El Niño North: El Niño Effects in the Eastern Subarctic Pacific Ocean" (W. S. Wooster and D. L. Fluharty, eds.), pp. 185–187. University of Washington Press, Seattle.

McHugh, J. L. (1952). The food of albacore (*Germo alalunga*) off California and Baja California. *Bull. Scripps Inst. Oceanogr. Univ. Calif.* **6**, 161–172.

Newell, R. E., and Gould, S. S. (1981). A Stratospheric Fountain? *J. Atmos. Sci.* **38**, 2789–2796.

Ochoa, N., and Gómez, O. (1987). Dinoflagellates as indicators of water masses during El Niño, 1982–1983. *J. Geophys. Res.* **92**, 14355–14367.

Parrish, R. H., and MacCall, A. D. (1978). Climatic variation and exploitation in the Pacific mackerel fishery. *Fishery Bull.* **167**, 1–110.

Parrish, R. H., Nelson, C. S., and Bakun, A. (1981). Transport mechanisms and reproductive success of fishes in the California Current. *Biol. Oceanogr.* **1**, 175–203.

Parrish, R. H., Bakun, A., Husby, D. M., and Nelson, C. S. (1983). Comparative climatology of selected environmental processes in relation to eastern boundary current pelagic fish production. Proceedings of a Joint FAO–IOC "Expert Consultation to Examine Changes in Abundance and Species Composition of Neritic Fish Stocks." San José, Costa Rica, April 18–29. FAO, Rome, pp. 1–45.

Pearcy, W. G., and Schoener, A. (1987). Changes in the marine biota coincident with the 1982–1983 El Niño in the northeastern subarctic Pacific Ocean. *J. Geophys. Res.* **92**, 14417–14428.

Pearcy, W. G., Hosie, M. J., and Richardson, S. L. (1977). Distribution and duration of pelagic life of larvae of Dover sole, *Microstomus pacificus*; rex sole, *Glyptocephalus zachirus*; and petrale sole, *Eopsetta jordani*, in waters off Oregon. *Fish. Bull. U.S.* **75**, 173–184.

Philander, G. S. (1989). "El Niño, La Niña and the Southern Oscillation." Academic Press, San Diego.

Pinkas, L., Oliphant, M. S., and Iverson, I. L. K. (1971). Food habits of albacore, bluefin tuna, and bonito in California waters. *Calif. Dept. Fish Game Fish. Bull.* **152**, 1–105.

Pomeroy, L. R. (1979). Secondary production mechanisms of continental shelf communities. *In* "Ecological Processes in Coastal and Marine Ecosystems" (R. L. Livingston, ed.), pp. 437–465. Plenum Publishing, New York.

Pyke, G. H. (1984). Optimal foraging theory: a critical review. *Annu. Rev. Ecol. Syst.* **15**, 523–575.

Ramage, C. S. (1969). Role of a tropical "maritime" continent in the atmospheric circulation. *Mon. Weather Rev.* **96**, 365–370.

Rasmusson, E. M., and Wallace, J. M. (1983). Meteorological aspects of the El Niño/Southern Oscillation. *Science* **222**, 1195–1202.

Redfield, A. C. (1958). The biological control of chemical factors in the environment. *Am. Sci.* **46**, 205–221.

Rothschild, B. J. (1986). "Dynamics of Marine Fish Populations." Harvard University Press, Cambridge, Massachusetts.

Rothschild, B. J., and Osborn, T. R. (1988). Small-scale turbulence and plankton contact rates. *J. Plankton Res.* **10**, 465–474.

Rothschild, B. J., and Osborn, T. R. (1990). Biodynamics of the sea: preliminary observations on high dimensionality and the effect of physics on predator–prey interactions. *In* "Large Marine Ecosystems: Patterns, Processes, and Yields" (K. Sherman, ed.), pp. 71–81. American Association for Advancement of Science, Washington, D.C.

Sanders, H. L., and Hessler, R. R. (1969). Ecology of the deep sea benthos. *Science* **163**, 1419–1424.

Scheltema, R. S. (1986). On dispersal and planktonic larvae of benthic invertebrates: an eclectic overview and summary of problems. *Bull. Mar. Sci.* **39**, 290–322.

Scheltema, R. S. (1988). Initial evidence for the transport of teleplanic larvae of benthic invertebrates across the east Pacific barrier. *Biol. Bull.* **174**, 145–152.

Smith, C. R. (1985). Food for the deep sea: utilization, dispersal and flux of nekton falls at the Santa Catalina basin floor. *Deep-Sea Res.* **32**, 417–442.

Smith, C. R. (1986). Nekton falls, low intensity disturbance and community structure of infaunal benthos in the deep sea. *J. Mar. Res.* **44**, 567–600.

Smith, R. L. (1983). Peru coastal currents during El Niño: 1976 and 1982. *Science* **221**, 1397–1399.

Strickler, J. R. (1985). Feeding currents in calanoid copepods: two new hypotheses. *In* "Physiological Adaptations of Marine Animals" (Symposia of the Society for Experimental Biology 23) (M. S. Lavarack, ed.), pp. 459–485. Pinder Group of Companies, Scarborough, UK.

Suess, E. (1980). Particulate organic carbon flux in the oceans—surface productivity and oxygen utilization. *Nature* **288**, 260–263.

Sundby, S., and Fossum, P. (1990). Feeding conditions of Arcto-Norwegian cod larvae compared with the Rothschild–Osborn theory on small-scale turbulence and plankton contact rates. *J. Plankton Res.* **12**, 1153–1162.

Sverdrup, H. U. (1955). The place of physical oceanography in oceanographic research. *J. Mar. Res.* **14**, 287–294.

Thiriot A. (1978). Zooplankton communities in the West African upwelling area. *In* "Upwelling Ecosystems" (R. Boje and M. Tomczak, eds.), pp. 32–61. Springer-Verlag, New York.

Trenberth, K. E. (1990). General characteristics of El Niño–Southern Oscillation. *In* "ENSO Teleconnections Linking Worldwide Climate Anomalies: Scientific Basis and Societal Impact" (M. Glantsy, R. Katz, and N. Nicholls, eds.), pp. 13–42. Cambridge University Press, Cambridge, England.

Volterra, V. (1926). Fluctuations in the abundance of a species considered mathematically. *Nature* **118**, 558–560.

Walsh, J. J., Whitledge, T. E., Kelley, J. C., Huntsman, S. A., and R. D. Pillsbury (1977). Further transition states of the Baja California upwelling ecosystem. *Limnol. Oceanogr.* **22**, 264–280.

Weickmann, K. M. (1983). Intraseasonal circulation and outgoing long wave radiation modes during northern hemisphere winter. *Mon. Weather Rev.* **111**, 1838.

Wiebe, P. H., Madin, L., Haury, L., Harbison, G. R., and Philbin, L. (1979). Diel vertical migration by *Salpa aspera* and its potential for large-scale particulate organic matter transport to the deep sea. *Mar. Biol.* **53**, 249–255.

Wyrtki, K. E. (1975). El Niño—the dynamic response of the equatorial Pacific Ocean to atmospheric forcing. *J. Phys. Oceanogr.* **5**, 572–584.

Part II: Climate Controls

C H A P T E R   3

Northern Hemisphere Climate Change: Physical Processes and Observed Changes

I. Introduction

Climate changes affect all the components of the climate system in varying degrees, but are manifest through the atmospheric changes. The atmospheric circulation forms the main link between regional changes in wind, temperatures, precipitation, and other climatic variables, and there is also a reasonably strong relationship between some of these even on monthly or longer time scales. These relationships are exploited in preparing long-range (monthly and seasonal) forecasts (Gilman, 1983; Epstein, 1988). In studies of climate variations, internal physical consistency between variables can add confidence to results of analyses of observed changes and thereby help indicate possible causes.

Many studies illustrate the links. Often, relationships are related to particular phenomena, such as the Southern Oscillation (Walker and Bliss, 1932; van Loon and Madden, 1981; Ropelewski and Halpert, 1987, 1989) or the North Atlantic Oscillation (Walker and Bliss, 1932; van Loon and Rogers, 1978; Rogers and van Loon, 1979; Meehl and van Loon, 1979), or to zonal and meridional indices, perhaps as part of an index cycle (Namias, 1950; Lamb and Johnson, 1966; Trenberth, 1976; Salinger, 1980a,b; Makrogiannis *et al.*, 1982).

We now consider the physical reasons why the observed relationships should

EARTH SYSTEM RESPONSES TO GLOBAL CHANGE
Contrasts between North and South America

Copyright © 1993 by Academic Press, Inc.
All rights of reproduction in any form reserved.

exist and go on to examine the observed large-scale multidecadal changes in climate, especially surface temperatures and their links to the circulation, with a special emphasis on insights into observed patterns of change and possible causes. Prospects for the future, with increased greenhouse gases, are also discussed.

II. Physical Reasons for Relationships

Local climates at the same latitude vary considerably around the globe because of the distribution of land and sea, the topography over land, and the associated planetary waves in the atmosphere. Variations in local temperatures on decadal time scales or from year to year are far from uniform but also occur in distinctive large-scale patterns (e.g., Lysgaard, 1949; Petterssen, 1949; Mitchell, 1963; van Loon and Williams, 1976a, 1976b, 1977; Williams and van Loon, 1976; Jones, 1988; Jones *et al.*, 1988; Folland and Parker, 1989); see Fig. 1. The large-scale coherence in these climate variations arises because they are associated with changes in the quasi-stationary planetary waves and other factors, as discussed in the following.

A. Heat Capacity

Generally, the variability of the temperatures over land is a factor of 2 to 6 greater than that over the oceans (Barnett, 1978). In part this can come about because of the formation of strong surface temperature inversions over land in winter, whereas over the oceans surface fluxes of heat into the atmosphere are a major limiting factor. This is really a reflection of the very different heat capacities of the underlying surface and the depth of the layer linked to the surface. The specific heat of dry land is roughly a factor of 4.5 less than that of seawater (for moist land the factor is probably closer to 2). Moreover, heat penetration into land is limited and only the top 2 m or so typically play an active role (as an *e*-folding depth for variations on annual time scales, say). In contrast, convection and wind-induced mechanical mixing within the ocean result in an active mixed layer typically 50 m or so in depth but ranging from less than that during the heating season (spring and early summer) to over 100 m in the fall and winter, when surface cooling helps trigger convection (e.g., Meehl, 1984). These values also vary considerably geographically. Over land, month-to-month persistence in surface temperature anomalies is greatest near bodies of water (van den Dool *et al.*, 1986). Consequently, it is clear that for a given heating anomaly, the response over land should be much greater than over the oceans; the atmospheric winds are the reason why the observed factor is only in the range of 2 to 6.

A further example of this is the contrast between the Northern Hemisphere (NH) (60.7% water) and Southern Hemisphere (SH) (80.9% water) mean annual cycle of surface temperature (Fig. 2). The amplitude of the first harmonic between 40 and 60° latitude ranges from less than 3°C in the SH to about 12°C in the NH. Similarly, from 22.5 to 67.5° latitude, the average lag in temperature response relative to the sun for the annual cycle is 32.9 days in the NH versus 43.5 days in the SH (Trenberth, 1983), again reflecting the difference in thermal inertia. Even the sea surface temperatures (SSTs) in the two hemispheres undergo quite different amplitudes in the annual cycle, and SSTs are always warmer in the NH than in the SH at each latitude (Shea *et al.*, 1990) (Fig. 3).

At high latitudes over land in winter, there is usually a strong surface temperature inversion whose strength is very sensitive to the amount of stirring in the atmosphere. Such wintertime inversions are also greatly affected by human activities; for instance, an urban heat island effect exceeding 10°C has been observed during strong surface inversion conditions in Fairbanks, Alaska (Weller, 1982). Increases in wind occur with increases in cyclonic activity, usually weakening the inversions and producing much warmer surface temperatures. Cyclonic activity is often accompanied by increases in low-level cloudiness, which also typically acts to warm the high latitudes in winter. Therefore, a modest increase in storminess and wind may produce a large warming in these regions (and conversely, of course, during a decrease in wind) (e.g., Hesselberg and Johannessen, 1958).

An important corollary of the preceding discussion is that changes in surface temperature may bear little relation to changes in heat content, especially in high latitudes over land in winter, where changes may be representative of only a very shallow layer.

It also follows that the oceans will act to delay the onset of any climate change. The heat capacity of the atmosphere corresponds to that of only 3.2 m of the ocean. The seasonal variations in heating penetrate below the mixed layer and on average involve about 90 m of ocean (Meehl, 1984). Using simple models, we can estimate the *e*-folding time for the temperature response to approach equilibrium due to an abrupt change in forcing of the climate (e.g., Ramanathan *et al.*, 1987). The thermal inertia of the 90-m layer would add a delay of about 6 years to the temperature response to a change. The total ocean, with a mean depth of 3800 m, if rapidly mixed would add a delay of 230 years to the response. In reality, the response depends on the rate of ventilation of water through the thermocline. Such mixing is quite uncertain, and an overall estimate of the delay caused by the oceans is 10 to 100 years. The slowest response should be in high latitudes, where deep mixing and convection occur, and the fastest response is expected in the tropics.

B. Planetary Waves

To understand the observed changes it is necessary to understand the changes in planetary waves and teleconnections within the atmosphere. In the NH in winter, the time-mean planetary waves have a strong baroclinic structure and slope westward with height, propagate into the stratosphere, and transport heat and momentum poleward (e.g., Chen and Trenberth, 1988). In particular, from the standpoint of surface climate, changes in the poleward heat transport have a pronounced effect on mean temperatures through changes in local temperature advection (van Loon and Williams, 1977). In other words, anomalous northerly winds often give rise to colder than normal conditions (in the NH), whereas anomalous southerlies are typically warmer than normal. Alternatively, in the NH in winter, winds from the ocean are usually milder while winds from the interior of the high-latitude continents may be considerably colder than normal. In addition, changes in the planetary waves may be

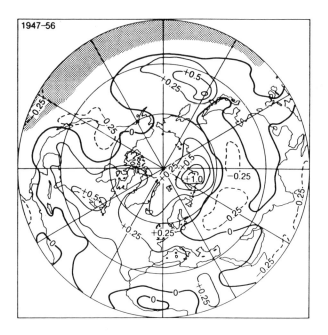

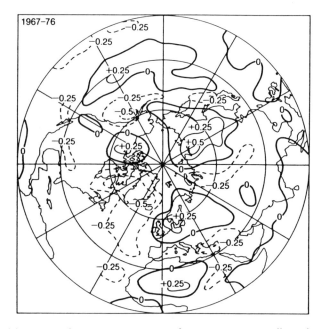

Fig. 1. Decadal average surface temperature or sea surface temperature anomalies as departures from the 1951–1980 mean, for 1947–1956, 1957–1966, 1967–1976, and 1977–1986. Contours every 0.25°C. (From Folland and Parker, 1989; reprinted by permission of Kluwer Academic Publishers.) (*Figure continues.*)

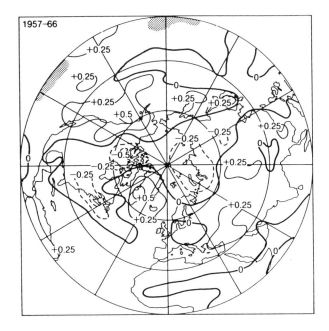

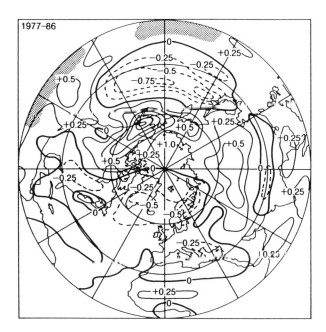

Fig. 1 (*Continued*)

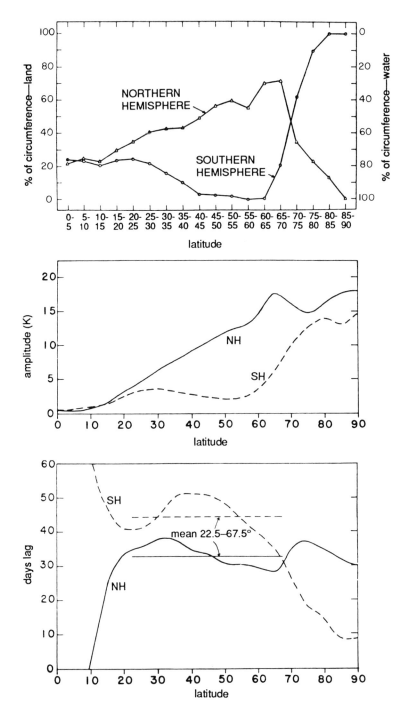

Fig. 2. Zonal averages for each hemisphere of percentage of land around each latitude circle (top); amplitude of the 12-month annual cycle harmonic of surface temperature, °C (center); and phase of the 12-month harmonic of surface temperature in terms of lag behind the sun (bottom). (From Trenberth, 1983; reprinted by permission of the American Meteorological Society.)

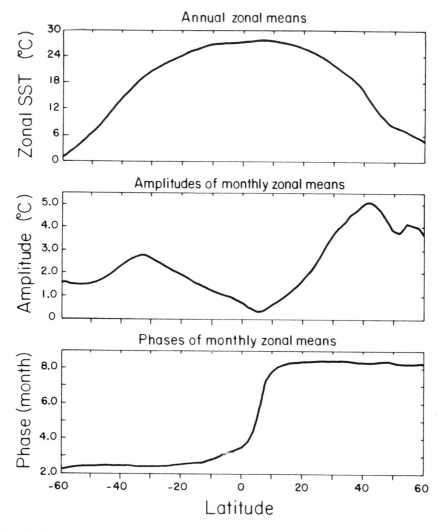

Fig. 3. Zonal average mean sea surface temperatures over the global oceans and zonal average of the amplitude of the annual cycle 12-month harmonic (K). (From Shea *et al.*, 1990.)

associated with periods of persistent blocking in both hemispheres (Namias, 1963; Austin, 1980; Trenberth and Mo, 1985; Trenberth, 1986; Folland, 1983).

In the NH in summer, temperature gradients are much weaker, so changes in winds do not produce much change in advection (van Loon and Williams, 1976b) and local effects become more important. The impact of local solar heating depends greatly on the availability of surface moisture through the partitioning of heat into sensible and latent components. In dry conditions, such as during droughts, more heat goes into raising surface air temperatures, thereby contributing to heat waves. Trenberth *et al.* (1988) suggest that this was a significant factor in maintaining the 1988 North American summer drought. It is therefore not surprising to find a highly significant negative correlation between temperature and precipitation totals in most of the United States and Europe in summer (Madden and Williams, 1978). In contrast, in winter the correlations between precipitation and temperature are lower and more patchy, with positive correlations over most of Europe and eastern parts of the United States but with negative correlations over the northern Great Plains of the United States. Positive correlations arise from the dynamical association between warm advection, increased moisture influx, and vertical motion ahead of a trough.

C. Links between the Circulation and Temperatures and Precipitation

It has been implied in the foregoing discussion that temperatures are strongly related to advection in the NH in winter. Strongest advection occurs in the lower troposphere from 850 to 700 mb and is related to that at the surface. Van Loon and Williams (1977) compare 700-mb and surface temperature changes and show, however, that differences exist in trends. Van Loon (1979) and van Loon and Williams (1980) have examined more systematically the relation at 700 mb between heat transports by the stationary planetary waves and transient waves and relationships with temperatures and temperature gradients. Of particular note is the remarkable negative correlation of less than −0.9 for 29 winters between the zonally averaged total poleward heat flux at 45°N and zonal mean temperatures at 40°N. This relationship comes about because of the stationary waves; strong negative correlations extend from 30 to 65°N, showing that *lower temperatures are accompanied by stronger poleward heat fluxes in the stationary waves.* Van Loon (1979) further shows that the transient heat flux varies inversely with the stationary flux and, whereas the transient flux is positively correlated with temperature gradients, as might be expected from baroclinic instability arguments, the total flux is negatively correlated with temperature gradients, implying that the temperature gradients respond to changes in the planetary waves and not the other way around.

The implications of this work have not been sufficiently appreciated. The stationary planetary waves are forced by orography and patterns of diabatic heating arising from the distribution of land and sea, both in the extratropics (Chen and Trenberth, 1988) and in the tropics (Nigam *et al.*, 1988). Therefore, changes in diabatic heating, for instance, can change the planetary waves and associated poleward heat flux (Chen and Trenberth, 1988), with many repercussions. First, reduced heat transports by the planetary waves favor increased temperatures by reducing the effectiveness of the Arctic cold sink in winter. Note that this mechanism does not operate in summer, nor does it work at any time of year in the SH, where heat transports are dominated by the transient component (van Loon, 1979). Consequently, not only the local changes in NH winter temperature (Fig. 1) but also the zonal means and, by extension, the NH mean are greatly influenced by changes in planetary waves.

Second, alterations of the temperature gradients that the transients feed on through baroclinic instability are accompanied by profound changes in the storm tracks (Trenberth, 1984; Lau, 1988). Storm tracks may be identified by the high-frequency (less than about a week in period) variances in geopotential height in the midtroposphere in both the NH (Blackmon et al., 1977; Lau, 1978, 1979) and SH (Trenberth, 1981, 1982). In this way, the circulation changes may also be systematically linked to changes in precipitation. In addition, the wind accompanying transient disturbances influences the amount of stirring, so the strength of winter surface temperature inversions and probably cloudiness and radiative losses to space would change.

D. Teleconnections

Because of the tendency for atmospheric motions to conserve absolute vorticity, atmospheric motions occur as Rossby waves. Local forcing of such waves sets up a wave train of disturbances and "teleconnections" downstream. In the context of the current topic, teleconnections are important because the anomalies in the circulation and associated temperature and precipitation anomalies may arise from forcing in remote regions. The best-known examples of global impacts of local forcing are changes in SSTs, such as in the El Niño–Southern Oscillation (ENSO) phenomenon, where coupling occurs between the tropical Pacific and higher latitudes via teleconnections. The 1988 North American drought appears to have been one example of this (Trenberth et al., 1988). General circulation modeling experiments have confirmed the causal link between the SST changes and remote atmospheric changes in the winter hemisphere (e.g., Shukla and Wallace, 1983).

E. El Niño–Southern Oscillation Influences

One of the most prominent sources of interannually variability in weather and climate around the world is the ENSO phenomenon. The Southern Oscillation (SO) is a global teleconnection pattern principally consisting of a seesaw (or standing wave) in atmospheric mass involving exchanges of air between the eastern and western hemispheres centered in tropical and subtropical latitudes. The centers of action are located over Indonesia and the tropical South Pacific Ocean. The influence of the SO extends to higher latitudes mostly in wintertime, as wavelike patterns that change the jet stream and storm track locations. El Niño is an anomalous warming of the eastern tropical Pacific Ocean, but in major "Warm Events" the warming extends over much of the tropical Pacific, and in these cases it is clearly linked to changes in the atmosphere, manifested as the SO. The "Cold Events" of the opposite phase are now sometimes referred to as "La Niña." The El Niño events occur every 2 to 10 years and have far reaching climatic and economic effects around the world (Fig. 4).

The mechanisms involved are broadly known (Trenberth, 1990b). ENSO results from an interaction between the changes in the tropical Pacific Ocean and the global atmosphere. Surface atmospheric winds drive tropical ocean currents and change the SSTs, which in turn alter the location and strength of atmospheric convection and precipitation in the tropics and so change the atmospheric heating patterns, atmo-

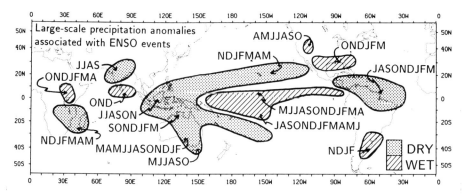

Fig. 4. Schematic of areas influenced by ENSO in terms of precipitation. The months indicated are those affected during the year of and the year following the ENSO event. (Based on Ropelewski and Halpert 1987, 1989; from Trenberth, 1990b.)

spheric waves (teleconnections), and winds in the tropics. It appears that the Pacific Ocean is large enough that the tropical atmosphere and ocean in the Pacific can evolve as a coupled system. Modeling studies reveal that a self-contained irregular oscillation can result (e.g., Zebiak and Cane, 1987).

Although ENSO is a natural part of the earth's climate variations, there are major questions concerning whether the intensity and/or frequency of ENSO events might change as a result of global warming (Trenberth, 1990a). Until recently, climate models used to examine changes due to increased greenhouse effects had such simplified oceans that ENSOs were not included. More recent models are simulating some ENSO-like, but not entirely realistic, variations, but the questions currently remain unanswered.

The observational record reveals that ENSO events have changed in frequency and intensity in the past. Figure 5 shows the Darwin pressure series as one homogenous index of the SO. High pressures at Darwin coincide with El Niño events (e.g., 1982–1983). The strong SO fluctuations from 1880 to 1920 led to the discovery and naming of the SO (Walker and Bliss, 1932) and strong SO events are clearly evident in recent decades, but a much quieter period occurred from about 1928 to 1950 with the exception of the very strong ENSO of 1939–1942 (Trenberth and Shea, 1987). Quinn *et al.* (1987) have documented ENSO events as seen on the northern west coast of South America for 450 years, and the potential exists for a longer paleorecord based on alluvial deposits in coastal Peru (Wells, 1987) and data from ice cores (Thompson *et al.*, 1984), coral cores (Shen *et al.*, 1987), and tree rings (Michaelsen, 1989).

During ENSO events, the heat stored in the warm pool in the tropical western Pacific is redistributed, resulting in the El Niño itself, and there is a net loss of heat by the ocean to the atmosphere, producing a short period of global warming. Consequently, the warmer individual years in the record of global temperatures are often associated with El Niños. Maxima in global temperatures tend to occur about 6 months after the El Niño itself (Pan and Oort, 1983). In one sense, therefore, this

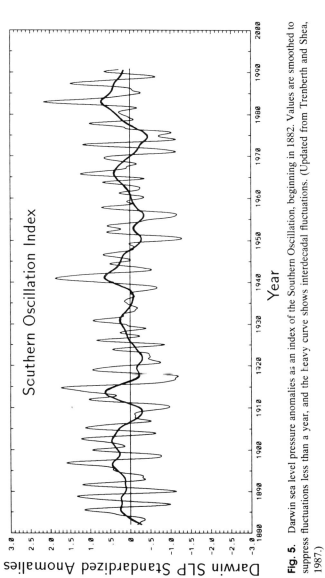

Fig. 5. Darwin sea level pressure anomalies as an index of the Southern Oscillation, beginning in 1882. Values are smoothed to suppress fluctuations less than a year, and the heavy curve shows interdecadal fluctuations. (Updated from Trenberth and Shea, 1987.)

natural variability represents a source of noise and high-frequency variations in global temperatures. It is therefore tempting to use linear regression to remove this noise statistically (Jones, 1989), and the result is certainly a smoother curve for global temperature change, with 20 to 30% of the high-frequency variance removed. However, this would assume that ENSO is not itself influenced by other mechanisms of global climate change, which is unlikely to be the case. Indeed, the fact that several of the warmest years on record in the 1980s were associated with ENSO events implies a reduction of the warming over the past 15 years from other sources because of the imbalance owing to the three sequential Warm Events (1976–1977, 1982–1983, 1986–1987), while there were no strong Cold Events between 1975 and 1988 (see Fig. 5). Because it is not clear how ENSO might change with other climate changes, it seems preferable to retain and acknowledge the ENSO variability in the record.

III. Large-Scale Observed Trends

Low-frequency temperature trends in the NH reveal a rise from about 1890 to 1940 of about 0.5°C, a downward trend until the early 1970s of a few tenths of a degree Celsius, and a further increase in the 1980s (Fig. 6). At any one place in the NH, there

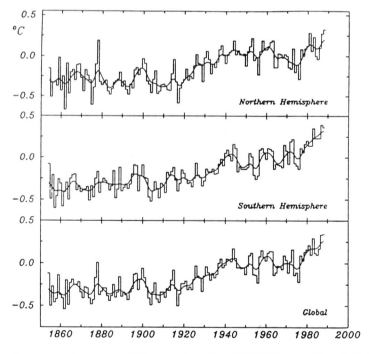

Fig. 6. Annual mean temperatures (°C) as departures from 1950–1979 mean. (From P. D. Jones; see Jones *et al.*, 1991.)

are major departures from this time sequence (Fig. 1), although van Loon and Williams (1976a) note that the biggest zonally averaged 15-year trends north of about 50°N exceed the biggest trends elsewhere by more than a factor of 7, so the sign of the hemispheric trends closely follows that of the subpolar and polar regions. Kelly *et al.* (1982) also note the broad similarity between the air temperature trends for the hemisphere and the Arctic (annual correlation 0.81), although with the Arctic variations much greater in magnitude (by a factor of 3 to 4) and more rapid. This amplification in the Arctic is less apparent in the most recent 20 years (Hansen and Lebedeff, 1987).

Of particular note in Fig. 6 in the NH is the fairly abrupt jump in temperatures from about 1920 to 1930, of about 0.3°C for the hemisphere while the increase was 1.1°C for the Arctic. This change has been referred to as a "climatic jump" (Yamamoto *et al.*, 1987). The change is most pronounced in winter but is clearly evident in the other seasons too.

The standard deviations of monthly mean NH surface temperatures in winter are roughly double those in summer, and for the Arctic the winter-to-summer ratio is more than a factor of 3 (Jones and Kelly, 1983). But Jones and Kelly show that individual monthly hemispheric means correlate best with the annual means in summer ($r \sim 0.75$), although correlations are still positive in winter ($r \sim 0.6$), perhaps reflecting larger month-to-month variability in winter. Most studies have focused on wintertime relationships, and the reason for the similarity in trends in all seasons is not clear, although it is probably related to heat storage through anomalies in sea temperatures and sea ice. The latter is strongly linked to circulation and surface temperature, and anomalies persist for up to 6 months (Walsh and Johnson, 1979).

A. The North Atlantic

The hemispheric warming that took place in the 1920s and 1930s was generally regarded by early workers as having resulted from an influx of heat from lower latitudes (Scherhag, 1936, 1939; Lysgaard, 1949). However, several studies have noted that the warming of Europe, which began somewhat earlier than the hemispheric warming, coincided with a period of stronger westerlies from about 1900 to 1938 (Lamb and Johnson, 1966; Lamb, 1972; Makrogiannis *et al.*, 1982; Parker and Folland, 1988), so that, especially in winter, there was an absence of very cold outbreaks. Rogers (1985) has noted that the best correlation between European temperatures and wind is with the westerly component, reflecting whether or not the encroaching air masses have had an oceanic moderating influence imposed.

Variations in the meridional pressure gradient across the North Atlantic from 40 to 60°N are associated with pronounced changes in the Icelandic Low. Such changes are strongly tied to the North Atlantic Oscillation (van Loon and Rogers, 1978). However, van Loon and Rogers show that there is a seesaw in winter temperatures between Greenland and northern Europe associated with this mode, whereby, most of the time, above-normal temperatures in Europe go hand in hand with below-normal temperatures in Greenland and the Canadian Arctic (referred to as GB, Greenland below), and vice versa (GA, Greenland above). Warm winter weather in

Europe goes along with a deeper Icelandic Low and a stronger Azores high, with a greatly enhanced westerly wind flow onto Europe. In Greenland, however, the reversals in temperature anomalies are associated with the meridional wind (Rogers, 1985).

This seesaw in temperatures has strong canceling effects for the Arctic as a whole and does not account for the Arctic warming of the 1920s. [Any cancellation of temperature anomalies presumes that there is adequate data coverage to resolve the opposite anomalies properly. Kelly *et al.* (1982) show that much of the Canadian Arctic and northern Greenland could not be analyzed until after 1935, whereas data were available before 1895 for northern Europe and the southern half of Greenland.] Van Loon and Rogers (1978) show that there were nine Januaries with Greenland below and Europe above normal by at least 4°C (GB) but only two with Greenland above and Europe below normal (GA) from 1900 to 1925, whereas the overall incidence of the two types is about equal. Iceland (Einarsson, 1984) and Spitzbergen began to warm after about 1917 (see Hesselberg and Johannessen, 1958), whereas Greenland and northern Canada warmed in the mid-1920s (Kelly *et al.*, 1982; Rogers, 1985). For the entire Arctic to warm, it seems necessary for both Greenland and Europe to be warm (see Jones and Kelly, 1983), cases considered by van Loon and Rogers (1978) as "both above" (BA), in contrast to the seesaw conditions. From 1926 to 1939, van Loon and Rogers found eight BA Januaries versus two GA and four GB cases and compared with only one January in this class from 1900 to 1925 (Fig. 7).

Sea level pressure composites for the "both above" versus "both below" cases (Fig. 8) (van Loon and Rogers, 1978) reveal higher than normal pressures over the Atlantic south of 55°N and a broad region of lower than normal pressures throughout the Arctic, so that there is a stronger westerly flow onto Europe but an absence of strong meridional flow anywhere from the polar regions. Van Loon and Madden (1983) show that the time of much colder mean temperatures (see Fig. 6), 1901 to 1916, was remarkable in the North Atlantic for the very low standard deviations of monthly mean sea level pressures north of 55°N, implying that there was less variability. In contrast, lower pressures in the Arctic (e.g., Fig. 8) are accompanied by an increased number of cyclonic disturbances (Rogers and van Loon, 1979) that increase stirring and cloudiness and thus break up the shallow surface temperature inversions in winter.

We conclude, therefore, that during the Greenland–northern Europe seesaw in temperatures the location of cold advection out of the Arctic is changed to affect principally either Greenland or Europe. We infer that for both Greenland and northern Europe to be above normal in temperature, the circulation is more zonal, there is less mean meridional flow out of the Arctic, there are fewer cold outbreaks anywhere, but there is increased transient cyclonic activity in the Arctic.

The relationship is not altogether straightforward, as pointed out by van Loon and Williams (1977), because of differences between what happens at the surface and at 700 mb. For the period 1949 to 1972, they show that when winter zonal mean 700-mb temperatures were generally decreasing but with a maximum cooling at 45°N, surface temperature trends were also down but with maximum cooling in the

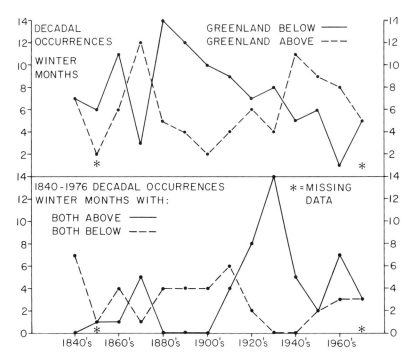

Fig. 7.   Number of cases per decade of the four temperature categories Greenland above (northern Europe below), Greenland below, both above, and both below for the northern winter (December, January, February). (From van Loon and Rogers, 1978.)

Arctic. Because there was enhanced poleward heat flux at 700 mb during this trend, thereby cooling midlatitudes, one might expect a compensating warming from this source at high latitudes. Such a warming is not found, evidently because of higher pressures in the Arctic and the associated shifts in storm tracks to lower latitudes following the changes in temperature gradients; see discussion in Section 2c.

Several studies provide clues about possible teleconnections and causes for changes in the North Atlantic. Rogers and van Loon (1979) and Meehl and van Loon (1979) have described associations of the seesaw in temperatures between Greenland and northern Europe and various other variables that may be relevant in determining causes. Because of changes in the location of the Intertropical Convergence Zone (ITCZ) in January, there are changes in rainfall over southern Africa between the two modes of the seesaw, but not in other circulation types (Meehl and van Loon, 1979). Moreover, anomalies in SSTs and sea ice develop locally, but significant SST anomalies in the tropical Atlantic are also associated with the seesaw. Ratcliffe and Murray (1970) and Palmer and Sun (1985) have associated changes in pressures over the North Atlantic and Europe with SSTs in an area near Newfoundland, and Palmer and Sun performed a modeling study with specified SST anomalies in the northwest Atlantic which produced results that compared well with the observations. Rowntree

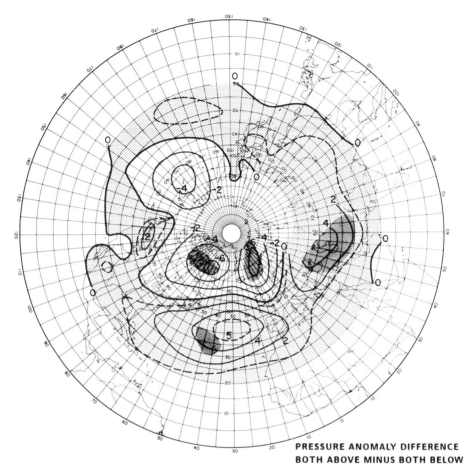

Fig. 8. Difference in pressure anomalies between January months when temperatures at Jakobshavn (Greenland) and Oslo (Norway) were both above and those when they were both below normal (mb). The dark shading indicates 5% significance. (From van Loon and Rogers, 1978; note: the original published figure stated incorrectly that it was for "both below minus both above.")

(1976) pursued the possibility that observed above-normal SSTs in the tropical North Atlantic helped to maintain the blocked, cold 1962–1963 winter in northern Europe using both observations and modeling and was able to show some promising, although not conclusive; support for this idea. Folland *et al.* (1986) have similarly used observed changes in SST extending over almost the entire global oceans in a climate model to show that they were a likely factor in circulation changes that led to the prolonged Sahel drought that began after the 1950s. Lau and Nath (1990) analyzed atmospheric responses to specified global SSTs in an atmospheric climate model and found North Atlantic and northern Europe changes to be strongly influenced by SSTs off Newfoundland and in the tropical South Atlantic.

B. The North Pacific

Van Loon and Rogers show that going along with the Greenland–Europe seesaw in temperatures is a distinctive pattern over North America whereby GB cases correspond to colder conditions in Alaska and along the West Coast but warmer than normal conditions over the eastern three-quarters of North America south of about 55°N. Temperatures in western North America appear better related to changes in the North Pacific and the strength of the Aleutian Low; however for the GA versus GB winters, van Loon and Rogers show a link between a weaker Aleutian Low and a stronger Icelandic Low. Namias (1970, 1972) and Dickson and Namias (1976) have noted the out-of-phase relationship in temperature anomalies across North America during the 1960s, when the east was colder and the west warmer than usual.

Examination of Fig. 1 for the decadal mean changes in temperatures shows an alternation in the Pacific, between cooling along the west coast of North America and Alaska from 1947 to 1956 and 1967 to 1976 and warming in the western or central Pacific, and vice versa from 1957 to 1966 and especially 1977 to 1986. Figure 9 shows time series of Pacific mean sea level pressure for the winter period November to March, averaged from 27.5 to 72.5°N, 147.5°E to 122.5°W, or virtually the entire North Pacific (from Trenberth, 1990a). This index is closely related to changes in the intensity of the Aleutian Low and is fairly strongly tied to the Pacific–North American (PNA) teleconnection pattern [see Wallace and Gutzler (1981), who show that the surface signature of the PNA is mostly confined to the Pacific in the area selected for Fig. 9]. This reveals a huge decrease in the pressures in the Aleutian Low for 1977 to 1988, with pressures lower by 2.0 mb averaged over the vast area of the North

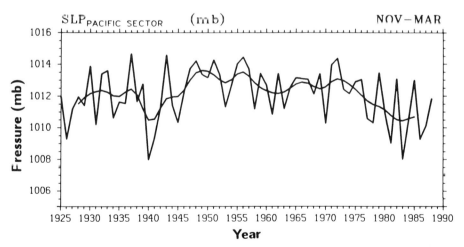

Fig. 9. Time series of mean North Pacific sea level pressures averaged over 27.5 to 72.5°N, 147.5°E to 122.5°W for the months November through March. A low-pass filter has been used to show decadal trends. (From Trenberth, 1990a.)

Pacific. These differences are present in all five winter months and are highly statistically significant (Trenberth, 1990a). They correspond to the center of the low farther east and deeper by 4.3 mb for the five winter months and deeper by 7 to 9 mb in January. A new climatology for surface wind stress based on 1980–1986 (Trenberth, 1991) revealed changes from previous climatologies that confirm the reality of the sea level pressure changes (Trenberth, 1990a). It is also consistent with and provides a partial explanation for the very large regional Pacific temperature anomalies for 1977–1986 in Fig. 1, with warming of over 1.5°C in Alaska and cooling of more than 0.75°C in the central and western North Pacific, as would be expected in association with a stronger Aleutian Low from considerations of thermal advection and increased ocean mixing.

When possible causes of changes and teleconnections are considered for the North Pacific, one prospect is *in situ* forcing. The question of the role of extratropical SST anomalies in the North Pacific and their influence on the circulation is pursued vigorously by Namias in a series of articles (e.g., Namias 1959, 1963) and by Wallace *et al.* (1990) and Lau and Nath (1990). Although it is likely that there is an effect, it probably depends greatly on the synoptic situation and other influences (as noted later) as well, so definitive answers do not yet exist. Also, insofar as the effects on planetary waves are concerned, enhanced land–sea contrasts would be needed to give a deeper Aleutian Low, whereas the observations show that the latter was accompanied by below-normal SSTs.

It is much more likely that the North Pacific changes are linked to interdecadal changes in the tropics (Trenberth, 1990a). In the NH winter, modeling results confirm that the observed deeper Aleutian Low during El Niño events is part of the teleconnection response to warmer tropical Pacific SSTs. In the 1977–1988 period both the tropical Pacific and tropical Indian Ocean had above normal SSTs (Nitta and Yamada, 1989). In the Pacific there were no strong Cold Events but several Warm Events (El Niños) (see Fig. 5). Whether this can be ascribed to any cause or is merely a part of natural variability is a very difficult question.

IV. Future Prospects: Links with the Greenhouse Effect

As well as all of the internal physical processes that we have discussed so far, several processes external to the climate system are known to be changing in ways that should affect climate. Although variations in solar activity are observed, the impact on the climate system is not known and is mostly thought to be small. The main external process of note in this context is the gradual buildup of several greenhouse gases, notably carbon dioxide, methane, nitrous oxide, and the chlorofluorocarbons (CFCs), due to human activities (IPCC, 1990). Because these gases become well mixed in relatively short times, they are globally distributed and the increases will undoubtedly cause climate change. The questions then are what form the climate change will take, how quickly it will happen, and how much will happen. One of the main prospects is for global warming.

There is considerable pressure on climate modelers to make projections into the future, and they have done so. In interpreting their results, however, it must be borne in mind that the models are far from perfect. In surveying the modeling results, looking for common features and assessing the reliability of different models (see IPCC, 1990), the following assessment of likely climate change due to increased greenhouse gases is made:

1. The models indicate global warming of 2 to 5°C for an equilibrium climate with doubled CO_2. Effective doubling of CO_2, taking into account the other greenhouse gases too, is likely around 2030 to 2050 with corresponding manifestations of Northern Hemisphere climate change occurring 20 to 30 years later (see Section 2a). The lag is likely to be greater over the SH; see Fig. 10 (Stouffer *et al.*, 1989). Over the past century, during which carbon dioxide has increased from 280 to 350 parts per million by volume (25% increase), the observed temperature increase has been fairly modest, about 0.5°C; see Fig. 6. Because of this and because of uncertainties in the models, particularly the role of clouds, it appears that the best estimate of future warming is at the low end of the values predicted in the climate models, perhaps 2 to 3°C for effective doubling of CO_2. The larger values do not appear to be tenable unless the oceans are playing tricks on us. Moreover, a big factor in producing the

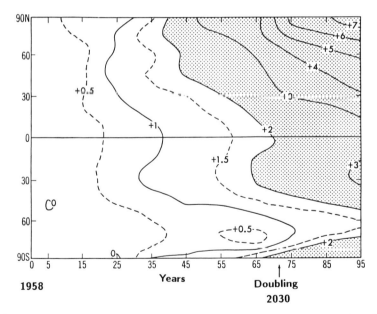

Fig. 10. Latitude–time section of the difference in zonally averaged decadal mean surface air temperature (°C) between a climate model run with CO_2 increasing at 1% per year (compounded) and a control run with constant CO_2. (Adapted from Stouffer *et al.*, 1989.)

larger values is positive feedback in clouds, which is not proved. Nevertheless, the error bars are significant.

2. The hydrological cycle is likely to speed up by about 10% with CO_2 doubling, giving increased evaporation and increased rainfall in general. Because more precipitation is apt to fall as rain in winter and there would be faster snowmelt in spring, soil moisture availability going into summer over midlatitude continents is likely to be less. Combined with increased evapotranspiration in summer, this implies that any natural tendency for a drought to occur is likely to be enhanced. However, there is not good agreement among the models on this aspect. Increases in rainfall in winter but drier conditions in summer imply that water management will become a prominent issue in the future.

3. Because warming results in expansion of the ocean and melt of snow and glacial ice, one real threat is a rise in sea level. However, there may be some compensation through increased snowfall on top of the major ice caps (Greenland and Antarctica), so they could increase in height even as they melt around the edges. Current rises in sea level of 1 to 2 mm/year are observed, and this should increase, but the main impacts will probably be felt much later next century.

4. Increases in water vapor (a greenhouse gas) and decreases in snow cover and sea ice (lower albedo) with warming provide positive feedbacks that should enhance the effects as time goes on. One result of this is that largest warming should occur in the Arctic.

5. Stratospheric cooling is another likely effect of increased greenhouse gases, and this has important implications for ozone because the kind of heterogeneous chemistry responsible for the Antarctic ozone hole is more effective at lower temperatures. The CFCs that are responsible for the ozone hole are also greenhouse gases.

6. Another possibility, because of increased SSTs, is increased frequency and/or increased intensity of tropical storms and hurricanes. This is because SSTs above about 27°C are needed before hurricanes can sustain themselves by feeding on the extra water vapor and thus latent heat available in such regions.

7. Greatest confidence exists on the global scale; regional climate changes are much more uncertain.

8. Coupled ocean–atmosphere general circulation models used to make projections have only very recently been able to simulate, as part of the overall variability within the model, very rudimentary ENSO cycles. Only one study exists so far of the possible effects on ENSO of climate change associated with the increasing greenhouse gases. It does not address the possible changes in frequency and magnitude of events; much longer model runs would be needed. However, it does find that ENSOs continue to exist in a warmer world (G. A. Meehl, personal communication, 1990). Because El Niños and La Niñas have the effect of creating droughts and floods in different parts of the world (Fig. 4) and global warming tends to enhance the hydrological cycle, there is a real prospect that future such events will be accompanied by more severe droughts and floods. In the tropics, in particular, because of the great dependence on rainfall and its tendency to fall at certain times of the year

(the dry season, the wet or monsoon season), the main prospect that looms is one of larger variability and larger extremes in weather events.

It is clear that natural variations are capable of producing low-frequency fluctuations in the climate system through the long-term memory of the ocean. In the case of the El Niño–Southern Oscillation, a self-contained low-frequency oscillation can result. Because the ocean is simultaneously affected by many climate forcings, including increasing greenhouse gases, there is great difficulty in separating out the real cause of any observed change. Indeed, generally multiple forcings will contribute to any observed change in differing amounts.

From the perspective of changes in the atmosphere, it may be possible to attribute a particular change in surface temperatures and circulation to SST anomalies, but this does not necessarily rule out other external mechanisms. In that case, it simply changes the focus of the question to the origin of the changes in SSTs. Because it is clear that changes in temperatures are not uniform over the globe, but rather contain considerable large-scale coherent structure with regions of opposite sign (Fig. 1), it is vital to establish the linkages between changes in circulation and temperatures and precipitation to build up understanding of the climate changes. Inevitably, all the different factors involved will complicate any interpretation of surface temperature changes in terms of global warming.

Acknowledgment

The National Center for Atmospheric Research is sponsored by the U.S. National Science Foundation.

References

Austin, J. F. (1980). The blocking of middle latitude westerly winds by planetary waves. *Q. J. R. Meteorol. Soc.* **106**, 327–350.

Barnett, T. P. (1978). Estimating variability of surface air temperature in the Northern Hemisphere. *Mon. Weather Rev.* **106**, 1353–1367.

Blackmon, M. L., Wallace, J. M., Lau, N.-C., and Mullen, S. L. (1977). An observational study of the Northern Hemisphere wintertime circulation. *J. Atmos. Sci.* **34**, 1040–1053.

Chen, S.-C., and Trenberth, K. E. (1988). Forced planetary waves in the Northern Hemisphere winter: wave-coupled orographic and thermal forcings. *J. Atmos. Sci.* **45**, 682–704.

Dickson, R. R., and Namias, J. (1976). North American influences on the circulation and climate of the North Atlantic sector. *Mon. Weather Rev.* **104**, 1255–1265.

Einarsson, M. Á. (1984). Climate of Iceland. *In* "Climate of the Oceans" (World Survey of Climatology, Vol. 15) (H. van Loon ed.), pp. 673–697. Elsevier, Amsterdam.

Epstein, E. S. (1988). Long-range weather prediction: limits of predictability and beyond. *Weather Forecasting* **3**, 69–75.

Folland, C. K. (1983). Regional-scale interannual variability of climate—a north–west European perspective. *Meteorol. Mag.* **112**, 163–183.

Folland, C. K., and Parker, D. (1989). Observed variations of sea surface temperature. "Proceedings of the NATO Advanced Research Workshop on Climate Ocean Interaction, Oxford, England; September 1988," pp. 31–52. Kluwer Academic Press, Boston.

Folland, C. K., Palmer, T. N., and Parker, D. E. (1986). Sahel rainfall and worldwide sea temperatures, 1901–85. *Nature* **320**, 602–607.

Gilman, D. L. (1983). Predicting the weather for the long term. *Weatherwise* **36**, 290–297.

Hansen, J., and Lebedeff, S. (1987). Global trends of measured surface air temperature. *J. Geophys. Res.* **92**, 13345–13372.

Hesselberg, Th., and Johannessen, T. W. (1958). The recent variations of the climate at the Norwegian Arctic stations. *In* "Polar Atmosphere Symposium," Pt. I. Meteorology Section (R. C. Sutcliffe, ed.), pp. 18–29. Pergamon Press, London.

IPCC. (1990). "Climate Change. The IPCC Scientific Assessment" (Report for IPCC by Working Group I.) (J. T. Houghton, G. J. Jenkins, and J. J. Ephraums, eds.). WMO, UNEP. Cambridge University Press, Cambridge.

Jones, P. D. (1988). Hemispheric surface air temperature variations: recent trends and an update to 1987. *J. Climate* **1**, 654–660.

Jones, P. D. (1989). The influence of ENSO on global temperatures. *Climate Monit.* **17**, 80–99.

Jones, P. D., and Kelly, P. M. (1983). The spatial and temporal characteristics of Northern Hemisphere surface air temperature variations. *J. Climatol.* **3**, 243–252.

Jones, P. D., Wigley, T. M. L., and Farmer, G. (1991). Marine and land temperature data sets: a comparison and a look at recent trends. DOE Workshop, 8–12 May 1989, Amherst, Massachusetts. *In* "Greenhouse-Gas-Induced Climate Change: A Critical Appraisal of Simulations and Observations" (M. Schlesinger, ed.), pp. 153–172. Elsevier, New York.

Jones, P. D., Wigley, T. M. L., Folland, C. K., and Parker, D. E. (1988). Spatial patterns in recent worldwide temperature trends. *Climate Monit.* **16**, 175–185.

Kelly, P. M., Jones, P. D., Sear, C. B., Cheery, B. S. G., and Tavakol, R. K. (1982). Variations in surface air temperatures: Pt 2. Arctic regions, 1881–1980. *Mon. Weather Rev.* **110**, 71–83.

Lamb, H. H. (1972). British Isles weather types and a register of the daily sequences of circulation patterns 1861–1971. *Geophys. Mem.*, **116**, London.

Lamb, H. H., and Johnson, A. I. (1966). Secular variations of the atmospheric circulation since 1750. *Geophys. Mem.*, **110**, London.

Lau, N.-C. (1978). On the three-dimensional structure of the observed transient eddy statistics of the Northern Hemisphere wintertime circulation. *J. Atmos. Sci.* **35**, 1900–1923.

Lau, N.-C. (1979). The structure and energetics of transient disturbances in the Northern Hemisphere wintertime circulation. *J. Atmos. Sci.* **36**, 982–995.

Lau, N.-C. (1988). Variability of the observed midlatitude storm tracks in relation to low-frequency changes in the circulation patterns. *J. Atmos. Sci.* **45**, 2718–2743.

Lau, N.-C., and Nath, M. J. (1990). A general circulation model study of the atmospheric response to extratropical SST anomalies observed in 1950–79. *J. Climate* **3**, 965–989.

Lysgaard, L. (1949). Recent climatic fluctuations. *Folia Geogr. Dan.* **5**, 86 pp.

Madden, R. A., and Williams, J. (1978). The correlation between temperature and precipitation in the United States and Europe. *Mon. Weather Rev.* **106**, 142–147.

Makrogiannis, T. J., Bloutsos, A. A., and Giles, B. D. (1982). Zonal index and circulation change in the North Atlantic area, 1873–1972. *J. Climatol.* **2**, 159–169.

Meehl, G. A. (1984). A calculation of ocean heat storage and effective ocean surface layer depths for the Northern Hemisphere. *J. Phys. Oceanogr.* **14**, 1747–1761.

Meehl, G. A., and van Loon, H. (1979). The seesaw in winter temperatures between Greenland and Northern Europe. Pt. III: Teleconnections with lower latitudes. *Mon. Weather Rev.* **107**, 1095–1106.

Michaelsen, J. (1989). Long-period fluctuations in El Niño amplitude and frequency reconstructed from tree-rings. In "Aspects of Climate Variability in the Pacific and Western Americas." *Geophys. Monogr.* **55** (D. H. Peterson, ed.), pp. 69–74. AGU, Washington, D.C.

Mitchell, J. M. (1963). On the world-wide pattern of secular temperature change. *Arid Zone Res.*, UNESCO, Paris. 161–181.

Namias, J. (1950). The index cycle and its role in the general circulation. *J. Meteorol.* **7**, 130–139.

Namias, J. (1959). Recent seasonal interactions between North Pacific waters and the overlying atmospheric circulation. *J. Geophys. Res.* **64**, 631–646.

Namias, J. (1963). Large-scale air–sea interactions over the North Pacific from summer 1962 through the subsequent winter. *J. Geophys. Res.* **68**, 6171–6186.

Namias, J. (1970). Climatic anomaly over the United States during the 1960s. *Science* **170**, 741–743.

Namias, J. (1972). Experiments in objectively predicting some atmospheric and oceanic variables for the winter of 1971–72. *J. Appl. Meteorol.* **11**, 1164–1174.

Nigam, S., Held, I. M., and Lyons, S. W. (1988). Linear simulation of the stationary eddies in a GCM. Pt. II: The "mountain" model. *J. Atmos. Sci.* **45**, 1433–1452.

Nitta, T., and Yamada, S. (1989). Recent warming of tropical sea surface temperature and its relationship to the Northern Hemisphere circulation. *J. Meteorol. Soc. Jpn.* **67**, 375–383.

Palmer, T. N., and Sun, Z. (1985). A modeling and observational study of the relationship between sea surface temperature in the north-west Atlantic and the atmospheric general circulation. *Q. J. R. Meteorol. Soc.* **111**, 947–975.

Pan, Y. H., and Oort, A. H. (1983). Global climate variations connected with sea surface temperature anomalies in the eastern equatorial Pacific Ocean for the 1958–1973 period. *Mon. Weather Rev.* **111**, 1244–1258.

Parker, D. E., and Folland, C. K. (1988). The nature of climatic variability. *Meteorol. Mag.* **117**, 201–210.

Petterssen, S. (1949). Changes in the general circulation associated with the recent climatic variation. *Geogr. Ann.* **31**, 212–231.

Quinn, W. H., Neal, V. T., and Antunez de Mayolo, S. E. (1987). El Niño occurrences over the past four and a half centuries. *J. Geophys. Res.* **92**, 14449–14462

Ramanathan, V., Callis, L., Cess, R., Hansen, J., Isaksen, I., Kuhn, W., Lacis, A., Luther, F., Mahlman, J., Reck, R., and Schlesinger, M. (1987). Climate–chemical interactions and effects of changing atmospheric trace gases. *Rev. Geophys.* **25**, 1441–1482.

Ratcliffe, R. A. S., and Murray, R. (1970). New lag associations between North Atlantic sea temperature and European pressure applied to long-range weather forecasting. *Q. J. R. Meteorol. Soc.* **96**, 226–246.

Rogers, J. C. (1985). Atmospheric circulation changes associated with the warming over the northern North Atlantic in the 1920s. *J. Climate Appl. Meteorol.* **24**, 1303–1310.

Rogers, J. C., and van Loon, H. (1979). The seesaw in winter temperatures between Greenland and Northern Europe. Pt. II: Some oceanic and atmospheric effects in middle and high latitudes. *Mon. Weather Rev.* **107**, 509–519.

Ropelewski, C. F., and Halpert, M. S. (1987). Global and regional scale precipitation patterns associated with the El Niño/Southern Oscillation. *Mon. Weather Rev.* **115**, 1606–1626.

Ropelewski, C. F., and Halpert, M. S. (1989). Precipitation patterns associated with the high index phase of the Southern Oscillation. *J. Climate* **2**, 268–284.

Rowntree, P. R. (1976). Response of the atmosphere to a tropical Atlantic ocean temperature anomaly. *Q. J. R. Meteorol. Soc.* **102**, 607–625.

Salinger, M. J. (1980a). New Zealand climate. I. Precipitation patterns. *Mon. Weather Rev.* **108**, 1892–1904.

Salinger, M. J. (1980b). New Zealand climate. II. Temperature patterns. *Mon. Weather Rev.* **108**, 1905–1912.

Scherhag, R. (1936). Die zunahme der atmosphärischen zirkulation in den letzen 25 jahren. *Ann. Hydrogr.* **64**, 397–407.

Scherhag, R. (1939). Die Erwärmung des polargebiets. *Ann. Hydrogr.* **67**, 57–67.

Shea, D. J., Trenberth, K. E., and Reynolds, R. W. (1990). A global monthly sea surface temperature climatology. NCAR Tech. Note NCAR/TN-345+STR.

Shen, G. T., Boyle, E. A., and Lea, D. W. (1987). Cadmium in corals as a tracer of historical upwelling and industrial fallout. *Nature* **328**, 794–796.

Shukla, J., and Wallace, J. M. (1983). Numerical simulation of the atmospheric response to equatorial Pacific sea surface temperature anomalies. *J. Atmos. Sci.* **40**, 1613–1630.

Stouffer, R. J., Manabe, S., and Bryan, K. (1989). Interhemispheric asymmetry in climate response to a gradual increase of atmospheric CO_2. *Nature* **342**, 660–662.

Thompson, L. G., Moseley-Thompson, E., and Morales-Arnao, B. (1984). El Niño–Southern Oscillation events recorded in the stratigraphy of the tropical Quelccaya Ice Cap, Peru. *Science* **226**, 50–53.

Trenberth, K. E. (1976). Fluctuations and trends in indices of the Southern Hemispheric circulation. *Q. J. R. Meteorol. Soc.* **102**, 65–75.

Trenberth, K. E. (1981). Observed Southern Hemisphere eddy statistics at 500 mb: frequency and spatial dependence. *J. Atmos. Sci.* **38**, 2585–2605.

Trenberth, K. E. (1982). Seasonality in Southern Hemisphere eddy statistics at 500 mb. *J. Atmos. Sci.* **39**, 2507–2520.

Trenberth, K. E. (1983). What are the seasons? *Bull. Am. Meteorol. Soc.* **64**, 1276–1282.

Trenberth, K. E. (1984). Interannual variability of the Southern Hemisphere circulation: representativeness of the year of the Global Weather Experiment. *Mon. Weather Rev.* **112**, 108–123.

Trenberth, K. E. (1986). The signature of a blocking episode on the general circulation in the Southern Hemisphere. *J. Atmos. Sci.* **43**, 2061–2069.

Trenberth, K. E. (1990a). Recent observed interdecadal climate changes in the Northern Hemisphere. *Bull. Am. Meteorol. Soc.* **71**, 988–993.

Trenberth, K. E. (1990b). General characteristics of El Niño–Southern Oscillation. *In* "ENSO Teleconnections Linking Worldwide Climate Anomalies: Scientific Basis and Societal Impact" (M. Glantz, N. Nicholls, and R. Katz, eds.), pp. 13–42. Cambridge Univ. Press, Cambridge, U.K.

Trenberth, K. E. (1991). Recent climate changes in the Northern Hemisphere. DOE Workshop, 8–12 May 1989, Amherst, Massachusetts. *In* "Greenhouse-Gas-Induced Climate Change: A Critical Appraisal of Simulations and Observations" (M. Schlesinger, ed.), pp. 377–390. Elsevier, New York.

Trenberth, K. E., and Mo, K.-C. (1985). Blocking in the Southern Hemisphere. *Mon. Weather Rev.* **113**, 3–21.

Trenberth, K. E., and Shea, D. J. (1987). On the evolution of the Southern Oscillation. *Mon. Weather Rev.* **115**, 3078–3096.

Trenberth, K. E., Branstator, G. W., and Arkin, P. A. (1988). Origins of the 1988 North American Drought. *Science* **242**, 1640–1645.

van den Dool, H. M., Klein, W. H., and Walsh, J. E. (1986). The geographical distribution and seasonality of persistence in monthly mean air temperatures over the United States. *Mon. Weather Rev.* **114**, 546–560.

van Loon, H. (1979). The association between latitudinal temperature gradient and eddy transport. I. Transport of sensible heat in winter. *Mon. Weather Rev.* **107**, 525–534.

van Loon, H., and Madden, R. A. (1981). The Southern Oscillation. I. Global associations with pressure and temperature in northern winter. *Mon. Weather Rev.* **109**, 1150–1162.

van Loon, H., and Madden, R. A. (1983). Interannual variations of mean monthly sea-level pressure in January. *J. Climate Appl. Meteorol.* **22**, 687–692.

van Loon, H., and Rogers, J. C. (1978). The seesaw in winter temperatures between Greenland and Northern Europe. I. General description. *Mon. Weather Rev.* **106**, 296–310.

van Loon, H., and Williams, J. (1976a). The connection between trends of mean temperature and circulation at the surface. I. Winter. *Mon. Weather Rev.* **104**, 365–380.

van Loon, H., and Williams, J. (1976b). The connection between trends of mean temperature and circulation at the surface. II. Summer. *Mon. Weather Rev.* **104**, 1003–1011.

van Loon, H., and Williams, J. (1977). The connection between trends of mean temperature and circulation at the surface. IV. Comparison of surface changes in the Northern Hemisphere with the upper air and with the Antarctic in winter. *Mon. Weather Rev.* **105**, 636–647.

van Loon, H., and Williams, J. (1980). The association between latitudinal temperature gradient and eddy transport. II. Relationships between sensible heat transport by stationary waves and wind, pressure and temperature in winter. *Mon. Weather Rev.* **108**, 604–614.

Walker, G. T., and Bliss, E. W. (1932). World weather V. *Mem. R. Meteorol. Soc.* **4**, 53–84.

Wallace, J. M., and Gutzler, D. S. (1981). Teleconnections in the geopotential height field during the Northern Hemisphere winter. *Mon. Weather Rev.* **109**, 784–812.

Wallace, J. M., Smith, C., and Jiang, Q. (1990). Spatial patterns of atmosphere–ocean interaction in the northern winter. *J. Climate* **3**, 990–998.

Walsh, J. E., and Johnson, C. M. (1979). Interannual atmospheric variability and associated fluctuations in Arctic sea ice extent. *J. Geophys. Res.* **84**, 6915–6928.

Weller, G. (1982). Urban climates in Alaska. *In* "Geophysical Institute Annual Report 1981–82." University of Alaska, Fairbanks.

Wells, L. E. (1987). An alluvial record of El Niño events from coastal Peru. *J. Geophys. Res.* **92**, 14463–14470.

Williams, J., and van Loon, H. (1976). The connection between trends of mean temperature and circulation at the surface. III. Spring and autumn. *Mon. Weather Rev.* **104**, 1591–1596.

Yamamoto, R., Iwashima, T., and Hoshiai, M. (1987). Climatic jump in the polar region. *Proc. NIPR Symp. Polar Meteor. Glaciol.* **1**, 91–102.

Zebiak, S. E., and Cane, M. A. (1987). A model El Niño–Southern Oscillation. *Mon. Weather Rev.* **115**, 2262–2278.

C H A P T E R 4

Climate along the Extratropical West Coast of South America

PATRICIO ACEITUNO   HUMBERTO FUENZALIDA
BENJAMÍN ROSENBLÜTH

I. Introduction

The growing concern about regional aspects of the global change that may result from an enhancement of the greenhouse effect is reflected in renewed interest in the mechanisms involved in climate anomalies at all scales. A general account of the most important factors determining climate along the extratropical west coast of South America is offered as background for a review of current knowledge of interannual climate variability in this region and for a description of the most significant changes in climate, inferred from the analysis of instrumental records of surface and upper-air temperature and rainfall.

II. Climate Background

The key atmospheric factors determining the climate characteristics along the extratropical west coast of South America are the subtropical anticyclone in the southeastern Pacific and the circumpolar belt of migratory low-pressure systems. These two atmospheric circulation features explain the predominant intense westerlies along the coast south of approximately 40°S. Other relevant factors are the influence of the ocean, which contributes to a remarkable temperature homogeneity along the coast (Fuenzalida, 1971), and the Andes Cordillera, which effectively isolates the west coast from the influence of air masses originating eastward from the mountains. The

61

Andes also contributes to the development of local wind systems, which in some cases represent an important ingredient of climate at a regional scale.

Seasonal changes in the intensity and position of the subtropical high during the year are determined by the annual cycle of meridional circulation in the tropics (Hadley cell). Thus, during austral winter the high is relatively more intense and displaced equatorward than in austral summer, when it is weaker. On the average, its center is located around 30°S and 90°W (Fuenzalida, 1971).

In the oceanic region under the influence of the subtropical high, the relatively cool and humid air in the boundary layer contrasts with the warmer and drier subsiding air above. The extremely stable conditions in the transition layer between both air masses, characterized by a sharp increase in temperature (temperature inversion layer), hinders the vertical development of clouds that originate in the boundary layer. Those clouds form a low-level stratus deck that usually covers a large portion of the subtropical southeast Pacific.

Consistent with the position of the subtropical high, the surface wind along the coast blows predominantly from the southwest. This atmospheric circulation forces a large-scale northward oceanic current (Humboldt or Peru current) and favors the cold water upwelling along the coast. These two oceanic processes explain the relatively cold waters along the west coast of the continent.

The subtropical high and the migratory low-pressure systems at midlatitudes are also the most important factors determining the different rainfall regimes along the coast. Northward from about 30°S the permanent anticyclonic conditions contribute to the extremely arid climate that characterizes the coastal and Atacama deserts. In the region from 30 to 40°S rainfall is associated with the occasional passage of cold fronts during the austral winter, when the subtropical high and the midlatitude westerlies are at their northernmost annual position. Southward of 40°S the westerlies prevail all year long and the migratory low-pressure systems and associated fronts produce rainfall episodes that are more evenly distributed through the year, so no dry season can be defined.

III. Climate Variability

Climate anomalies in Chile have been related to global-scale phenomena and regional anomalies in the atmospheric circulation. In particular, the strength of the subtropical high is closely related to the Southern Oscillation (SO), which is characterized by a tendency for atmospheric pressure changes in the east and west of the tropical Pacific to be out of phase. During the negative SO phase, which is associated with El Niño episodes, pressure is anomalously low over the eastern sector, particularly in the domain of the subtropical high (Aceituno, 1988). In contrast with this anomalous pressure pattern, the positive SO phase is characterized by a strengthened anticyclone.

With respect to anomalies in rainfall, a tendency for anomalously wet winters during El Niño episodes, or in general during the negative SO phase, has been documented in several studies (Quinn and Neal, 1983; Aceituno 1988; Rutllant and Fuenzalida, 1991). The excess rainfall is consistent with a relatively weak subtropical anticyclone during those events and with a tendency to atmospheric blocking toward the southwest of South America (Rutllant and Fuenzalida, 1991) that seems to be part of a hemispheric teleconnection pattern during the El Niño episodes (Karoly, 1989).

In addition to remote factors determining the interannual climate variability, some regional anomalies in the atmospheric circulation contribute to the seasonal climate variability in the region. Among these are low-pressure systems, not associated with extratropical disturbances, that develop in the lower troposphere at subtropical latitudes and move southward, trapped along the coast (Rutllant, 1981, 1983). These coastal lows, also observed along the subtropical west coasts of North America, Africa, and Australia (Mass and Albright, 1987; Reason and Jury, 1990), produce some remarkable changes in the atmospheric boundary layer. The continental warm air advection in the southern half of the low is conducive to a positive temperature anomaly at the surface; a decrease in altitude of the base of the temperature inversion layer, which in some extreme cases reaches the ground; and elimination of the marine stratus. On the other hand, the westerly flow in the northern portion of the low and the associated rise of the temperature inversion base bring cool and humid air inland, usually accompanied by low-level stratus or fog. Changes in the frequency of this phenomenon, whose origin and generating mechanism are still uncertain, may explain some intraseasonal anomalies in temperature, as well as in other variables such as cloudiness, relative humidity, and insolation.

IV. Long-Term Climate Changes

Long-term climate evolution along the extratropical west coast of South America may be associated with changes in the intensity and position of large-scale circulation features (i.e., the subtropical anticyclone) or with changes in the frequency of atmospheric and oceanic phenomena influencing regional climate (i.e., El Niño events). The following discussion summarizes the observed changes in surface and upper-air temperature and rainfall at selected chilean stations (Table I) for the period with available historical records.

A. Surface Temperature

A group of 19 series, along the west coast of southern South America, were initially selected from stations with relatively long temperature records. The latter were subjected to detailed scrutiny for inhomogeneities. Specifically, monthly time series of temperature differences between a minimum of three stations with similar temperature regimes were analyzed for consistency. Figure 1 shows the changes in annual mean air temperature at 5 of the 19 stations selected to characterize the different evolution patterns along the coast. In order to assess low-frequency changes, a smoothing exponential filter (Essenwanger, 1986) with a parameter value of 0.12, rather than straight line fitting, was applied to annual mean value series. Most of the comments that follow refer to the smooth variations.

The southernmost station (Evangelistas, Fig. 1e), located well in the westerlies, shows a total warming of 0.7°C in the period 1901–1988. However, this upward trend

TABLE I
List of Stations and Observations

Station	Latitude (°S)	Longitude (°W)	Temperature Surface	Temperature Upper air	Rainfall
Iquique	21	70	*		
Antofagasta	23	70		*	
La Serena	30	71			*
Quintero	33	72		*	
Valparaíso	33	72	*		
Santiago	33	71			*
Puerto Montt	41	73	*	*	
Puerto Aisén	45	73	*		
Evangelistas	52	75	*		

does not appear to be steady. In fact, a warming tendency was observed on the average during the 1920s, 1930s, and 1970s and a weak negative trend prevailed between approximately 1940 and 1970. The changes observed at Evangelistas are also apparent at other Chilean stations in the region, particularly at Punta Dungenes (52°S) and Punta Arenas (53°C), and are similar to those reported for the Southern Hemisphere by Jones *et al.* (1990), although with a larger amplitude. Farther north, at Puerto Aisén (Fig. 1d), still located in the westerlies, a small increase in temperature is seen during the past 60 years.

The temperature records of Puerto Montt and Valparaíso (Fig. 1c,b) illustrate changes in the transitional region between the westerlies and the subtropical anticyclonic domain. Changes differ appreciably from those for stations located more to the south (Fig. 1d,e). A significant feature is a well-defined cooling between the early 1950s and the mid-1970s. This change is best defined at Puerto Montt (Fig. 1c) and at other nearby stations (Punta Corona, 42°S; Valdivia, 40°S; and Temuco, 39°S) and seems to reflect a generalized hemispheric cooling in the belt at 30–60°S during that period (Angell and Korshover, 1983; Angell, 1988). A seasonal analysis of the series indicates that this change was relatively less important during the austral winter. Moreover, the analysis of the upper-air data at Puerto Montt indicates that this remarkable cooling period was restricted to the lower troposphere, pointing to the sea surface temperature (SST) as a likely forcing mechanism. This relation is supported by data for Valparaíso, where SST and air temperature records are available.

In the northern part of the Chilean coast, dominated in all seasons by the arid anticyclonic regime, the interannual temperature variability is closely associated with the El Niño and La Niña episodes. Thus, the annual temperature record at Iquique (Fig. 1a) clearly shows the warm events of 1905, 1913, 1925, 1941–1942, 1957, 1982–1983, and 1987. The analysis of the smoothed record suggests that the warming in process since the early 1970s was preceded by a long cooling period beginning around 1915.

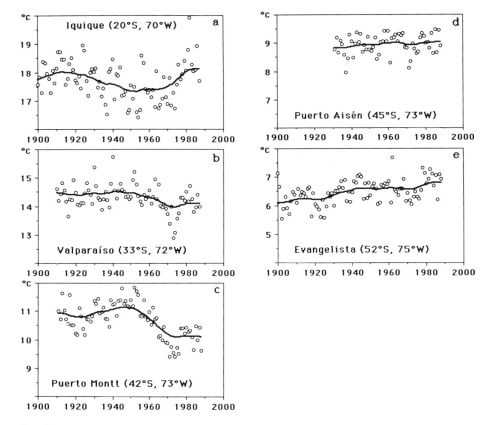

Fig. 1. Annual mean temperature (circles) and smoothed values (heavy line) obtained with an exponential filter (coefficient 0.12) for five stations along the extratropical west coast of South America: (a) Iquique (20°S, 70°W); (b) Valparaíso (33°S, 72°W); (c) Puerto Montt (42°S, 73°W); (d) Puerto Aisén (45°S, 73°W); (e) Evangelistas (52°S, 75°W).

B. Upper-Air Temperature

From the analysis of a sparse global network of upper-air stations, Angell (1988) has reported a tendency for warming at tropospheric levels and cooling above the tropopause from the early 1960s to the late 1980s. This vertical structure in the temperature change agrees with that expected from an intensified greenhouse effect. Figure 2 shows the linear trend in temperature at 700, 500, 300, 150, 100, 50, and 30 hPa at three radiosonde stations along the west coast at approximately 23°, 33°, and 42°S for the period 1958–1988. Separate analyses were performed for each season of the year.

In general terms, a warming trend is apparent in the troposphere (700, 500, and 300 hPa) and cooling prevails at levels above the tropopause (100, 50, and 30 hPa). The stratospheric cooling is particularly well defined at Antofagasta (Fig. 2a) during

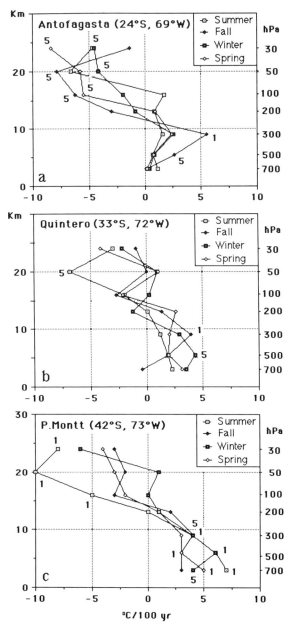

Fig. 2. Linear trend in upper-air temperature, expressed in °C/100 years, calculated at standard isobaric levels. Corresponding altitudes for a standard atmosphere are indicated: (a) Antofagasta (24°S, 69°W); (b) Quintero (33°S, 72°W); (c) Puerto Montt (42°S, 73°W). Labels 1 and 5 refer to values significant at the 1% and 5% levels, respectively, using a Student one-tail test for a zero trend null hypothesis.

all seasons. At this station, the warming trend at lower levels seems more pronounced in the higher part of the troposphere (300 hPa). At Puerto Montt (Fig. 2c) a tropospheric warming is apparent during all seasons but the cooling trend at higher levels reaches significance only during the austral summer. The significant 50-hPa cooling at this station during summer may be partly related to the springtime ozone depletion over Antarctica, which may affect this latitude by horizontal mixing associated with the polar vortex collapse in November.

The tropospheric warming over Antofagasta during the period 1958–1988 (Fig. 2a) agrees with a simultaneous positive trend in the surface temperature, although this is at least twice as large as the upper air trend. At Puerto Montt, the significant surface cooling between the 1950s and 1970s is present at the 850-hPa level but not from 700 hPa upward, where a positive trend prevails.

C. Rainfall

Identification of coherent changes in rainfall is complicated by the large spatial variability of this element. In central and southern Chile this limitation is mitigated by the fact that precipitation is mostly associated with frontal activity. This guarantees a relatively high spatial coherence in rainfall distribution that favors the definition of indices by clustering the data from a group of stations. Analysis was restricted to the subtropical domain, where a relatively larger number of stations with long records allow the definition of reliable indices. Figure 3 shows the rainfall evolution at La Serena (30°S, 71°W) and Santiago (33°S, 71°W), where records are available for more than a hundred years. Decadal mean values of two rainfall indices based on a principal component analysis of standardized precipitation during the rainy season, April–September, at six stations around 30°S and 33°S for the period 1911–1990 (Del Río, 1989) are also included.

A prevailing negative trend followed a period of relatively wet years at the end of the nineteenth century. This behavior was also detected in the analysis of time series of river discharge in the Andean region from approximately 30 to 33°S. As a distinctive feature of a semiarid region, the interannual rainfall variability in subtropical Chile is relatively large, and long-term positive or negative trends are mostly associated with changes in the frequency of rainy or dry years. On the other hand, wet years are generally associated with El Niño episodes (Quinn and Neal, 1983; Rutllant and Fuenzalida, 1991). Thus, the high decadal values observed at La Serena and Santiago during the last part of the nineteenth century and the increase in rainfall during the past two decades may be associated with the relatively more frequent warm episodes in the central Pacific during those periods. Furthermore, the analysis of the atmospheric pressure time series at Santiago (33°S) revealed a positive trend during the available homogeneous record (1861–1958), suggesting that a strengthening of the subtropical high is a factor in the decrease in rainfall.

V. Concluding Remarks

In general terms, surface temperature has varied differently along the extratropical west coast of South America during the present century. At midlatitudes south of

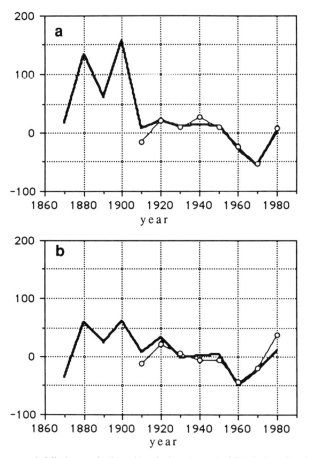

Fig. 3. Long-term rainfall changes in the subtropical west coast of South America: (a) Decadal mean values of standardized rainfall in hundredths (heavy line) for April–September at La Serena (30°S, 71°W) with respect to 1911–1987. Open circles indicate decadal mean values of a rainfall index (Del Río, 1989) using principal components analysis on six rainfall records at around 30°S. (b) Decadal mean values of standardized rainfall in hundredths (heavy line) for April–September at Santiago (33°S, 71°W) with respect to 1911–1987. Open circles indicate decadal mean values of a rainfall index (Del Río, 1989) using principal components analysis on six rainfall records at around 33°S.

approximately 45°S, temperature has been increasing in a stepwise fashion. In the area spanning about 35 to 45°S, the most significant feature is a well-defined cooling of 1 to 2°C from the mid-1950s to the mid-1970s. This change, restricted to low tropospheric levels, seems to be related to negative sea surface temperature anomalies in the southern Pacific. At lower latitudes, in the region from 20 to 30°S, the series are characterized by a relatively large interannual variability that is closely related to the occurrence of warm and cold episodes in the central Pacific.

Temperature evolution for the past 30 years at higher levels in the atmosphere exhibits a vertical structure that is compatible with an intensification of the greenhouse effect. In fact, a positive trend was detected consistently at tropospheric levels, in contrast with a negative trend at higher elevations, although only a few of the trends reach statistical significance.

Rainfall evolution in the subtropical domain is characterized by a prevailing negative trend following a period of relatively frequent rainy years at the end of the nineteenth century and ending in the early 1970s. Long-term changes in the intensity and position of the subtropical anticyclone and in the frequency of El Niño events may have contributed to the observed rainfall evolution in this region.

Acknowledgments

Most of the surface climatic data and part of the radiosonde data were provided by the Chilean Meteorological Service. This contribution was partially supported through FONDECYT grant 1177-89 and DTI grant E2830-8812. The authors appreciate the valuable comments by Dr. K. E. Trenberth and Dr. J. Rutllant.

References

Aceituno, P. (1988). On the functioning of the Southern Oscillation in the South American sector. I. Surface climate. *Mon. Weather Rev.* **116**, 505–523.

Angell, J. K., and Korshover, J. (1983). Global temperature variations in the troposphere and stratosphere, 1958–82. *Mon. Weather Rev.* **111**, 901–921.

Angell, J. K. (1988). Variations and trends in tropospheric and stratospheric, 1958–87. *J. Climate* **1**, 1296–1313.

Del Río, A. (1989). Estudio de algunos aspectos físicos que contribuyen a la variación interanual de la precipitación en Chile Central. Tesis de Ingeniería, Engineering School, University of Chile.

Essenwanger, O. M. (1986). "Elements of Statistical Analysis" (World Survey of Climatology, Vol. 1B) (H. E. Landsberg, ed.). Elsevier, Amsterdam.

Fuenzalida, H. (1971). Climatología de Chile. Department of Geophysics, Univ. of Chile.

Jones, P. D., Wigley, T. M. L., and Wright, P. B. (1990). Global and hemispheric global temperature variations between 1861 and 1988. NDP-022/R1, Carbon Dioxide Information Analysis Center, Oak Ridge National Laboratory, Oak Ridge, Tennessee.

Karoly, D. J. (1989). Southern Hemisphere circulation features associated with El Niño–Southern Oscillation events. *J. Climate* **2**, 1239–1252.

Mass, C. F., and Albright, M. D. (1987). Coastal southerlies and alongshore surges of the west coast of North America: evidence of mesoscale topographically trapped response to synoptic forcing. *Mon. Weather Rev.* **115**, 1707–1738.

Quinn, W. L., and Neal, V. T. (1983). Long-term variations in the Southern Oscillation, El Niño and the Chilean subtropical rainfall. *Fish. Bull.* **81**, 363–374.

Reason, C. J. C., and Jury, M. R. (1990). On the generation and propagation of the southern African coastal low. *Q. J. R. Meteorol. Soc.* **116**, 1133–1151.

Rutllant, J. (1981). Subsidencia forzada sobre ladera andina occidental y su relación con un episodio de contaminación atmosférica en Santiago. *Tralka* **2**, 57–76.

Rutllant, J. (1983). "Coastal lows" in Central Chile. First International Conference on Southern Hemisphere Meteorology, American Meteorological Society, Sao Jose dos Campos, Brazil, July 31–August 6, pp. 344–346.

Rutllant, J., and Fuenzalida, H. (1991). Synoptic aspects of the central Chile rainfall variability associated with the Southern Oscillation. *Int. J. Climatol.* **11**, 63–76.

CHAPTER 5

North–South Comparisons:
Climate Controls

KEVIN E. TRENBERTH

The nature of the material presented for the Southern and Northern Hemispheres was rather different. Trenberth discussed global aspects and processes, and Fuenzalida discussed the Chilean temperature records showing some warming trends but also conflicting evidence in central Chile, such as at Puerto Montt. Trenberth showed that there has been little or no trend in temperatures for the 48 contiguous states in the United States over the past century, but there is evidence for some warming in the west versus slight cooling in the east. There has been noteworthy warming in Alaska, early in this century prior to 1930 and again after 1975. More detailed regional trends for the United States are available in Boden *et al.* (1990). In the western United States upward temperature trends are most distinct along the South Pacific coast and in the Southern Desert, regions where temperatures have been much warmer than average since about 1975. Generally, in the West temperatures were cooler from about 1905 to 1925, especially in the coastal regions. Trends of any sort are much less in evidence inland in the North Cascades and Great Basin regions.

A central question is, how might the climate change in the future? The IPCC (1990) has provided an overall assessment, and further comments are contained in Chapter 3 of this volume. In the past, the main guidance provided by climate models as to what might be expected in the future due to increased greenhouse gases has been for equilibrium climate conditions with doubled CO_2 concentrations. As noted in discussing Fig. 10 of Chapter 3, the real climate is not in equilibrium and it will take some decades to approach equilibrium values, with a much greater lag likely over the southern oceans. In the Stouffer *et al.* (1989) model, about 70% of the equilibrium

EARTH SYSTEM RESPONSES TO GLOBAL CHANGE
Contrasts between North and South America

response is realized in the Northern Hemisphere but as little as 20% is realized near 60°S when CO_2 is increased by 1% per year cumulatively.

The climate models used to date to make such predictions differ in sensitivity and give values for global warming ranging from 2 to 5°C for doubling CO_2. A primary source of these differences is the way in which clouds are handled. With no cloud feedbacks, the sensitivity is at the low end of the scale, 2 to 3°C, whereas in many models clouds act to produce a positive feedback. However, these cloud parameterizations are all crude and quite unrealistic. Because the hydrological cycle is likely to increase with global warming, leading to increased moisture in the atmosphere, it seems intuitively possible that clouds might increase in such a way as to offset part of the warming. In fact, the outcome depends greatly on the relative changes in high versus low cloud. At present there is no reliable evidence as to how clouds might actually change.

In the case of the Stouffer *et al.* (1989) model (Fig. 10 of Chapter 3), the global equilibrium value for doubled CO_2 is 4°C. Accordingly, if we discount cloud feedback and follow Trenberth's arguments concerning the observational record so that we take 2.5°C as more reasonable for doubling, it would be appropriate to divide the numbers in Fig. 11 by nearly 2 (in practice, the response time of a less sensitive model may not be the same, however). Accordingly, around the year 2030, when effective doubling of CO_2 (combining effects of all greenhouse gases) is anticipated, warming of about 1 to 1.5°C could be expected along the West Coast of North America and warming would be about 1°C in most of Chile and even less over southern Chile.

In the Northern Hemisphere the large-scale circulation is apt to become less active with weaker westerlies, and the reverse or little change might be expected in the Southern Hemisphere, where the presence of Antarctica and cold southern oceans guarantees a continuing strong equator-to-pole temperature gradient. Because the land is apt to warm faster than the oceans in general, the overall slowdown in the westerlies in the Northern Hemisphere may be offset by enhanced monsoons driven by enhanced land–sea contrasts. These aspects highlight the rather different trends in the two hemispheres that might be expected with climate change due to increased greenhouse gases.

References

Boden, T. A., Kanciruk, P., and Farrell, M. P. (1990). "Trends '90: A Compendium of Data on Global Change." Carbon Dioxide Information Analysis Center, Oak Ridge National Laboratory, Oak Ridge, Tennessee.

IPCC. (1990). "Climate Change. The IPCC Scientific Assessment. "Report prepared for IPCC by Working Group I (J. T. Houghton, G. J. Jenkins and J. J. Ephraums, eds.). WMO, UNEP. Cambridge University Press, Cambridge.

Stouffer, R. J., Manabe, S., and Bryan, K. (1989). Interhemispheric asymmetry in climate response to a gradual increase of atmospheric CO_2. *Nature* **342**, 660–662.

Part III: Hydrology and Geomorphology

C H A P T E R   6

Regional Hydrologic Responses to Global Change in Western North America

R. G. LAWFORD

I. Introduction

There is a need to gain a better understanding of the response of regional hydrological systems to global change. In the coming decades water availability will become a major international concern. Already, lack of water is a major constraint on industrial development and aggravates the water pollution problems responsible for the lack of good drinking water and spread of disease in many parts of the world. Climate changes arising from the buildup of greenhouse gases could have significant effects on regional water resources. Local land use changes caused in part by human responses to global warming will also affect the hydrologic cycle. For example, deforestation and urbanization affect runoff, and effluents from heavy industries frequently lead to deterioration of water quality on the local scale.

This chapter describes the relationships between climate and hydrology in western North America. It also explores some of the possible impacts of global change on hydrologic and geomorphic processes along the west coast of North America. In addition, it provides some baseline information about hydrologic processes that could be used in identifying the links between processes in this region and those in South America. This information should also be of assistance in identifying links between hydrology and other characteristics of regional ecosystems. As a result of its exploratory nature, this chapter poses more questions and challenges than definitive answers. Before exploring the hydrology of the west coast of North America, the space and time scales of hydrologic processes and the factors that control them will be considered.

II. Nature of the Hydrologic Cycle

A. The Hydrologic Cycle

The hydrologic cycle is summarized by the relationship:

$$R = P - E + S,$$

where R is runoff, P is precipitation, E is evaporation or evapotranspiration, and S is water released from storage.

In most areas the P and E terms are relatively large on an annual basis. The runoff (R) tends to be smaller and is sometimes considered a residual term. In some regions the storage (S) term is very complex. It includes storage of water in lakes, snowpacks, glaciers, and shallow and deep aquifer systems and storage in the soil in both liquid and ice forms. The natural storage "reservoirs" operate on a range of time and space scales; consequently, they affect the transient responses of the hydrologic cycle to global change on regional and basin scales.

Each parameter in the hydrologic balance will be affected by global change. In particular, most precipitation scenarios produced by global climate models (GCMs) indicate that P will increase west of North America's Continental Divide. They also project an increase in temperature and, by inference, an increase in potential evaporation. The storage term cannot be easily assessed using GCM scenarios alone. However, one could project that the winter snowpack will not last as long and that glaciers may diminish in size with global warming. Assuming that changes in the storage term will average out to zero for long enough periods of time, it is reasonable to expect that the net long-term effect of global change on annual runoff will be governed by the changes in precipitation and evaporation.

B. Spatial Variability

A number of factors influence spatial scales in the hydrologic cycle. In order to analyze and predict river flows, hydrologists use the concept of a river basin as a geographic unit. River basins vary in size from a few square kilometers to tens of thousands of square kilometers. Because river basins are land based, their size and shape are determined by the geomorphologic characteristics of the region. The spatial dimensions of river basins are to a large extent a function of topography, with larger watersheds tending to occur on flatter terrain, often with mountains on one boundary. In addition, the topographic features within a watershed control the spatial dimensions of the storage terms. Although the spatial scales of precipitation are controlled primarily by atmospheric processes, they are also influenced by the scale of variations in the underlying topography.

The spatial scale of evaporation is less variable than that of precipitation and is strongly influenced by the distribution of the underlying vegetation cover and water bodies. Lakes and wetlands vary in size and tend to cover larger areas where the climate is humid, soils are only moderately permeable, and the terrain is relatively flat. The water-holding capacities of soils and underground aquifers also depend on terrain features such as the slope, depth, and type of soil.

The amount of snow on the ground also varies spatially. Snowfall amounts associated with synoptic storms tend to be uniform over relatively large areas. Orographic influences often produce local snowfall maxima. However, wind transport of fallen snow tends to concentrate it in preferred areas. The snowpack tends to be more variable spatially in undulating or mountainous terrain.

C. Temporal Variability

Temporal variability in the hydrologic cycle ranges from millennia on one extreme to minutes on the other. Most geomorphologic processes operate on longer time scales, and meteorologic processes operate on much shorter time scales. The annual cycle is the dominant periodicity in precipitation and evaporation processes, particularly at more northerly latitudes. However, the timing of the annual precipitation maximum is dependent on geographic location and the atmospheric processes producing precipitation (e.g., synoptic processes or localized convection).

Potential evaporation has a strong annual signal, with the highest rates occurring in the summer. Longer-term variability has also been observed in many of these hydrologic parameters. This longer-term variability leads to difficulties in separating long-term hydrologic variations from possible changes forced by global warming and other factors such as land use change. In addition, phenomena such as the El Niño, which occurs every 4 to 8 years, lead to variations in hydrology worldwide. Runoff also shows strong seasonal signals, with peak flows tending to occur in the spring in many areas of Canada and the northern United States. In coastal areas, peak flows may occur in different seasons and extreme events may be initiated suddenly. Farther south, in Mexico, precipitation and runoff maxima occur in the summer months. Nearer the equator in Central America, two precipitation maxima occur each year.

Some storage terms such as snowpack have strong annual cycles, whereas others tend to vary on longer time scales (e.g., ground water and lake levels). Different water reservoirs reach their maximum storage values at different times of the year. Storage terms with an ice phase, such as glaciers and snowpacks, reach their maxima in the early spring and their minima in the autumn. At higher latitudes the maxima occur later in the spring and the minima occur earlier in the autumn. Mountain snowmelt makes its maximum contribution to runoff during the late spring and early summer, and glaciers contribute throughout the summer. In summary, the temporal variability of the regional hydrologic cycle cannot be understood without a knowledge of the spatial and temporal characteristics of the regional temperature and precipitation patterns and the local and regional geomorphology.

III. Regional Hydrology

A. Atmospheric Controls

The mountains along the west coast of North America and their influence on atmospheric flow determine the region's hydrology. A significant portion of this area is at elevations greater than 1000 m. For example, in British Columbia 54.8% of the province is at elevations greater than 1000 m and 5.1% at elevations greater than 2000 m (Leith, 1990). In Mexico more than 50% of the land is located above 1000 m (Aleman and Garcia, 1974). At latitudes north of 35°N, moist air from the Pacific Ocean moves eastward over land where the mountains force it to rise, resulting in condensation and precipitation. The effectiveness of the mountains in "squeezing" moisture from the atmosphere depends on the speed at which the air is moving, the air's angle of incidence with the coastal mountains, and the vertical distribution of moisture and temperature in the air. A significant proportion of the moisture removed from the atmosphere in this way is either transformed to streamflow within the next few hours or days (if it falls as rain) or stored in the snowpack to be released in the following spring or summer (if it falls as snow).

Figure 1 shows an estimate of the amount of streamflow that moves to the different oceans after originating in the Canadian portion of the western Cordillera. Although the percentage that flows to the east toward Hudson Bay is small (approximately 2%), it is critical because it is the main water supply for the semiarid areas of the Canadian prairie provinces. A similar situation exists in the northern United States, where most of the water removed from the atmosphere by the mountains flows westward into the Pacific Ocean and a relatively small, but economically valuable, amount flows eastward into the Mississippi River network.

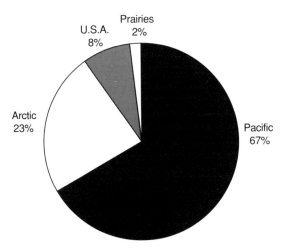

Fig. 1. Percentage of streamflow moving in different directions from the Canadian portion of the western Cordillera.

The amount of moisture removed from the air can be estimated from the moisture flux divergence. Using data from the National Meteorological Centre's daily global analysis of mixing ratios for the period from January 1979 to September 1987, Sargent (1988) derived monthly integrated moisture fluxes for the layer between the pressure levels of 1000 and 100 millibars. Moisture flux convergence implies that precipitation exceeds evaporation, and flux divergence implies the opposite condition.

The strongest net annual moisture flux convergence values are centered over the Queen Charlotte Islands on the west coast of British Columbia (Fig. 2). This maximum area of convergence occurs in winter and stretches down the coast to Washington State and north toward Alaska. With the onset of spring this flux convergence maximum in British Columbia tends to weaken, and by July a weak divergence maximum exists on the west coast of British Columbia. In September, a local flux convergence maximum reemerges in the vicinity of the Queen Charlotte Islands. In contrast, a moisture flux divergence maximum that retains its intensity all year round occurs off the coast of the Baja. In summary, the moisture flux patterns indicate that removal of moisture is most effective along the west coastal range of North America at latitudes north of 49°N during the fall and winter months, an observation that is confirmed by seasonal precipitation patterns in the mountainous areas.

The circulation patterns giving rise to these moisture flux patterns are influenced by the hydrometeorologic processes occurring over the land mass between 30 and 70°N. For example, the seasonal variations in snow cover at higher latitudes result

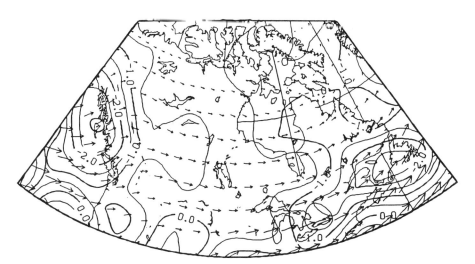

Fig. 2. Mean annual divergence flux calculated from the National Meteorological Centre's daily global analysis. (From Sargent, 1988.)

in varying albedos, and the wide range of soil moisture values leads to variability in both albedo and moisture fluxes in the planetary boundary layer. It follows that the Northern Hemisphere may be more sensitive to global change as a result of changes in snow cover, soil moisture, and other hydrometeorologic parameters that are highly dependent on surface temperature and local precipitation regimes.

Temperature exerts an important control on the hydrologic cycle, particularly at northern latitudes. In nearly all areas where the mean annual surface temperature is $-7°C$ or colder, the ground is underlaid with continuous permafrost. Because the ground never thaws out completely during the summer, both the thermal and hydrologic characteristics of these regions are significantly different from those farther south. Above the permanently frozen ground is the active layer where melt occurs in the summer. Because of its frozen state, water below this active zone does not move laterally and only a small fraction of the summer rain and meltwater infiltrates the permafrost layer. In areas farther south where the ground is frozen for 3–5 months of the year, infiltration in the spring is limited by the frost in the ground. Streamflow generation can be very efficient during spring melt in these areas provided a well-developed drainage systems exist.

B. Regional Geomorphologic Processes and Patterns

Geomorphic processes play a critical role in the hydrology of western North America. The geomorphic characteristics of this area have resulted from diastrophism and the glaciations that shaped this region's river basins and channel orientations in the past. They are also affected by fluvial processes, which are currently shaping the landscape, and by wave and eolian processes that occur along the coast. In certain localities fluvial erosion and mass wasting are also important processes.

It is generally agreed that at latitudes north of 40°N the Wisconsin ice sheet was a major factor in shaping the landscape and determining the distribution of lake and river systems in the prairie regions of North America. As a result of the glaciation, the soil layers are relatively shallow in intermontane regions north of 40–45°N. South of 40°N the ice sheet did not have the same effect on the distribution of soils, hence much deeper soil layers and finer sediments exist. The residual effects of glaciation are still contributing to a dynamic landform as isostatic rebound continues to raise the northern parts of the continent relative to the southern parts.

The western coast of North America can be divided into a number of geomorphic regions. Figure 3 shows the major geomorphic regions based on a synthesis of work by Hunt (1974) and Slaymaker (1972) and an atlas of Mexico published by the University of Texas at Austin (1975). In these regions the sensitivity of water resources to global change is partially determined by their climate and partly by water demand for industrial activities.

In Canada and the United States, the spatial scale of river basins west of the Continental Divide is limited by the distance between the Continental Divide and the oceans. River basins in this area range in size from 2.0×10^4 to 1.0×10^6 km². Farther south, in Mexico and Central America, river basin size is limited by the size of the

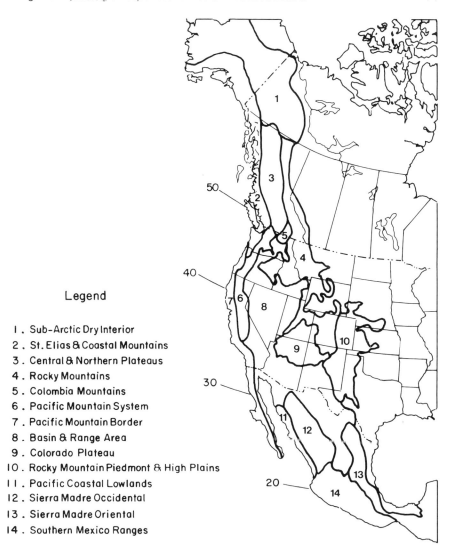

Legend

 1 . Sub-Arctic Dry Interior
 2 . St. Elias & Coastal Mountains
 3 . Central & Northern Plateaus
 4 . Rocky Mountains
 5 . Colombia Mountains
 6 . Pacific Mountain System
 7 . Pacific Mountain Border
 8 . Basin & Range Area
 9 . Colorado Plateau
10 . Rocky Mountain Piedmont & High Plains
11 . Pacific Coastal Lowlands
12 . Sierra Madre Occidental
13 . Sierra Madre Oriental
14 . Southern Mexico Ranges

Fig. 3. Geomorphic provinces of North America.

region. In spite of the relatively small size of some coastal basins, river discharges remain high north of 45°N because they are the areas of high runoff generation.

C. Hydrology to the West of the Continental Divide

In many ways hydrologic processes transform the effects of atmospheric phenomena operating at hemispheric and regional scales to hydrologic processes at local and river basin scales. The efficiency of runoff production depends on the shape of the land, its vegetative cover, geographic location, and climate. Feedback effects can also occur where variations in precipitation and runoff alter vegetative covers and, on a longer time scale, the shape of the land surface.

In broad terms, the mountains on the west coast of North America consist of two major chains: a western one composed of a number of small ranges including the coastal range in Canada and the Cascade and Sierra Nevada ranges in the United States, and an eastern range composed of the Rocky Mountains, which stretch from Alaska to southern Mexico. In Canada and the United States the separation distance between these two mountain chains generally increases with decreasing latitude (from 250 km at 55°N to 600–750 km at 30°N). Although the western mountain chain is generally less rugged than the eastern one that forms the Divide, it removes the largest amount of moisture from the atmosphere.

As shown in Fig. 4, there are significant variations in precipitation and temperature along the west coast of North America (MacDonald, 1977). Precipitation increases from south to north at latitudes south of 60°N. The rapid decrease north of 60°N can be explained by the change in the orientation of the mountains from north–south to east–west. Precipitation amounts also increase south of 20°N to values of 1500 to 2000 mm annually in southern Mexico and Central America.

Along the west coast of North America the amount of runoff generated per unit area is dependent on the location of the watershed. As shown in Fig. 5, coastal runoff tends to be lowest in California and increases with latitude to a maximum in the Queen Charlotte Islands. Although precipitation amounts are highest along the west coast, the Rocky Mountain range is also characterized by a secondary maximum in precipitation amounts. Small coastal basins are often confined to one climatic regime, whereas larger basins are usually influenced by two or three climatic regimes. For example, the Fraser River basin occupies three geomorphic zones and experiences a wide range of climatic conditions. The mean annual runoff per unit area generated in the Fraser River's headwaters is four times the average runoff generated in the intermontane plateau.

Dreyer *et al.* (1982) reported that the average annual streamflow for different basins in the Rocky Mountains increased with basin elevation. However, the effects of elevation on precipitation are more difficult to assess. Climatologic stations are frequently located in valleys, so their data generally underestimate the snow that has accumulated at higher elevations. In spite of these limitations, Steinhauser (1974) has shown that the duration of a snowpack is determined by the height of the freezing level, whereas depth of snow is dependent on temperature and atmospheric moisture conveyance. More diagnostic and modeling studies are required to understand fully the effects of climatic change on orographic precipitation.

In the United States, most of the precipitation in the Coastal and Cascade Mountains falls in the autumn and early winter. Peak flows in the coastal range are produced by heavy winter rains. Farther inland, the peak flows occur with snowmelt in the months of April and May. Farther south, in central and southern California, the snow is more transient, occurring only irregularly at higher elevations. Peak flows in this area tend to occur in March.

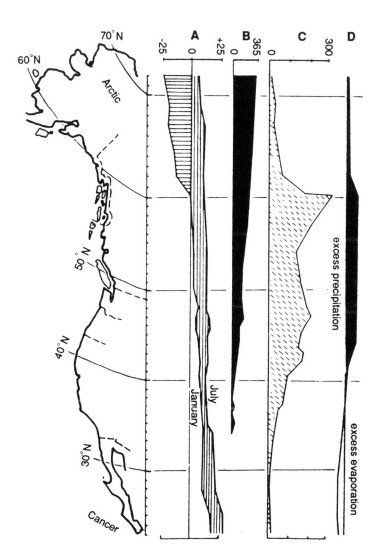

Fig. 4. Variation in temperature and precipitation at latitudes along the west coast. *A*, mean monthly air temperatures, °C; *B*, day of frost; *C*, precipitation, cm; *D*, evaporation. (Adapted from MacDonald, 1977.)

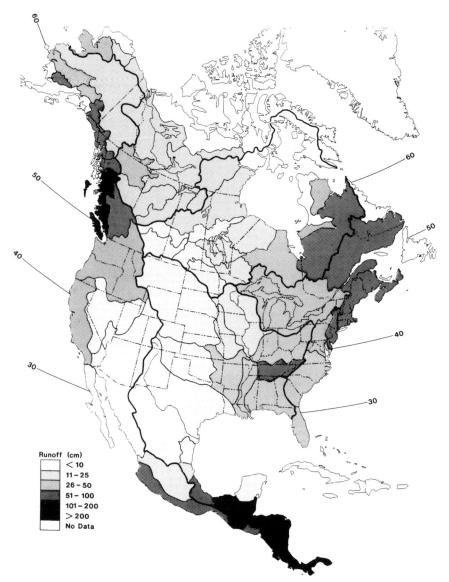

Fig. 5. Runoff per unit area for basins in North America. National averages are shown for Central American countries.

The area west of the Continental Divide at latitudes north of 35°N produce high streamflows, with the largest occurring on the west coast of British Columbia. Figure 5 shows the sharp discontinuity between annual streamflow per unit area for basins to the west and east of the Continental Divide. According to UNESCO (1978), the average drainage for North America is 339 mm. With the exception of those in the Basin and Range Province (see Fig. 3), river basins west of the Continental Divide and north of 40°N exceed the continental average.

In general, runoff is highest in areas of highest precipitation, although the runoff-to-precipitation ratios vary with latitude and topography. To the west of the mountains in the latitude band from 45 to 60°N annual precipitation amounts exceed the global average of 2.74 mm/day.

The ratio of precipitation to runoff has been reviewed for European countries by Falkenmark (1989a) using data from Chernogaeva (1971). A version of the graph, shown in Fig. 6, indicates that the proportion of precipitation forming runoff in Europe is much higher in mountainous regions than it is in flatter terrain. Similar findings would also apply in Alaska, British Columbia, Washington, and Oregon. At latitudes between 40 and 25°N the circulation pattern is not as conducive to the formation of precipitation and the production of runoff; hence these regions, which include mountains, are either semiarid or arid. Farther south in central and southern Mexico, the Intertropical Convergence Zone (ITCZ) generally brings heavy rains in June to September, leading to higher runoff rates.

D. Hydrology to the East of the Continental Divide

On the eastern side of the Continental Divide nearly all of the U.S. rivers and streams with their headwaters in the foothills of the Rocky Mountains drain into the Mississippi River and then run into the Gulf of Mexico. The streamflow generated on the eastern slopes of the Rocky Mountains in Canada flows northward in the Mackenzie River to the Arctic Ocean and eastward in the Saskatchewan–Nelson river system to Hudson Bay.

In general, the areas immediately to the east of the Rockies are characterized by limited precipitation and runoff. Knox and Lawford (1990) have shown that springs and summers with above-average precipitation to the east of the Canadian Rockies are associated with distinctive circulation patterns over western North America. In most years potential evaporation east of the Continental Divide is well in excess of precipitation. Monthly precipitation amounts are highest in the summer. Strong (1986) has suggested that the mountains enhance the occurrence of summer convective events in the eastern foothills of the Rocky Mountains.

Precipitation is highly variable from year to year and in some years vegetation can become partially desiccated by the high evaporative demands. Many U.S. and Canadian farmers in these areas rely extensively on irrigation to assist in crop production during the dry summers. Most of the irrigation water in Canada comes from surface sources. In the United States, where the demand is more developed, aquifers are used to meet a significant proportion of the irrigation demand.

Snowmelt, rain, and glacier melt all contribute to the streamflow of rivers flowing

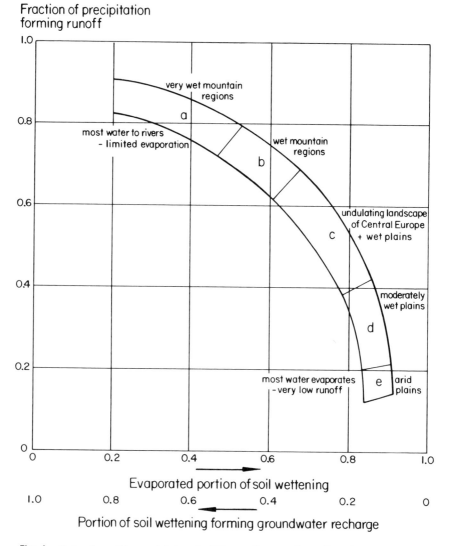

Fig. 6. Ratio of runoff to precipitation for different climate regimes. (From Falkenmark 1989a.)

eastward from the eastern slopes of the Rocky Mountains. For a typical northern river, such as the North Saskatchewan River at the border between Alberta and Saskatchewan, much of the flow, particularly in the spring, is derived from snowmelt (Fig. 7). In the summer, local rain events contribute to this flow, particularly during widespread thunderstorms. The base flow that continues through the fall is maintained by the ground water inputs and melting glaciers. In the United States, the contributions of glaciers to runoff are smaller. Snowmelt also becomes less important

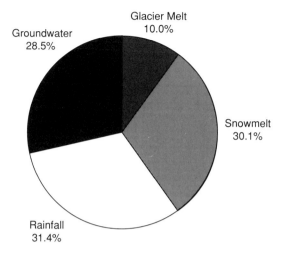

Fig. 7. Estimated flow at the Alberta–Saskatchewan border in the North Saskatchewan River. (From Lawford, 1988.)

and begins earlier in the southern United States, having limited influence in southern California.

IV. Ground Water Systems

If ice caps and glaciers are excluded, ground water accounts for 97% of the earth's total "fresh water" (van der Leeden et al., 1990). Many watersheds have large, deep aquifers under them. The great age of the water in some of these aquifers suggests that significant amounts of water do not readily move between the surface and these aquifers. The boundaries of these deeper aquifers may differ markedly from the boundaries of the surface watersheds. In undulating terrain, hills tend to be recharge areas and valleys frequently constitute discharge areas. In these discharge areas, water in the soil is released into a wetland or a river channel. The precise locations of ground water recharge areas are often not fully known and depend on climatic extremes. Deeper aquifers may undergo episodic recharging when an unusually heavy precipitation or snowmelt event leads to a large flood and a sudden recharge event (Benson and Klieforth, 1988). After such a major recharge event, the ground water may be slowly released to the surface water system over a long period of time. Falkenmark (1989b) has shown how dating studies of water lens can be used as a means of obtaining information about the wet period leading to the formation of lens.

In the contiguous United States, the major factor affecting the ground water regime is the excessive pressure for the development and utilization of ground water reserves. In British Columbia the distribution of ground water reserves is similar to that in the United States, with limited reserves in the coastal range and good storage in the intermontane plateaus. In the northern parts of the Yukon and Alaska the ground water is found below the permafrost or in the sand and gravel deposits under the floodplains and channels of major streams. Ground water movement is restricted by permafrost and the longer time each year that water remains in a frozen state.

On the western coast of North America, the ground water regimes are highly dependent on the geomorphology. As noted by Heath (1984), the thin soils and rock fractures in the coastal mountains have very little capacity for recharge and are quickly and regularly filled by the melting snow and rainfall in these mountains. Farther south in California and Arizona, particularly in the Basin and Range Province (see Fig. 3), valleys have thick alluvial deposits and hence contain large ground water reserves. The water supply for these aquifers comes from runoff originating on the mountain slopes. In the industrialized areas of California and Arizona the extensive use of these reserves has resulted in declining ground water levels and, in a few cases, land subsidence. However, according to L'vovich (1973), the actual values of infiltration and soil moisture on the west coast are governed by water availability. Consequently, values are highest in southwestern British Columbia and western Washington State (800–900 mm) and lowest in Baja, California (less than 100 mm).

Farther north in the Columbia Mountains and south in the Columbia Plateau, the ground water storage capacity is variable, with the Snake River being underlain by one of North America's largest aquifers. This ground water reservoir has been used extensively for agriculture and large declines in ground water levels have been experienced. However, it is generally felt that regulations have brought "ground water mining" under control in this area.

The interchange between surface and ground water systems in mountain basins is dependent on both precipitation and the morphologic characteristics of the soils and rocks (Mihalik and Kajin, 1990). Some types of rocks, such as karst-fissured permeable rocks, result in higher ground water recharge rates than other materials. In mountains, ground water movement is frequently caused by a hydraulic head that forces water down gradient. Variations in hydraulic head can lead to increased runoff where precipitation gradients occur.

The role of ground water in transporting contaminants is an increasingly important aspect of global change. As water moves through the soil system, many of the soil's minerals and organic chemicals are absorbed. Contaminants, fertilizers, pesticides, and other chemicals added to the soil can also be absorbed by the water. The changes that take place in water chemistry depend on the nature of the soil, which in turn determines how fast the water moves and which chemicals are leached into the water. Permafrost slows the migration of contaminants in northern areas.

V. Hydrologic Phenomena in the Mountains

Mountain hydrology is highly variable, at times leading to sudden extremes that can put life and property at risk. Generally, hydrologic and geomorphologic hazards in the mountains are related to precipitation events. Changes in the distribution of precipitation arising from climatic change could be expected to influence the frequency and intensity of these phenomena. The following sections describe some of these hazards as well as other hydrologic phenomena in the mountains.

A. Floods

Floods occur when intense and unstable atmospheric circulation systems interact with the underlying terrain to produce excessive runoff. Figure 8 shows the months with peak flows and hence the most probable time of flooding for North America based on a publication by UNESCO (1978). In the coastal mountains in western Canada and the northwestern United States, the largest floods occur in the late autumn, when rains fall on relatively shallow snowpacks (0.5–1.0 m deep) to produce

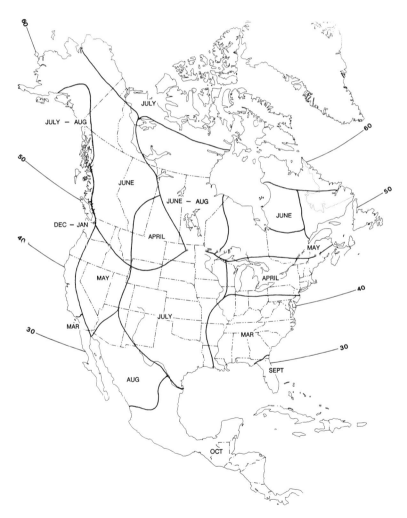

Fig. 8.   Timing of peak flows in North America. (Adapted from UNESCO, 1978.)

rapid melting and large runoffs. Snowmelt is the principal cause of flooding in the interior of British Columbia, Washington, and Oregon (Coulson, 1988). The timing and magnitude of peak flows in these areas depend on the water equivalent of the winter snowpack and the rate of spring melt. Farther north, ice jams during river breakup are a major cause of extreme peak flows (National Research Council, 1989).

Hirschboeck (1987) reported that flash floods are much more common than large-scale floods in the western United States. With the exception of one major regional flood event in northern California, all devastating floods west of the Continental Divide have been flash floods. According to Hirshboeck's analysis, the regional flood resulted from an unusual atmospheric circulation pattern in which a strong blocking high pressure ridge over the Pacific Ocean shifted a persistent upper air jet stream over the northwest coast of California. Records do not provide a reliable indication of long-term trends in flood events except to show that there has been a decrease in the number of floods in the 1980s.

B. Glaciers

Glaciers cover the peaks of a number of mountains in the Rocky Mountains at latitudes north of 45°N. Approximately 45% of North America's estimated 276,100 km² of glaciers (excluding Greenland) are located in Alaska and the western Cordillera (Haeberli et al. 1989). Alpine glaciers undergo a process of summer melt and winter snow and ice accumulation. Trends in glacier movements suggest that over the past 40 to 50 years most alpine glaciers have been retreating. Although there are exceptions to this trend, the few recorded cases of advances can be explained by the unique physical characteristics of the glacier or by uncertainties arising from data collection procedures. The predominance of retreating glaciers is consistent with the short-term warming trends that have been observed in northwestern Canada. Glacier trends suggest that the warming is occurring at higher elevations as well.

Glaciers can sometimes create their own lakes when ice dams up the water. These lakes may continue to grow until the water is suddenly released, leading to catastrophic floods. The formation of these floods, known as *jökulhlaups*, has been described by Björnson (1988) and Liestöll (1956). These events are most likely to occur in Alaska, the Yukon, or northern British Columbia. Large jökulhlaups can lead to the erosion of gorges and canyons.

C. Avalanches

In areas of the western Cordillera with steep slopes and meteorologic conditions leading to fractures in the snowpacks, avalanches frequently occur. In North America, avalanches are responsible annually for the loss of more than 20 lives and more than $1 million in damage (Perla, 1990). Besides death and injuries, they cause many inconveniences, including blocked roads and train tracks and damaged buildings.

D. Mudslides

Mudslides and landslides often occur during rainy seasons. On relatively steep mountain slopes, particularly along the west coast, rock and mud slides increase rapidly when the soils are saturated by rain. VanDine (1984) has shown that, for the coastal areas of British Columbia, debris torrents are most likely to occur in the period from September to February. In the interior mountains, similar events are likely to occur during heavy summer convective rains in June or July. During the fall of 1990, when mudslides and floods were unusually frequent in western British Columbia, concerns were raised that clear-cutting may have increased the frequency of these events.

VI. Lakes and Wetlands

North America holds 29.6% of the 91,000 km^3 of the world's fresh water stored in lakes. In contrast, South America contains only 1.1% (UNESCO, 1978) of the world's total. Fish populations are relatively diverse in U.S. lakes and considerably less diverse in Canadian lakes (Scott and Crossman, 1973). The reduced diversities at northern latitudes may reflect the effects of relatively recent glaciations. They may also result from the severe winters in the North, where there are thick ice covers, long seasons of ice cover, and large annual temperature ranges.

Examining more than 20 years of intensive measurements in northwestern Ontario during the 1970s and 1980s, Schindler *et al.* (1990) observed that over this 20-year period air and lake temperatures warmed by 2 degrees, the ice-free season increased by 3 weeks, and annual precipitation amounts decreased. The catchment basins were exposed to forest fires during this 20-year period, possibly because of these warmer, drier climatic conditions. During the same period the lakes had deeper thermoclines, increased diversity in phytoplankton, and decreased habitat for some cold-water organisms such as lake trout and opossum shrimp. In spite of these shifts in community composition, there were no discernible trends in primary production. It is postulated that the changes that occurred in these lakes with a short-term warming may reflect the types of changes that could be expected in western North America with a warming climate.

VII. Runoff and Coastal Oceanography

Manak and Mysak (1989) have shown that runoff influences the formation of ice in the Beaufort Sea. There is some evidence that runoff may play an important role in coastal oceanographic processes and could be an important factor in the circulation of the Arctic Ocean. McBean *et al.* (1991) report that runoff can influence the circulation of the coastal areas of the northeastern Pacific Ocean. This relationship is significant for fish migration, especially around Vancouver Island. For example, Xie and Hsieh (1989) found that they could use the Fraser River discharge as one of several predictors of the route that the largest proportion of salmon would take as they returned from the mid-Pacific to spawn in the Fraser River. River runoff along the west coast of North America also plays an important role in supplying nutrients to coastal biological communities.

Seasonal and interannual variations in runoff are also important. Festa and

Hansen (1976) have shown that vertical stratification and estuarine circulation increase and fresher water is found farther from the coast when terrestrial runoff increases. In summary, runoff provides an important link between terrestrial and oceanographic systems. The extent of these linkages has not been fully documented, but the evidence indicates that they include physical, chemical, and biological processes.

VIII. Climatic Change Effects

A. Climatic Change Scenarios

Scenarios based on a number of equilibrium global climate models show how the world's climate could change with a doubling of atmospheric carbon dioxide. As reported by the intergovernmental Panel on Climate Change (IPCC, 1990), these scenarios have some common features for the Northern Hemisphere, such as increasing warming with latitude during the winter months and larger warming over land than over oceans. They also project an increase of mean global temperature of 3.0 to 4.5°C and a 5 to 12% increase of the global annual precipitation. The Canadian Climate Centre's global climate model is one of three recommended in the IPCC report because of its ability to reproduce the existing climate.

The Canadian Climate Centre (CCC) model indicates that for the winter months precipitation along the west coast will increase by 1 to 2 mm/day between the latitudes of 40 and 55°N. Farther north there will be small decreases in Alaska and to the south larger decreases of up to 2 mm/day on the west coast of northern and central Mexico. These projected changes are larger than the precipitation changes projected by the Geophysical Fluid Dynamics Laboratory (GFHI) and the United Kingdom Meteorological Office (UKHI) models. However, with the exception of Alaska, where these last two models project increases in precipitation, the changes are in the same direction for all three models.

The models also predict that the summer precipitation regime will change significantly. The CCC model indicates that less rain will occur along much of the west coast, with the largest decreases in precipitation on the west coast of Mexico. It indicates increases of more than 1 mm/day in parts of Alaska. The GFHI and UKHI models also indicate small decreases in rainfall along the coasts of British Columbia and California and, in the case of the UKHI model, Oregon and Washington. One major difference between these last two models is the large increase of more than 2 mm/day projected for the west coast of central and southern Mexico. This suggests that the models are in good agreement along the west coast of North America except for central Mexico and to a lesser extent Alaska, where significant differences also occur.

The models all project that a warming will take place along the west coast, although it will not be as intense as that projected for the interior of the continent. With a doubled CO_2 scenario, surface temperatures along the coast from Alaska to Mexico are expected to increase by 1.5 to 4.5°C in the winter. Although a significant north–south gradient in the amount of warming over the land is generally expected, this is not the case along the west coast, where the temperature changes show stronger east–west gradients. In the summer, temperature increases will range from 2.5 to 5°C, indicating that the amplitude of the annual temperature cycle will be very similar to the current cycle's amplitude, although the annual and monthly mean temperatures will be 2 to 5°C warmer than the present values. The changes in temperature and precipitation along the coast of North America and South America are summarized in Fig. 9.

Precipitation scenarios are expected to become more reliable as global climate models improve. At present, the complex topography of the western Cordillera is not incorporated effectively into these models because of the coarse spatial resolution (~350 km per side for a grid square) of the models. The effects of oceans on the greenhouse warming are only now being investigated. According to Manabe's transient model runs reported in the IPCC Scientific Assessment of Climate Change (IPCC, 1990), oceans in the Southern Hemisphere will be able to absorb most of the greenhouse heat in the next century, while the Northern Hemisphere will warm more

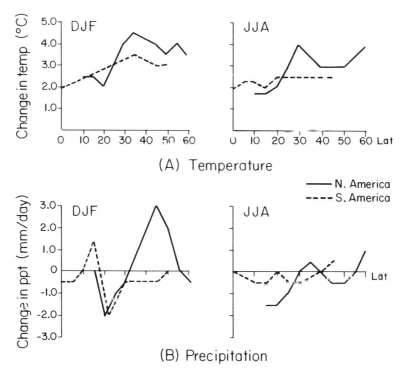

Fig. 9. Estimated changes in temperature (A) and precipitation (B) along the west coast of North and South America associated with a doubling of atmospheric CO_2. (Data taken from IPCC, 1990.)

rapidly. If large oceans such as the North Pacific are effective in absorbing excess heat resulting from increasing levels of atmospheric carbon dioxide at the regional scale, the augmentation of the hydrologic cycle on the west coast of North America may not be as vigorous as in other areas of the Northern Hemisphere. In general, it appears that along the west coast of North America climatic change will enhance the current patterns of precipitation, with the months and areas with precipitation amounts well above the global average receiving more precipitation and those with dry conditions receiving even less precipitation.

B. Climatic Change and Streamflow

The implications of changes in temperature and precipitation for hydrology at the regional level are likely to be quite significant. Different approaches have been used to assess the hydrologic impacts of climatic change. Gleick (1987) and Cohen *et al.* (1988) have used water balance models to assess the significance of temperature and precipitation scenarios for streamflow. Cohen *et al.* (1988) showed that the streamflow in the Saskatchewan River could either increase or decrease for a doubling of CO_2, depending on the GCM scenario chosen. in cases in which precipitation changes are small, runoff decreases are expected because of higher temperatures and hence higher evaporation. However, decreased water use by plants arising from an enriched CO_2 atmosphere may compensate for the effects of changing precipitation and temperatures on runoff. For example, Wigley and Jones (1985) showed that for river basins with low runoff ratios, the magnitude and even the direction of changes in runoff could be reversed if the effect of atmospheric CO_2 on plant evapotranspiration is increased.

More work is needed to assess how climatic warming could affect extreme events such as floods and record low flows. Because monthly mean temperature and precipitation scenarios are the easiest to manipulate, few studies have gone beyond them to consider the variability of precipitation and temperature in GCM outputs and the implications of this variability for the frequency of the extremes. As noted earlier, this is an important area for further study because the hydrologic and geomorphologic hazards of importance to society result from extreme events rather than changes in annual averages.

Another approach has been used by Gleick (1990), who assessed the sensitivity of water resource systems in the United States to climatic change. Key factors in determining the stress on the regional water resources were the current and projected demands for water. Gleick found that water resources in the Great Basin located between the coastal and Rocky Mountain ranges in the southern United States are highly vulnerable to changes in the climate. Other critical areas are the Missouri Basin and the State of California, both of which will be strongly affected by climatic change in the Cordillera. His analysis confirms the economic importance of the effects of climatic change on the hydrology of the Cordillera region.

The timing of peak streamflows is also expected to be affected by warmer temperatures. A preliminary assessment suggests that the snowpack may start accumulating later in the autumn and may melt earlier in the spring. In turn, this would produce earlier spring freshets in the northwestern United States and British Columbia and lower peak flows because of reduced snowpacks.

C. Climate Change and Other Hydrologic Effects

1. Impacts on Snow and Ice. Warmer temperatures would have a number of hydrologic impacts in western North America, particularly at higher elevations and more northern latitudes. With global warming in the mountains, the snow line is expected to retreat to higher elevations. Alpine glaciers may also retreat, although higher precipitation rates, particularly in the autumn, make net glacier responses to higher temperatures somewhat uncertain. However, the work of Lauman and Tvede (1989) suggests that movements of glacier fronts may not occur immediately but may show a lag time of 30 to 40 years. Increased glacier melt rates will probably increase glacial contributions to streamflow in the short term but may decrease them in the long term.

The confinement of snow and ice to higher elevations may also shift some alpine geomorphologic processes to higher elevations (Slaymaker, 1990). The most significant impacts of climatic change are likely to occur in the dry interior mountains of northern British Columbia, Alaska, and the Territories because of the effects of warmer temperatures on the distribution of permafrost. In addition, there is likely to be a reduction in discontinuous permafrost and periglacial activity in the Rocky Mountains.

2. Permafrost. Global warming would have a very significant effect on the distribution of permafrost throughout the north. As temperatures increase, the boundary for continuous permafrost would shift northward. Active layers would also deepen and exposed ice lenses would melt, leading to site-specific land slumping. The reduction in the area of permafrost could lead to large-scale ecological changes if wetlands and small lakes in Alaska, the Yukon, and northern British Columbia are drained as the permafrost melts.

3. Soil Moisture. GCM outputs indicate that soil moisture values are likely to decrease in areas where precipitation is projected to decrease or remain constant, because of the higher potential evaporation arising from higher temperatures. Along the west coast of North America summer soil moisture values are expected to decrease by 1 to 3 cm in most areas north of 45°N. Farther south and in selected areas of Alaska, slight increases in soil moisture are projected. In the winter the models project soil moisture increases along the coast (IPCC, 1990). The effects of warmer temperatures on evaporation and hence soil moisture are sensitive to the method used for computing evaporation. Complementary relationships for computing evaporation are less responsive to temperature than empirical formulas that relate temperature and evaporation directly. More work is needed to develop better relationships for assessing the effects of climate warming on evaporation.

4. Ground Water. The effects of climate change on ground water regimes have not been studied extensively. However, the decreasing precipitation in the U.S. southwest will reduce the water available to recharge aquifers. In areas where ground water levels are already decreasing because of overutilization, climatic change will place them even more at risk. The processes leading to the formation of playas will expand, and these phenomena, which tend to characterize the southern part of the Range and Basin Province, will spread northward in areas where alluvial basins exist.

5. Sea Level Rise. Rising sea levels are another expected consequence of global warming. Higher sea levels would accelerate erosional processes along the coast, particularly in Mexico and southern California. Combined with changes in wind regimes, rising sea levels could alter significantly the distribution of beaches and sand dunes along the coast of California and the west coast of Mexico. Rising sea levels would also increase the potential for flood damage to structures located on the coast and lead to saltwater intrusion into coastal ground water systems. Although sea levels have been rising in some areas, observations indicate they have been decreasing in other locations. In areas where significant isostatic rebound is taking place, sea levels are falling relative to the land. Projections for sea level rise must incorporate local geological trends if they are to be accurate.

6. Aquatic Ecosystems. It is anticipated that global change will affect the aquatic biology of lakes. In Canada, particularly in the north, temperatures and hours of sunlight set limits on the biological productivity of lakes and the diversity of their fish populations. Warmer temperatures are expected to result in thinner lake ice covers and lead to earlier overturning in lakes in which this process is important for cycling nutrients.

Climatic change may eliminate the habitats and populations of some freshwater fish types and encourage others as temperature regimes change. If migration routes are not available for dislocated fish species, some populations could become extinct. This is not expected to be such a problem for saltwater fish communities because they will be able to shift northward as air and water temperatures change. The lack of quantitative models that can link climate and hydrology to fish habitats and populations makes it difficult to assess the full range of impacts possible from climatic change.

7. Water Demand. One of the major factors influencing hydrologic patterns in western North America is the increasing demand placed on water resources. As populations grow, more water is required to meet their domestic needs. Water requirements increase as industrial activity expands. Communities in the southwestern United States, particularly California, already experience limits to growth as a result of sparse water supplies. Global warming will intensify demands for water for irrigation and hydroelectricity production, as well as industrial and domestic use. The implications of global warming for regional water demand have not been fully detailed in this chapter, however, in water-sparse areas, they may be as important a consideration as water supply.

8. General. Climate models produce projected values for a range of hydrologic outputs, but it is difficult to assess how well many of these parameters are simulated because very few observational data are available for validating the model outputs. Steps should be taken to obtain measurements of soil moisture, snow water equivalent, and other modeled hydrologic parameters for use in validating current and future GCMs.

IX. Land Use and Hydrology

Industrial activities also have local impacts on the hydrologic cycle. Although the individual effects of such activities may be small, their collective impacts can be very significant. Changes in the vegetation cover can influence interactions between the atmosphere, land cover, and surface and ground water systems. In a completely natural ecosystem precipitation supports vegetative growth and contributes to both the surface and ground water systems. If systems are left alone, the long-term average conditions of natural hydrologic systems, biomes, and landforms will reflect a near equilibrium between the climatic regime and vegetative cover, runoff, erosion, and weathering.

Manipulation of vegetation cover can have a significant impact on local hydrology. Industrial development, agriculture, and forestry lead to land cover changes. For example, the deforestation of tropical rainforests affects the hydrology of these areas. Clear-cutting is a favored means of tree harvesting in the boreal forests. Clear-cutting can affect the quality and quantity of water entering general river systems. Kovacs et al. (1989) confirmed that changes in vegetation cover can affect a basin's response to precipitation. Forests tend to transpire more than grasslands or crops. They also affect the amount of water held in storage. For a given amount of winter precipitation, snowpacks in forested areas are frequently less than those in open areas, particularly in coniferous stands.

Packer (1962) reported that, for a western white pine basin, a 1-cm increase in snow cover would occur for a 10% decrease in the density of the canopy cover. One explanation for these reduced snowpacks within a dense forest is that trees hold snow in their branches until it sublimes back into the atmosphere. Meiman (1987) confirms that in the first 3 to 4 years immediately after clear-cutting and before the stand has regrown, there is an increase in the amount of snow accumulated in a clear-cut area. The major factor in these higher accumulations appears to be the reduced snow interception by tree branches. The interception effect also takes place in summer as the decreased canopy in a clear-cut area results in lower interception losses through evaporation and less soil water depletion due to the decreased moisture uptake by trees. As a result, the mean, peak, and base flows are likely to increase in logged areas. Stottlemeyer (1987) reported that tree removal also affects the chemistry of the stream water, most notably increasing the NO_3^- ion loss. Other forestry practices such as road building may also increase the potential for debris flows, mudslides, and erosion, particularly on steep slopes, and thereby cause short-term increases in sedimentation.

During the past century, as grasslands have been converted to agricultural lands, many prarie potholes or shallow wetlands have been drained and plowed to expand agricultural production. The removal of these ephemeral wetlands has had an impact on ground water recharge, although the magnitude of this impact has not been quantified. Most areas affected by drainage have been waterfowl habitats; consequently, the number of ducks and other waterfowl bred in the prairie region has been reduced. It is suggested by some agencies that the trends toward higher evaporation rates that have been identified at some prairie locations between 1930 and 1980 (Martin, 1988) could also be attributed to wetland drainage.

X. Conclusions

In considering global change it is important to recognize that there will be changes arising from the effects of global change on ecosystems and their hydrology. However, these changes will also lead to different responses by society to opportunities for resource development and the expansion of agricultural, industrial, and forestry activities. The effects of these changed land use patterns and industrial activities on hydrology may be greater than those produced directly by fluctuations in the climate. In view of the foregoing discussions, the following conclusions can be stated:

1. Regional hydrology depends on processes such as evaporation, precipitation, and infiltration operating on a number of time and space scales. The relative importance of these processes depends on the large-scale temperature and precipitation patterns and the characteristics of the underlying topography. This is particularly evident in the hydrologic patterns of western North America.

2. The hydrology of western North America will be affected by global change in a number of ways. Both precipitation and evaporation regimes are likely to change, generally enhancing both the present characteristics of the region (wet becoming wetter, dry becoming drier) and a general trend toward moister conditions in the North. Land–atmosphere interactions must be better understood to anticipate how local land use changes may affect hydrologic regimes.

3. Assessments of the hydrologic response to climate change require a better understanding of

(a) The implications of temperature increases for evaporation;
(b) The variability of precipitation and temperature and the occurrence of extreme values computed at regional levels;
(c) The linkages between meteorologic inputs and runoff at regional scales;
(d) The sensitivity of the spatial and temporal characteristics of the storage components in the hydrologic cycle to global change;
(e) The effects of changes in water quantity on water quality and on fluvial geomorphologic processes; and

(f) The chemical and biological consequences of changes in meteorologic and hydrologic inputs for aquatic ecosystems.

4. The future changes in water demand will be driven mainly by population growth, industrial development, and technology. However, any change in global climate will be a confounding factor because it will alter the demand–supply relationships at regional levels.

References

Aleman, P. A. M., and Garcia, E. (1974). The climate of Mexico. *In* "Climates of North America, World Survey of Climatology," Vol. 11 (R. A. Bryson and F. K. Hare, eds.), pp. 345–409. Elsevier Science Publishing, New York.

Benson, L., and Klieforth, H. (1988). Stable isotopes in precipitation and ground water in the Yucca Mountain region, southern Nevada: paleoclimatic implications. *In* "Aspects of Climate Variability in the Pacific and Western Americas" American Geophysical Union (Geophysical Monograph No. 55) (D. H. Petersen, ed.), pp. 41–59. Washington, D.C.

Björnson, H. (1988). "Hydrology of Ice Caps in Volcanic Regions" (Reykjavik, Societas Scientarium Islandica. University of Iceland. 139 p.

Chernogaeva, G. M. (1971). "Water Balance of Europe." Academy of Sciences of the USSR, Inst. of Geography–Soviet Geophysical Committee, Moscow.

Cohen, S. J., Welsh, L. E., and Louie, P. Y. T. (1989). Possible impacts of climatic warming scenarios on water resources in the Saskatchewan River subbasin. Canadian Climate Centre Report No. 89-9. Saskatoon, Saskatchewan.

Coulson, C. H. (1988). Manual of operational hydrology in British Columbia. Ministry of the Environment Internal Report, Victoria, B.C.

Dreyer, N. N., Nikolayeva, G. M., and Tsigelnaya, I. D. (1982). Maps of streamflow resources of some high-mountain areas in Asia and North America. "Proceedings of a Symposium on Hydrology Aspects of Alpine and High Mountain Areas, July 19–30, 1982. Exeter, U.K." International Association of Hydrological Sciences, No. 138, pp. 11–20.

Falkenmark, M. (1989a). Global-change-induced disturbances of water related phenomena—the European perspective. Collaborative Paper No. 89-1. International Institute for Applied Systems Analysis, Laxenburg, Austria.

Falkenmark, M. (1989b). Hydrological phenomena in geosphere–biosphere interactions—outlooks to past, present and future. International Association of Hydrological Sciences, Monographs and Reports, No. 1, Wallingford, U.K.

Festa, J. F., and Hansen, D. V. (1976). A two-dimensional model of estuarine circulation: the effect of altering depth and river discharge. *Estuarine Coastal Marine Sci.* **4**, 309–323.

Gleick, P. H. (1987). The development and testing of a water-balance model for climate impact assessment: modeling the Sacramento Basin. *Water Resour. Res.* **23**, 1049–1061.

Gleick, P. H. (1990). Vulnerability of water systems. *In* "Climate Change and U.S. Water Resources" (P. E. Waggoner, ed.), pp. 223–240. Wiley, New York.

Haeberli, W., Bösch, H., Scherler, K., Ostrem, G., and Wallén, C. C. (1989). "World Glacier Inventory, Status 1988" World Glacier Monitoring Service, IAHS-UNEP-UNESCO, Switzerland, pp. C131–C171.

Heath, R. C. (1984). Ground-water regions of the United States. Geological Survey Water-Supply Paper 2242. U.S. Department of the Interior, Washington, D.C.

Hirschboeck, K. K. (1987). Catastrophic flooding and atmospheric circulation anomalies. *In* "Catastrophic Flooding; Binghamton Symposia in Geomorphology," International Series (GBR), (L. Mayer and D. Nash, eds.), Vol. 18, pp. 23–56. Miami.

Hunt, C. B. (1974). "Natural Regions of the United States and Canada." W. H. Freeman, San Francisco.

IPCC. (1990). "Climate Change—The IPCC Scientific Assessment" (J. T. Houghton, G. J. Jenkins, and J. J. Ephraums, eds.). Cambridge University Press, Cambridge.

Knox, J., and Lawford, R. G. (1990). The relationship between Canadian prairie dry and wet months and circulation anomalies in the mid-troposphere. *Atmosphere–Ocean.* **28**(2), 189–215.

Kovacs, G., Zuidema, F., and Marsalek, J. (1989). Human interventions in the terrestrial water cycle. *In* "Comparative Hydrology: An Ecological Approach to Land and Water Resources" (M. Falkenmark and T. Chabmand, eds.), pp. 105–130. UNESCO.

Lauman, T., and Tvede, A. M. (1989). Simulation of the effect of climate changes in western Norway. *In* "Conference on Climate and Water," Vol. I, 11–15 September 1989, Helsinki, Finland, pp. 339–352. Valtion Painatuskeskus, Helsinki.

Lawford, R. G. (1988). Towards a framework for research initiatives involving the impacts of climatic variability and change on water resources in the Canadian praries. *In* "Symposium on the Impact of Climate Variability and Change on the Canadian Praries, September 9–11, 1987 (B. L. Magill and F. Geddes, eds.), pp. 275–307. Edmonton, Alberta.

Leith, R. (1990). Personal Communication.

Liestöll, O. (1956). Glacier dammed lakes in Norway. *Norsk Geogr. Tidskr.* **15** (3–4), 122–149.

L'vovich, M. I. (1973). *EOS* **54**(1), American Geophysical Union. Map published in van der Leeden, F. L. Troise, and D. K. Todd. (1990). "The Water Encyclopedia," 2nd ed., p. 69. Lewis Publishers, Chelsea, Michigan.

Manak, D. K., and Mysak, L. A. (1989). On the relationship between Arctic sea-ice anomalies and fluctuations in northern Canadian air temperature and river discharge. *Atmosphere–Ocean* **27**(4), 682–691.

MacDonald, K. B. (1977). Plant and animal communities of Pacific North American salt marshes. *In* "Ecosystems of the World, Vol. I, Wet Coastal Ecosystems," pp. 167–191. Elsevier Science Publishing, New York.

McBean, G. A., Slaymaker, O., Northcote, T., LeBlond, P., and Parsons, T. S. (1991). Review of models for climate change and impacts on hydrology, coastal currents and fisheries in British Columbia. Contract report to the Atmospheric Environment Service for the University of British Columbia.

Martin, F. R. J. (1988). Determination of gross evaporation for small to moderate-sized water bodies in the Canadian prairies using the Meyer formula. Hydrology Report 113, Prairie Farm Rehabilitation Administration, Regina, Saskatchewan.

Meiman, J. R. (1987). Influence of forests on snowpack accumulation. Proceedings of a Technical Conference on Management of Subalpine Forests: Building on 50 Years of Research, July 6–9, Silver Creek, Colorado. U.S. Department of Agriculture, General Technical Report RM-149.

Mihalik, F., and Kajin, J. (1990). Groundwater runoff in mountainous areas of Slovakia and its relations to precipitation and hydrogeological conditions. *In* "Hydrology of Mountainous Areas." Proceedings of the Strbske Pleso Workshop (L. Molnar, ed.), pp. 313–327. Czechoslovakia, June, UNESCO, No. 190.

National Research Council. (1989). "Hydrology of Floods in Canada: A Guide to Planning and Design." Ottawa, Ontario.

Packer, P. E. (1962). Elevation, aspect and cover effects on maximum snowpack water equivalent in a western white pine forest. *Forest Sci.* **8**, 225–235.

Perla, R. 1990. Personal Communication.

Sargent, N. (1988). The Variability of Vertically Integrated Moisture Flux Divergence. *In* "Proceedings of the Prairie Drought Workshop," (D. Bauer, ed.), pp. 273–282. National Hydrology Research Institute Symposium No. 2. Saskatoon, Sask.

Schindler, D. F. W., Beaty, K. G., Fee, E. J., Cruishank, D. R., DeBruyn, E. R., Findlay, D. L., Linsey, G. A., Shearer, J. A., Stainton, M. P., and Turner, M. A. (1990). Effects of climatic warming on lakes of the central boreal forest. *Science* **250**, 967–970.

Scott, W. B., and Crossman, E. J. (1973). Freshwater fishes of Canada. Bulletin 184, Fisheries Research Board of Canada, Ottawa.

Slaymaker, O. (1972). Physiography and hydrology of six river basins. *In* "Studies in Canadian Geography" (J. L. Robinson, ed.), pp. 32–68. University of Toronto Press.

Slaymaker, O. (1990). Climate change and erosion process in mountain regions of western Canada. *Mountain Res. Dev.* **10**(2), 171–182.

Steinhauser, P. (1974). Die Schneeverhältnisse Ostereichs and ihre ökonomische Bedeutung. 70–71. *Jahresbericht des Sonnblock–Vereines.* 1972–1973; Vienna, pp. 1–55. [Quoted by Barry, R. G. (1990). Changes in mountain climate and glacio-hydrological responses, 1990. *Mountain Res. Dev.* **10**(2), 161–170.]

Stottlemeyer, R. (1987). National and anthropic factors as determinants of long-term streamwater chemistry. "Proceedings of a Technical Conference on Management of Subalpine Forests: Building on 50 Years of Research," July 6–9, 1987. Silver Creek, Colorado, U.S. Department of Agriculture, General Technical Report RM 149, pp. 86–100.

Strong, G. (1986). Synoptic to mesoscale dynamics of severe thunderstorm environments: a diagnostic study with forecasting applications. Ph.D. thesis, Univ. of Alberta, Edmonton, Alberta.

UNESCO. (1978). "World Water Balance and Water Resources of the Earth" (Studies and Reports in Hydrology). UNESCO, USSR.

University of Texas at Austin. (1975). "Atlas of Mexico." Bureau of Business Research, Austin, Texas.

van der Leeden, F., Troise, F. L., and Todd, D. K. (1990). "The Water Encyclopedia," 2nd ed., p. 58. Lewis Publishers, Chelsea, Michigan.

VanDine, D. F. (1984). Debris flows and debris torrents in the Southern Canadian Cordillera. *Can. Geotech. J.* **22**(1), 44–68.

Wigley, T. M. L., and Jones, P. D. (1985). Influences of precipitation changes and direct CO_2 effects on streamflow. *Nature* **314**, 149–152.

Xie, L., and Hsieh, W. W. (1989). Predicting the return migration routes of the Fraser River sockeye salmon (*Oncorhynchus merka*). *Can. J. Fish. Aquatic Sci.* **46**(8), 1287–1292.

CHAPTER 7

Chilean Geomorphology
and Hydrology:
Response to Global Change

BELISARIO ANDRADE J.   HUMBERTO PEÑA T.

I. Introduction

Changes in climatic conditions affect the intensity and even the nature of morpho-genetic and hydrologic processes. In this chapter, the problem is analyzed considering the systems of basins of the western slope of the Andean Range between 17°S and 56°S. The present geomorphologic and hydrologic patterns are briefly described, and the possible effects of the global change on the most sensitive systems of the region are considered.

II. Main Geomorphologic and Hydrologic Characteristics

According to Inman and Nordstrom's classification (1971), the structural character-istics of the western coast of South America are controlled by its collision coast nature. Parallel alignments to the general coastline are established, defining four main morphostructural units; these are ordered from east to west as follows: coastal plains, coastal range, central depression, and the Andes range (Fig. 1).

The altitude factor, which controls glacial, periglacial, and pluvial areas (Fig. 2), is another important aspect of the patterns of distribution.

101

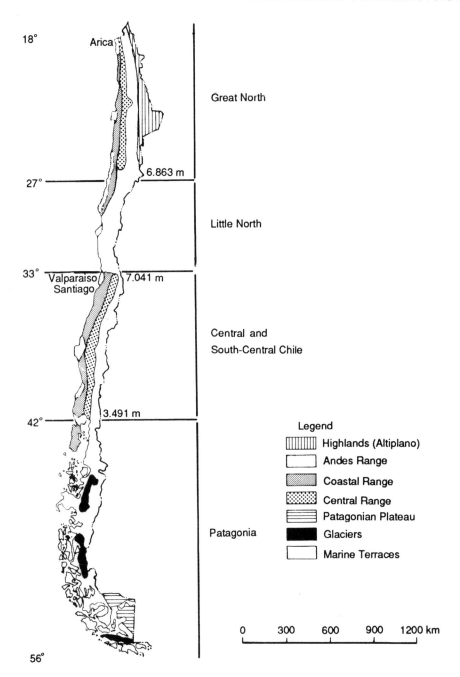

Fig. 1.   Main Chilean morphostructural units. (Modified from Borde and Santana, 1980.)

Altitude (x1000 m)

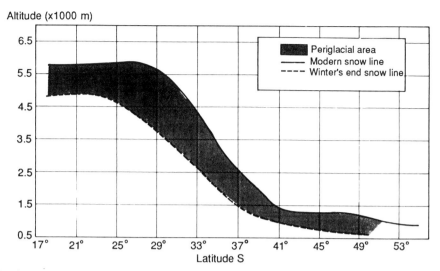

Fig. 2. Altitude of the modern snow line and the snow line at the end of the winter season. (Data from Nogami, 1976.)

A. The Great North (18–27°S)

The distinctiveness of this region is due to its hyperarid climate. This climate type can be dated to the second half of the Tertiary.

Abele (1981) describes the following altitude zonation at the latitude of Arica (18°S):

1. Nival Stage (>5800–5900 m). Bounded by the heights of some stratovolcanoes, it corresponds to the permanent snow environments.

2. Periglacial zone (5800–4800 m): The geomorphologic influence of gelifraction tends to drown forms under a detritic layer, a smoothing of escarpments, and a development of solifluction terraces.

3. Erosion impeded by vegetation zone (4500–3200 m): Moderate erosional processes, due to the existence of vegetation and a decrease of freezing frequencies. In addition, because freezing winter temperatures occur during a water shortage season, not much ice accumulates. Linear erosion is scarce in comparison with the following zone.

4. Intense linear erosion zone (3000–2000 m): Vegetation limited by a decrease of precipitation; however, an increase in the intensity of rain, together with steep slopes, causes the development of intense pitfalls because of linear erosion. Frequent mudflows alternate with muddy streams.

5. Weak linear erosion zone (2000–1100 m): Decrease of the dissection of the slopes resulting from a decrease in precipitation.

6. Coastal fog zone (<1100 m): Almost nonexistent pluvial precipitation but

abundant high-elevation fog that moistens the surface of the outcrops and regoliths and, together with a high salt content in the environment, favors the phenomena of salt weathering and salt-induced creeping. These cause rock alignment formations, solifluction terraces, and grooved soils resembling those of periglacial environments. There is a general trend to escarpment smoothing and development of flat forms. In this region, the geomorphologic units of the Coastal Range and the Intermediate Depression have no runoff, for they are part of the Atacama Desert. Only the highlands and the adjacent piedmont are hydrologically active.

In the basins of the highlands, rainfall is mainly of convective origin. Precipitation occurs from December through March, with an average of 150–300 mm/year. The mean annual temperature of the highlands is on the order of 4°C. Regarding the source, base flows must be distinguished from floods.

Base flows are very steady throughout the year, their main sources being the springs that emerge at the foothills of volcanoes. The springs are fed by the water accumulated as snow at high elevations; this water filters through lava and other volcanic materials covering the slopes. Base flows often represent a large fraction of the total runoff.

Floods are very violent and short-lived. They originate in the surface runoff generated by the active catchment below the snowline, specially in sectors with impermeable rocks or covered by a thin layer of soil.

Many of the highland basins in northern Chile are endoreic. The exoreic basins draining to the Pacific Ocean across the Atacama Desert are the exception. Net outflow is nonexistent in closed basins. Hence, in the case of wide basins, a long-term equilibrium between precipitation and evaporation is reached. The basin can be divided into two parts: one formed by the main depression toward which the whole system flows (lakes, salt lakes, marshes, etc.) and the other formed by subbasins that flow to the main depression. A partial balance may be established for a sufficiently long period between the inflow to the depression plus *in situ* precipitation and evapotranspiration from the water surfaces and the moist soil of the depression.

The importance of closed basins in the highlands is evident if one considers that 90% of the water resources available from latitude 17° to 26.30°S is lost by evaporation.

B. The Little North (27–33°S)

Aridity begins to decline in this region, compared to the great north, and a system of exoreic rivers appears, corresponding to the transition between the desert and more humid regions in the center and south of the country. Its morphostructure is characterized by absence of highlands, lack of a central depression, and disappearance of Quaternary volcanism.

The prevailing climatic conditions of the high mountains at this latitude favor periglacial processes that tend to cover the glacial modeling of the late Quaternary under a layer of debris. Debris talus slopes are very frequent.

At midelevations, torrential phenomena occur, explaining the development of small alluvial fans over those of ancient heritage. These storms also trigger mass wasting where the weathering products and slopes are conducive to it.

On the coastal zones, the essential morphoclimatic processes are related to eolian actions that produce coastal dune fields nourished by rivers carrying sediments from the steep inland areas (Castro, 1984–85), as well as the development of tafoni aided by salt weathering (Grenier, 1968).

The current geomorphologic development on the coast and midmountains reveals active intervention of humans in triggering morphogenetic processes and loss of soil by erosion as well as in the remobilization of dunes that had been stabilized by vegetation.

The hydrologic behavior of basins in this region varies over a wide range determined by rainfall regimes, rainfall–snow regimes, and snow–ice regimes, with climatic conditions ranging from extreme aridity to abundant rainfall.

Between 27 and 38°S, 12 basins are fed from the high Andes, all exhibiting a mixed snow–rain behavior. Small basins controlled by a rainfall regime are present in the interfluvials. Between 38 and 42°S the hydrologic regime is also pluvial.

The importance of the snow component in the basins is determined mainly by the elevation of the Andes and the amount of precipitation in mountains and valleys. Between 27 and 33°S, scant valley precipitation and the high altitude of the mountains result in a better-defined snow regime than occurs between 33 and 38°S, where the elevation of the Andes decreases and there is less difference in precipitation between valleys and mountains. In the same way, the small basins close to the Andes have a larger snow component than the basin as a whole when the winter maximum occurs. Figure 3 shows the latitudinal change of the hydrologic regime (A) and the variation along a typical river from the Andes to the ocean (B).

These regions have a mediterranean type of rainfall regime, and precipitation that accumulates as snow remains on the ground until the spring thaw. The duration of this process depends on the basin and may be up to 6 months. On the other hand, floods produced by rainfall are typical of wintertime.

Specific discharges increase substantially with latitude while their interannual variability decreases (Figs. 4 and 5), consistent with the rainfall trend. Regarding interannual discharge variability, it is worth noting the reduction produced by the presence of a large area covered by glaciers (800 km^2) between 32 and 35°S and by lakes from 38 to 42°S.

C. Central and South-Central Chile Region (33–42°S)

Before discussing the distribution of the current processes, it must be noted that the limit of permanent snows varies with latitude, descending southward. Thus, the modern snowline at 32.5°S is found at 4600 m altitude and at 42°S is found at 1400 m altitude (Nogami, 1976) (see Fig. 2).

In north-central Chile, the semiarid conditions are attributed to deficient nival nourishment and numerous rock glaciers are observed, with apparent weakening of this type of action in more recent ages. The situation changes toward the south, where active glacial processes can be observed.

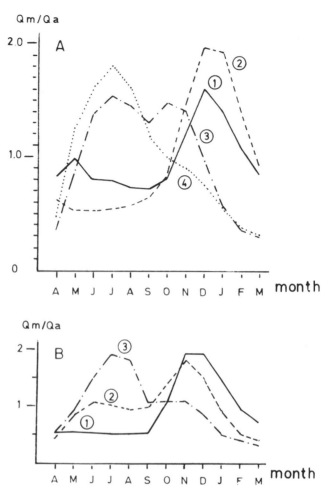

Fig. 3. (A) Latitude variation of hydrologic regime. ① Rio Elqui, 30° lat.; ② Rio Maipo, 34° lat.; ③ Rio Bio-bio, 38° lat.; ④ Rio San Pedro, 40° lat. (B) Change in the hydrologic regime along the river from the Andes to the ocean. ① Inflow to Invernada Lake, 1320 m.a.s.l.; ② Rio Maule at Colbun, 400 m.a.s.l.; ③ Rio Maule at Pichaman, 10 m.a.s.l. Q_m/Q_a = ratio of mean monthly discharge to mean annual discharge measured in the river.

Below the permanent snow limit, there is active periglacial morphogenesis in which gelifraction permits the development of large debris talus. The action of landslides in winter also contributes to the modeling of the mountains. Mudflows, triggered by very intense rainfall, frequently occur in the mountains and in the contact zone with the central depression (Hauser, 1985).

The longitudinal depression is dominated by fluvial processes receiving significant sediments from Andean basins.

Surface materials of the Coastal Range are strongly weathered, allowing for the

q

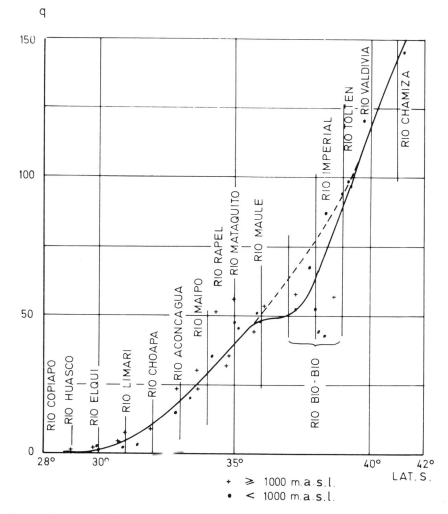

Fig. 4. Variation with latitude of specific discharge in the mountain basins. (From Benitez and Vidal, 1984.) q is the specific discharge (runoff yield, in liters per second, divided by basin area, in km^2).

development of intense anthropic erosion. This is reflected in the abundant ills, gullies, and badlands that have dramatic proportions toward 36°S (Endlicher, 1988). Southward, the frequencies of mass wasting processes increase with precipitation, also affecting urban areas.

In the coastal terraces extensive dune fields are developed where eolian actions predominate. Many generations of ancient dunes are stabilized by vegetation but are still subject to intense linear erosion and eolian remobilization (Castro and Vicuña, 1986). In the ancient dunes that cover abandoned cliffs, mass wasting processes have been induced by exceptionally intense rainfalls.

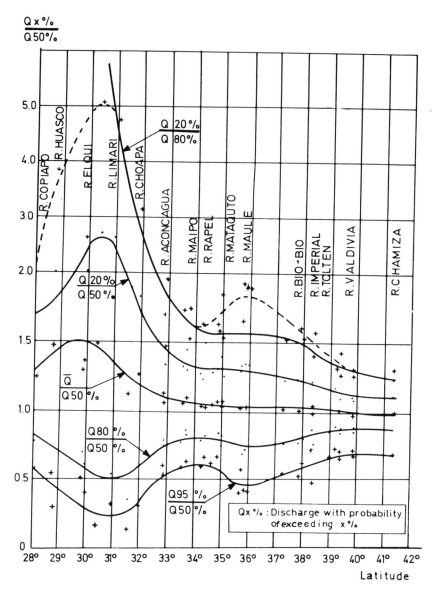

Fig. 5. Change with latitude of interannual variability of the discharge in Chilean basins. The ratio of the annual discharge that is exceeded with different levels and discharge with 50% probability of being exceeded is used as an index of the spread of annual discharge time series. (From Benitez and Vidal, 1984.)

D. Patagonia and Tierra del Fuego (42–56°S)

In this region it is difficult to distinguish the present from the ancient patterns because of the still fresh signs of intense glaciation. The most remarkable feature is the current glaciation; it has been estimated that ice bodies cover 24,000 km². Up to 46.5°S, valley glaciers predominate. The great ice caps are found to the south: the North Patagonian Ice Field with an estimated surface area of 4400 km², and the South Ice Field with an estimated area of 13,500 km². Southward, great valley and summital glaciers reach Tierra del Fuego in the Darwin Range.

The current morphologic patterns of this southern sector are poorly documented. However, a relationship reappears between current processes and anthropic erosion. Active deforestation has led to serious soil erosion, enlarging some deltas and increasing the sedimentation of estuarine ports. Mass wasting phenomena have been observed in the native forests.

Rainfall is steady throughout the year. On the western slope is ranges from 3000 to 8000 mm/year and on the eastern slope it decreases drastically to values as low as 400 mm/year, giving rise to a semiarid plain toward the Argentinean pampa. River discharges are under strong glacial influence and are well correlated with mean monthly temperatures. Rivers with a smaller glacial contribution show a secondary maximum from May to June (Fig. 6).

III. Global Change and Its Influence on Geomorphologic and Hydrologic Patterns

The present status of morphologic processes controlled by climate, climatic change models, and geomorphologic patterns is poorly documented in Chile. However, some

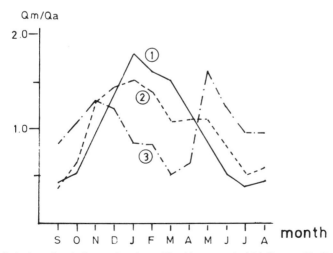

Fig. 6.   Hydrologic regime in Patagonian rivers: (1) with severe glacial influence, Rio Pascua; (2) with low glacial influence, Rio Ibañez; (3) without glacial influence, Rio Cisnes. Q_m/Q_a = ratio of mean monthly discharge to mean annual discharge measured in the river.

quantitative analyses relevant to projected climatic changes can be made. The following considerations will be made on the basis of the general circulation model predictions developed by Princeton University for a doubling of CO_2. In all cases the direct relationship between morphogenetic and hydrologic processes will be considered, regardless of complex feedbacks like the increase of vegetation coverage in some sectors of the Big North.

A. The Great North

If we assume a temperature increase of 2–3°C, together with an increase of precipitation in the highlands of order 50%, the following projections can be made:

Nival actions could increase above 5900 m because they greatly depend on an adequate snow input and altitude contributes to maintaining the cold conditions.

Gelifraction could also increase, and the greater availability of melting snow would contribute to better evacuation of the debris, probably intensifying linear erosion and solifluction in the altitude range 5800–4000 m.

In the vegetation-impeded erosion zone (4500–3200 m) the erosive activity of gelifraction could be intensified, but better development of vegetation could counteract this effect. Erosive activity would be intensified between 3200 and 2000 m in the intense linear erosion stage.

Playas or *salares* are a special case. They show evidence of ancient shorelines, suggesting that with higher precipitation they could be transformed into lakes or at least into permanent water bodies. The increase in water input should generate an increase in sediment nourishment.

The typical precipitation–runoff function for arid zones means that higher runoff is expected with increasing rainfall. On the other hand, the warmer air produces more evaporation (runoff deficit) with a consequent decrease in discharge. According to Turc (1954), if the discharge is estimated as a function of temperature and rainfall in the basin, the climatic scenario for the highlands predicts an increase of the flow on the order of 100%.

The empirical relations commonly used to relate evaporation rate to air temperature indicate 10 to 20% of this rate for an average warming of 3°C.

A reduction of snow areas and an increase in areas receiving rainfall are expected as an outcome of the rise of the snow line. This effect is moderate because of the large slope of the elevation–area curve along the level of the snowline.

The large runoff increase together with the moderate evaporation increase might produce a significant change in the present hydrologic balance of endoreic basins. In the remote past these systems were highly sensitive to climatic changes.

B. Little North and Central and South-Central Chile

A temperature increase of 3 to 4°C and a decrease in precipitation are expected for these regions. In central Chile particularly, a trend toward aridity could accentuate the intensity of meteorologic events.

In the high mountains of the Little North, the snow line is expected to become higher and fluvial transportation is expected to increase with winter rainfall, the disappearance of glacial actions, and the predominance of periglacial processes.

In the coastal zone of the Little North, further development of the present dune field is foreseen as a result of better sediment supplies.

In the high mountains of central and south-central Chile, a decrease in glacial processes is expected. Periglacial actions will increase in altitude, increasing the production of debris. At the same time, the area of the pluvial reception basins will increase. Thus, the winter discharge would intensify the transport of debris to the central depression, as occurred in the meteorologic events of 1982, 1986, and 1987.

In the longitudinal depression, sedimentary inputs from Andean rivers would be intensified, with a probable accentuation of winter floods. Alluvial fans would undergo more intense evolution.

The coastal ranges of central Chile will experience greater linear erosion because of a longer dry season, and the vegetation coverage will diminish accordingly. Probably, mass wasting phenomena will occur less frequently. Conditions for the development of dune fields would be accentuated as the rivers discharged sediments to the coastal area.

The most significant change expected in hydrologic pattern is a marked increase in the area of pluvial reception basins due to an elevation of the 0°C isotherm. This would generate a dramatic increase in the river discharge produced by winter rainfalls.

The response of the upper basins of central Chile rivers to a hypothetical climatic change, typified by a temperature increase of 3°C and unaltered rainfall, was analyzed by means of a snow and ice hydrologic model specially developed for this region (Peña, 1989; Peña and Nazarala, 1987). The model was run for the time series 1978–1979 to 1987–1988.

The series of daily flows, modified by this hypothetical climatic change, shows a new hydrologic regime with higher discharge in winter and spring and lower discharge in summer and fall. During some winter months, increments of up to 70–80% are expected, while the decrease in the summer months is of order 20–30%. This change would be due to the rise of the snowline in winter and the more intense snowmelt in spring as a result of warmer air. Similar effects have been suggested for California (Gleick, 1987).

C. Patagonia and Tierra del Fuego

A 4–5°C increase in temperature, with no change or a slight increase of precipitation, is a future scenario for this region. Consequently, withdrawal of the snow line is possible due to higher temperature, with a decrease of glacial processes, an intensification of periglacial processes, and an increase of the sediment load of the rivers. However, the high amount of present precipitation may maintain current conditions in the lower areas.

The rise of temperature may have a substantial impact on the ice-covered areas. However, the Patagonian glacial response could be extremely complex and markedly different from one case to another. At present, some glaciers show important retreat (withdrawal) rates while others are advancing.

Hydrologic changes associated with a reduction of the ice-covered area can be visualized by comparing them to the present behavior of rivers, either fed by glaciers or not (see Fig. 6).

The current marine processes on the coast have a low degree of zonality (Paskoff, 1990). However, they are liable to be affected by global change because of the greenhouse effect on the sea, i.e., a level rise. Andrade and Castro (1990) have analyzed the possible effect of a sea level rise in different scenarios, considering two types of coastline.

Because of the steepness of the rocky coastline along its 30,000-km length, a rise in sea level of 20 cm would not have any noticeable effect. A 50- to 100-cm rise would probably affect a few human installations on the coast but only during stormy weather. In the central zone, a rise of 200 cm might cause an acceleration of the cliff recession in soft rocks (Tertiary sandstones). South of 41°S latitude, the sea would attack the cliffs created in soft glacial outwash.

Although sandy coastlines constitute only 2.1% of the total coastline, they support important economic activities. All pocket beaches are located in front of cliffs, so they cannot retreat toward the inland in case of a sea level rise.

Due to Bruun's rule, which states that a sea level rise in sandy coasts triggers a regression of the shore by erosion (Bruun, 1962), a sea level rise of 50 cm or more would substantially diminish the area of pocket beaches and a rise of 2 m would totally eliminate them. In the oceanic temperate climate zone, a sea level rise of 20 cm or more would severely affect tidal salt marshes.

IV. Summary and Conclusions

The geomorphologic and hydrologic characteristics of the South American territory of Chile have been described in this chapter. The country is divided into natural regions that show clear differences according latitude and altitude. This are the Big North, which is hyperarid; the Little North, semiarid; south-central Chile, mediterranean and humid temperate; and Patagonia and Tierra del Fuego, temperate hyperhumid.

The influence of the Andes Mountains, especially from the point of view of their altitude, is shown along the territory because they affect the distribution of snow precipitations, the source of hydrologic reservoirs, and runoff in the Andean basins.

Hydrologically, three areas are especially sensitive to global climatic change: the northern Altiplano with its playas, the rain and snow basins of the central mountain range, and the Patagonian ice fields.

In the same sense, the variations of geomorphologic processes in the altitude zone of the country have been discussed, especially glacial and periglacial processes in the mountain range and increase in sedimentary inputs in the longitudinal depression. The changes in coastal behavior have been mentioned, particularly in relation to the development of coastal sand dune fields and beach erosion and tidal flats due to a probable sea level rise.

Even though our knowledge of the behavior of the Chilean natural system and the models of climatic change do not permit accurate predictions of the variations of geomorphologic and hydrologic patterns, it is feasible to suppose that a territory with such clear differences of zones and altitudes may show modifications like the ones discussed in this chapter.

The particular characteristics of the latitudinal extension and altitudinal zonation of Chile define it as an excellent laboratory for future research on possible consequences of global climatic change. This challenge must involve both the national and international scientific communities.

References

Abele, G. (1981). *Rev. Geogr. Norte Grande* **8**, 3–25.

Andrade, B., and Castro, C. (1990). *In* "Changing Climate and the Coast" (J. G. Titus, ed.), Vol. 2, pp. 399–418. U.S. Environmental Protection Agency.

Benitez, A., and Vidal, F. (1984). *In* "Jornadas de Hidrología de Nieves y Hielos en América del Sur," Vol. 1, pp. 1.0–1.22. PHI-UNESCO, Santiago, Chile.

Borde, J., and Santana, R. (1980). "Le Chili: La Terre et les Hommes." CNRS, Paris.

Bruun, P. (1962). *J. Waterways Harbors Div.* (ASCE) **1**, 117–130.

Castro, C. (1984–1985). *Rev. Geogr. Chile Terra Australis* **28**, 13–32.

Castro, C., and Vicuña, P. (1986). *Thalassas* **4**, 17–21.

Endlicher, W. (1988). *Rev. Geogr. Norte Grande* **15**, 11–27.

Gleick, P. (1987). *In* "The Influence of Climate Change and Climatic Variability on the Hydrologic Regime and Water Resources."

Grenier, P. (1968). *Bull. Assoc. Geogr. Fr.* 364–365, 193–211.

Hauser, A. (1985). *Rev. Geol.* **24**, 75–92.

Inman, D. L., and Nordstrom, C. E. (1971). *J. Geol.* **79**(1), 1–21.

Nogami, M. (1976). *Geogr. Rep. Tokyo Metropolitan Univ.* **11**, 71–86.

Paskoff, R. (1990). *Essener Geogr. Arbeit.* **18**, 237–267.

Peña, H. (1989). *In* "IX Congreso nacional de ingeniería hidráulica," Vol. 2, pp. 585–599. Sociedad Chilena de Ingenieríca Hidráulica, Santiago, Chile.

Peña, H. (1990). Internal paper, Dirección General de Aguas, Santiago, Chile.

Peña, H., and Nazarala, B. (1987). *In* "VIII Congreso nacional de ingeniería hidráulica," Vol. 2, pp. 45–60. Sociedad Chilena de Ingeniería Hidráulica, Santiago, Chile.

Turc, L. (1954). *In* "Tercera Jornada de la Hidráulica." Argel, pp. 36–43.

.

CHAPTER 8

North–South Comparisons: Hydrology–Geomorphology

R. G. LAWFORD

I. Introduction

The geomorphology and hydrology of the western parts of North America and South America have more similarities than most, if not all, other sets of landmasses in the Northern and Southern Hemispheres. However, significant differences also exist between these two regions. As a result of these similarities and differences, the responses of ecosystems in these two areas to global change should serve as useful indicators of the extent of global change in each hemisphere.

This preliminary assessment of the similarities and differences of the hydrologic characteristics of these areas is based on comparisons of:

1. Macroscale climatic, hydrologic, and geomorphologic controls
2. Geomorphologic processes
3. Hydrologic processes and patterns

The potential effects of global change on the patterns are also briefly discussed.

II. Macroscale Effects

A. Similarities

Both North America and South America have a chain of mountains along their western coasts. In both hemispheres the intensity of the upper air circulation is influenced by variations in the heat source intensity at the equator and the heat sink intensity variations at the poles. Near the equator the air generally moves from east to west, whereas south of 30–35°S and north of 35°N it flows from west to east. The low-level air interacts with the local topography to produce high precipitation rates and amounts on the windward sides of the mountains and lower rates and amounts on the leeward sides. This results in high precipitation rates and amounts in the latitudes poleward from 40° along the west coasts of the two continents. The latitudes between 20°N and 20°S are affected by the north–south migration of the Intertropical Convergence Zone (ITCZ), which produces high rainfall rates for one or two periods each year, depending on location.

B. Differences

The major difference between the two hemispheres is the much larger landmass in the Northern Hemisphere. Whereas ocean surface temperatures are the major determining factor in the Southern Hemisphere's climate, the Northern Hemisphere's climate is also influenced by land surface climate processes associated with snow cover and soil moisture. Consequently, energy and moisture inputs to the atmosphere are more variable in the Northern Hemisphere. The large areas of ocean in the Southern Hemisphere frequently result in lower temperatures for corresponding latitudes in that hemisphere. However, according to a preliminary comparison based on results presented by MacDonald (1977) and the Dirección General de Aguas (1987) (see Fig. 1), it appears that temperatures on the west coast of Chile at a given latitude can be as warm than those on the west coast of North America. Precipitation on the west coast of North America is also lower than amounts for Chile at latitudes between 30° and 45°, according to MacDonald (1977). However, it should be noted that values estimated from climatologic maps by the author indicate that MacDonald's values may be underestimates for the midlatitudes.

Differences in landforms affect the geomorphologic and hydrologic processes along the western coasts of the two American continents. The Andes Mountains are very rugged, rising rapidly to more than 6000 m above sea level at a relatively short distance (150–200 km) from the coast. In some parts of Chile there are indications of a second mountain chain next to the coastline, but this chain is not well developed. By contrast, there are two well-developed mountain chains in North America: one along the coast, next to the ocean, and a second chain, the Rocky Mountains, 500–750 km farther inland. In addition, there are intervening ranges between these principal chains. The landmass in Chile terminates at approximately 54°S, while the landmass and mountains in North America extend north of 70°N.

Other differences between the west coast of South America (particularly Chile) and western North America are the size and orientation of the river basins. As a result of the orientation of the Andes and their proximity to the ocean, most rivers run east–west and the dimensions of the river basins are relatively small. In North America, the dual mountain chains and their faulted structures result in larger river basins because some rivers flow north–south for a portion of their course in the valleys and trenches between these mountain ranges. Rivers with their headwaters on

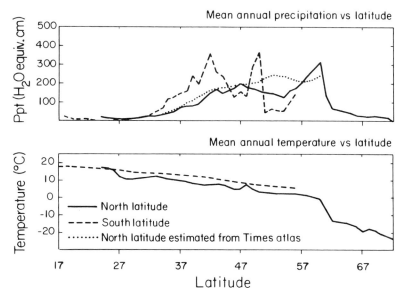

Fig. 1. Latitudinal variations in precipitation and temperature along the west coasts of North and South America. The dotted curve for precipitation was estimated from the Times Atlas and shows a significant difference between coastal precipitation amounts reported by MacDonald (1977) for latitudes from 50° to 55°N and those reported on standard precipitation maps. (South American data from Dirección General de Aguas, 1987.)

the eastern side of the Rocky Mountains flow east to the Hudson Bay, north to the Beaufort Sea, and south to the Gulf of Mexico.

III. Geomorphologic Processes

A. Similarities

In both North America and South America a range of geomorphologic processes, including coastal processes and fluvial processes, are at work. Mountain regimes on the western coasts of both continents have glaciers, but the higher incidence of glaciers in North America results in more widespread glacial erosion. The mountains in Canada have been shaped by glacial action. Slopes in both the Andes and the Rocky Mountains are relatively steep, leading to alluvial processes in both hemispheres. However, more dry geomorphic processes, such as landslides, take place in the arid regions of northern Chile, where mountain slopes are steeper.

Coastal processes are important in both North America and South America. Each western coast is affected by ocean currents, which flow eastward toward the coasts at 40°N in North America and 40–42°S in South America. The shape of the coastline determines where erosion and sediment accumulation will take place as these currents flow along the coast. Fluvial processes are also important in both hemispheres.

Through the processes of erosion and downcutting, rivers have changed the landscape extensively. The steeper slopes of the Andes result in higher stream velocities and erosion rates, while the longer paths of rivers to the ocean in North America lead to larger river flows with greater potential for bank cutting.

Eolian processes operate in both North and South America. They tend to be important in dune restructuring and movement in more arid climates; hence the northern parts of Chile are strongly affected. The longer rivers, higher sediment loads, and lower slopes associated with the mountain rivers in North America lead to the formation and maintenance of deltas at the mouths of a number of the major rivers, such as the Fraser and the Mackenzie Rivers.

B. Differences

Geomorphologic patterns in North America and South America are strongly affected by the location of the mountains. The location of mountains in the Southern Hemisphere closer to the equator results in dominance of arid region geomorphologic processes, that is, processes with episodic high-runoff events causing erosion. Average erosion rates are higher for the west coast of North America. Consequently, under global warming, a possibly more intense hydrologic cycle combined with warmer temperatures in the Northern Hemisphere could accelerate erosion processes.

The more poleward extension of mountains in North America results in more cold region geomorphic processes in this continent. For example, permafrost is more important in North America. It follows that climatic warming could be expected to have more impact on geomorphic processes in North America.

IV. Hydrology

A. Similarities

The hydrology of the western parts of the two continents is controlled by the interactions of the mountains with the atmosphere. Precipitation is highest on the western coasts north of 40°N and south of 38°S. Because of their elevation, a large proportion of the precipitation in these mountains at these latitudes falls as snow. In general, glaciers become important at higher latitudes. The main glacial contributor to Chile's streamflow is the Patagonia glacier, situated between 48 and 52°S. Its northern field covers 4400 km^2 and its southern field covers 13,500 km^2. In North America, alpine glaciers are more numerous. However, individually they are considerably smaller in dimension than the Patagonia. As a result, their contributions to river runoff are spread over a much larger geographic area.

B. Differences

The hydrology of the Northern and Southern Hemispheres differs, primarily because of the dominant atmospheric circulation patterns at corresponding latitudes. The large moisture convergence fluxes in flows in autumn at latitudes of 50 to 55°N lead to very high precipitation rates on the west coast of British Columbia, Canada.

Although relatively large annual precipitation amounts occur in southern Chile, in general they are not as large as those encountered on the west coast of British Columbia and Washington State. In part, this occurs because the ruggedness of the mountains in Chile decreases south of 50°S. The short-term precipitation rates along the west coast of North America also tend to be larger than those experienced in Chile. In addition, the quantities of sediment transported to the Pacific Ocean from Chilean rivers are much smaller than the quantities in North American rivers.

Moving from south to north in the Northern Hemisphere, snowmelt becomes a more significant contributing factor to runoff. Peak flows in the interior basins of North America at latitudes north of 50°N are clearly associated with snowmelt. In Chile, the influence of glaciers may also be important at latitudes between 48 and 52°S. Peak flows in rivers originating in these areas would be expected to be correlated with summer temperatures, suggesting that glacier melt could contribute most significantly to runoff south of 48°S. In North America, smaller glacier contributions occur over the 50–70°N latitude band.

V. Responses to Global Change

The responses of both hemispheres to global change will be highly dependent on the influence of global warming on the atmospheric circulation. According to equilibrium global climate models, the greenhouse warming will increase with latitude but will be less over oceans and coastal areas. In particular, the Canadian Climate Centre's GCM (IPCC, 1990) projects changes of 2.5–4.5°C along the west coast of North America with only minor latitudinal amplification. Increased sea surface temperatures will lead to increased moisture in the low-level air flowing inland from the oceans and higher precipitation rates on the mountains (provided wind intensities do not decrease significantly). Increased precipitation rates will lead to larger runoff values, deeper winter snowpacks, and greater glacier ice accumulation at higher latitudes. Conversely, the warmer temperatures will lead to a later start for the snow accumulation season, an earlier start for snowmelt, and a general decrease in the size of many glaciers (unless warming trends are compensated for by the effects of higher precipitation). In general, these trends will be the same in both hemispheres, with moist areas having increased runoff due to increased precipitation and arid regions having decreased runoff resulting from higher temperatures and evaporation rates.

With global warming, fluvial geomorphic processes will be enhanced in areas of high runoff and glacial melt. Local differences will likely occur, but the resolution of the climate model outputs is not yet adequate to address these differences.

An overriding influence on the hemispheric responses to global change could be heat absorption by the oceans. Oceans may effectively absorb the excess heat energy arising from the increasing atmospheric CO_2 concentrations for many decades. According to transient model results presented by the IPCC (1990), the Southern Hemisphere may warm much more slowly than the Northern Hemisphere as CO_2 concentrations gradually increase. Differences between the hydrologic responses of the two hemispheres to global change will be determined in large measure by the way in which atmospheric circulation patterns and regional temperatures change in response to global warming. Cloud formation and land surface feedback processes are other confounding influences that could regulate the rate of global warming.

VI. Summary

In summary, the western coasts of North and South America have many similar hydrologic and geomorphic patterns and processes. However, there are some significant differences in landform and land distribution that could lead to significant differences in their overall responses to global warming. Without further research on these atmospheric controls and feedbacks it will not be possible to determine whether the similarities or differences will prevail as individual ecosystems respond to global change.

References

Dirección General de Aguas. (1987). *Balance Hidrico de Chile.* Inscripcion No. 70115, Litografia Marinetti SA.

Intergovernmental Panel on Climate Change (IPCC). (1990). "Climate Change—The IPCC Scientific Assessment" (J. T. Houghton, G. J. Jenkins, and J. J. Ephramus, eds.), p. 339. Cambridge University Press, Cambridge.

MacDonald, K. B. (1977). Plant and animal communities of Pacific North American salt marshes. *In* "Ecosystems of the World," Vol. 1, "Wet Coastal Ecosystems," pp. 167–190. Elsevier Science Publishing, New York.

Part IV: Biogeochemistry

C H A P T E R 9

Response of Major North American Ecosystems to Global Change: A Biogeochemical Perspective

BARBARA I. KRONBERG

I. Introduction

The natural distribution and composition of vegetation zones are determined by interactions between climate, biota, and soils. During the past 2 million years, planetary-scale climate oscillations have been the predominant factor forcing dramatic changes in the distribution of continental ecosystems. However, in the most recent interglacial period, beginning approximately 10,000 years ago, human activities have become increasingly significant in modifying landscapes.

In this overview of possible responses of major North American ecosystems to global change, emphasis is given to biogeochemical processes that affect carbon exchanges between these systems and the atmosphere. In forested terrains the major pathway for these transfers is the fixation of carbon dioxide in life systems via photosynthesis and its release via respiration. Tundra and wetland systems contribute significant amounts of carbon as methane to the atmosphere. Because of the relatively low carbon and water storage capacities of grasslands and arid regions, climate–landscape interactions are more closely coupled than in systems with larger carbon and water budgets.

A broader scope is maintained in this overview primarily because of the paucity of information available on biogeochemical interactions. Also, in future planning of field studies, this information may be useful for guiding the selection of variables to be measured.

II. Tundra Ecosystems

The characteristics features of tundra ecosystems are (1) the absence of tree canopy, (2) the importance of organic soils relative to those dominated by clay and rock-forming minerals, and (3) the presence of permafrost, which limits decomposition and microbial activity (MacLean *et al.*, 1983).

The importance of these systems in the global carbon cycle is attributed to the high capacity of tundra soils for carbon storage (13–17 kg C/m^2; Post *et al.*, 1982; Prentice and Fung, 1990), accounting for about 15% of the global soil carbon budget, and their significant contributions (~10%) of methane (CH_4) to the global atmosphere (Cicerone and Oremland, 1988). The difficulty in predicting the response of these areas to climate warming is underscored by field and laboratory studies of methane fluxes across the soil–atmosphere interface. Studies by Whalen and Reeburgh (1990a) indicate significant variation in methane flux rates between sites. Mean CH_4 fluxes (mg/m^2 per day) ranged from 90 (wet tundra) to 0.6 (alpine tundra). These emission rates were only weakly correlated with soil temperature, water table depth, thaw depth, and depth of the organic soil layer. A study of tundra soils in the Aleutian Islands showed methane consumption rates exceeding those for emission in non-waterlogged soils. Surface vegetation consumed no methane, and the highest rates of methane consumption occurred in surface soil layers (0–5 cm). Strong biological controls on methane consumption in tundra soils were indicated by ^{14}C tracer experiments. These showed that about 50% of consumed $^{14}CH_4$ was assimilated into the microbial biomass and organic and inorganic matter, and about 50% was respired as $^{14}CO_2$. In artificial atmosphere experiments CH_4 oxidation increased with methane increases, at least at ambient temperatures of about 7°C, and final methane concentrations were up to 10 times lower than ambient atmospheric concentrations (Whalen and Reeburgh, 1990b).

Consumption of atmospheric CH_4 by tundra soils depends on transport to zones of consuming activity. Transport in waterlogged tundra soils is by aqueous molecular diffusion, and maximum methane consumption rates near the water table (oxic–anoxic boundary) are sustained by upward diffusion of dissolved CH_4. Transport in porous tundra soils is by gas diffusion (~10^5 times faster than aqueous), so the supply of methane is not limited. The consequences of atmospheric warming (such as a lower water table and increased seasonal thaw depths or permafrost melting) could increase the extent of oxidized tundra soil and CH_4 consumption. This possible negative feedback on atmospheric CH_4 concentration may be mitigated by some vascular plants, which transport CH_4 and bypass the CH_4-oxidizing zone.

Gas hydrates stored at high latitudes are among the largest potential sources of CH_4 emission to the atmosphere. Most of these hydrates (10^{15}–10^{18} m^3; Kvenholden, 1988) are stored in shelf and continental slope environments. However, the much smaller land reservoir (3–6 × 10^{13}; Makogon, 1982) could be the more important source of CH_4 addition to the atmosphere, if simulated increases in Arctic air temperatures (10–15°C) are realized. There is tentative evidence that the permafrost surface has warmed by as much as 2–4°C during the last few decades (Lachenbruch and Marshall, 1986). Another study of a high Arctic ice core indicates that summers over the past century have been the warmest for more than 1000 years, but not as warm as those of the early Holocene (Koerner and Fisher, 1990).

III. Wetlands

Between latitudes 50 and 70°N are vast regions in which, because of high water tables, plant production has exceeded decay and at least 40 cm of peat has accumulated. Most of the world's peatlands are located in Canada (150 Mha or 54%) and Eurasia (30%). The United States contains approximately 3% of global peatlands. In North America most peatlands occur in the central and northwest region of the boreal forest, where summers are cool. The effects of climate on wetland dynamics and peat accumulation are not well understood. Peatland development seems to be constrained by rough terrain, cold arctic summers, and summer moisture deficits. Along the northern border of the boreal forest, the period of greatest peat accumulation was the early to middle Holocene, when summers were warmer than present; along the south and central borders peatland development was delayed until the middle to late Holocene, when summers became cooler and wetter. In northern wetlands permafrost occurs in about 60% of peatland and has a major influence on vegetation patterns, hydrology, and carbon balance. In the discontinuous permafrost zone, peatlands show signs of increased thawing during the past 40 years, possibly reflecting climate warming since the Little Ice Age (Ovendene, 1989).

It is considered that wetland terrains between 50 and 70°N contribute approximately 60% (63×10^{12} g/year) of total CH_4 wetland emissions to the atmosphere (Matthews and Fung, 1987). There is also evidence for seasonal fluctuations in methane levels. Measurements show an annual minimum followed by a rapid climb to an annual peak in late autumn. This is unexpected, given the intense destruction of CH_4 by OH from summer to autumn and the transport of CH_4 during the northern summer from the Northern to the Southern Hemisphere (Blake and Rowland, 1988). The implication of CH_4 autumn rises at northern latitudes is that there are seasonal sources of CH_4 that produce strongly enough to outweigh summer losses. A most likely source of the large late summer–autumn increases in global methane is peatland. These ecosystems have very high summer biological productivity because of their high temperature (maximum averages 21–26°C, maxima ranging to 40°C) and light levels. The autumn methane signal could be enhanced by overturn and exclusion of CH_4 as northern forest wetlands and lakes freeze (Nisbet, 1989).

Recent increases in wetland area may be attributed to increasing populations of beaver *(Castor canadensis)* since 1950 as a result of conservation measures, the decline in beaver as a food source, and the low price for their pelts. Beavers cut trees and remove these to ponds and streams. In so doing, each beaver may introduce into aquatic ecosystems up to a metric ton of wood every year. Carbon from these sources in beaver-influenced aquatic ecosystems is converted to CH_4 much more efficiently than in waters without beavers. Methane production from beaver ponds has been shown to be, per unit length, about 100 times greater than that from an equivalent length of unmodified stream (Naiman *et al.,* 1986). Rates of methane emission from surfaces of beaver ponds are up to 33 times those from aquatic surfaces not influenced by beavers. In drainage networks with beavers, some 90% of methane evasion may be related to their activities.

IV. Forests

North American forests are estimated to cover 0.7×10^9 ha (WRI, 1990–1991). Throughout the temperate latitudes of North America there is a series of forests, which vary from mixed broadleaf to pine and boreal coniferous forests as altitude and/or latitude increases. Along these gradients primary productivity, respiration, and decomposition rates decline and physical and chemical soil processes and turnover of matter and energy become less intense, so there are larger accumulations of soil carbon as latitude and altitude increase. The largest forests are the boreal systems, which cover vast circumpolar regions between 48 and 60°N. These forests are transitory and have been repeatedly disrupted by the major ice sheet advances of the past 2–3 million years, during which these ecosystems were relegated to niches with very different light and climate conditions (Delcourt and Delcourt, 1987).

Forests act as semitransparent-porous exchange layers between the ground and atmosphere. In northern forests, the insulation keeps soil temperatures low during summer, limiting biological activity. The boreal forests are structurally adapted to the climatic environment. The many daylight hours during the vegetation period, the high absorption and low reflection of global shortwave radiation by the conifers (albedo 5–7%), and crown shapes adapted to intercept sunlight at low angle make high daily rates of net photosynthesis possible during midsummer in boreal regions (Bruenig, 1987). The main feature of boreal soils is that large amounts of nutrients, particularly nitrogen, are held in the litter and organic soil pools, for example, 1100 kg/ha in boreal forest litter (MacLean *et al.,* 1983). The accumulation of humus in northern climates removes nutrients from circulation and nutrient supply may become limiting to plant growth.

In temperate and boreal forests, spontaneous fire is an important ecological regulating mechanism. In these ecosystems fire results in a major reorganization of nutrients because of the redistribution of nutrients during fire and postfire changes in nutrient cycling. Redistribution processes during fire include (1) oxidation of biomass to gaseous products (e.g., carbon, nitrogen, and sulfur oxides) and particulate matter (products of incomplete combustion) and (2) accumulation of nutrient-rich debris on the soil surface. In temperate forests nitrogen losses may begin at 200°C and volatilization increases to about 60% at 700°C. Few temperature measurements are available for boreal regions. One study indicates mean temperatures of 300–400°C (Smith and Sparling, 1966). Nitrogen losses of 500–800 kg/ha have been reported (Grier, 1975). Losses of inorganic nutrients (phosphorus, potassium, calcium, magnesium are less (<100 kg/ha for calcium and potassium and <40 kg/ha phosphorus and magnesium). Few data are available on the oxidation of soil organic matter during fires. In boreal forests, field observations include losses of one-third to all of organic matter in soil in jack pine stands and 24–62% reduction in forest floor thickness. With large quantities of nutrients tied up in soil organic matter in northern forests, these effects could conceivably be among the most important consequences of fire (MacLean et al., 1983).

Indirect effects of fire include changes in plant growth and species composition, which may be dramatic, due to increased soil nutrient and light availability. After fire, the energy budget of the soil surface changes. These effects are especially important in permafrost terrains for enlarging nutrient pools. There is little information on how postfire conditions affect nitrogen fixation and other processes associated with the plant–microbiological–litter–soil organic matter interface. Volatilization losses of nitrogen for the entire ecosystem may be large. Increases in temperature and availability of cations may result in higher decomposition and nitrogen fixation rates.

It is difficult to forecast the potential for climate-induced alteration of fire frequency and intensity (Clark, 1988). The limited information available indicates that fire patterns responded to climate changes of the past few centuries. Fire cycles appear to be driven by climate and dynamics of fuel loading, consisting of short-term less intense fires superimposed on less frequent intense burns. It is considered that increased evapotranspiration and decreased precipitation trends observed in this century would probably have resulted in at least a 25% increase in fire frequency in the absence of fire suppression. With continued fire suppression and further warming, fuel buildup could result in more intense and/or more frequent fires.

Carbon exchanges between forests and the atmosphere are evident from the annual oscillations superimposed on the atmospheric CO_2 concentration curve (Woodwell, 1987). These oscillations are attributed to the net decrease (in autumn) and increase (in spring) in productivity in the northern temperate and boreal forests, which are major terrestrial carbon reservoirs. The influence of forests on the atmospheric carbon budget is underscored in the rapid (<1 month) response, in spring and autumn, of atmospheric CO_2 levels (Moore and Bolin, 1987). The impact of boreal forests on this signal is evident from the larger amplitude (8 ppm) measured at Alaska (77°N) than that (~3 ppm) measured at Hawaii (20°N) (D'Arrigo et al., 1987).

Throughout the 1980s a consistent year-by-year increase in amplitude has been measured (Woodwell, 1987) and presumably is caused by changes in rates of photosynthesis and respiration in the seasonal northern forests. The ratio of gross photosynthesis to total respiration is affected by several factors, including temperature, availability of energy, water, and nutrients; however, respiration appears to be especially responsive to changes in temperature (Woodwell and Dykeman, 1966). A significant fraction (20–30%) of global respiration on land takes place in the forest and tundra systems of the middle to high latitudes, where warming is expected to be greatest (Houghton and Woodwell, 1989). A shift in rates of photosynthesis and respiration could also be due to the conversion of old-growth forests to young forests (Harmon et al., 1990), which would decrease carbon storage capacity. More than 65% of Canadian boreal forests (McLaren, 1990) and about 95% of original U.S. forests (Deighton, 1990) have been logged. To replace the biomass and nutrients removed by this activity may take 60 to 80 years (Likens *et al.*, 1978). Deforestation combined with a general warming trend in this century has been offered as an explanation for the changes in the annual accumulation of atmospheric CO_2 from 1.5 to 2.4 ppm since 1988 (Houghton and Woodwell, 1989).

V. Grasslands and Deserts

Biogeochemical cycles in semiarid and arid lands are particularly susceptible to spatial and temporal disruptions due to climate changes as well as human exploitation. A key difference between semiarid grasslands and desert shrublands is the more heterogeneous distribution of water and nutrients in deserts. Here the soils supporting shrubs have higher nutrient and soil moisture contents than the more easily degraded intershrub soils (Schlesinger *et al.*, 1990). The close coupling between ecosystem components in deserts is underscored by a study showing the important contribution to soil nitrogen budgets of snails, which ingest rock and lichens and deposit their digested nitrogen-rich wastes in soils (Jones and Shachak, 1990).

If the warmer and drier climate conditions predicted for much of North America are realized, dryland areas will expand. A review of potential effects of climate change on North American agriculture highlights the uncertainties in current knowledge (Adams *et al.*, 1990). National Ocean and Atmosphere Administration (NOAA) and Goddard Institute for Space Studies (GISS) models predict precipitation changes that would leave some major grasslands drier and others wetter. An important effect of climate change may be an increase in the frequency and severity of fires in grasslands. The reasons are complex: (1) some ranges will be prone to fire because they have less rainfall; (2) CO_2-enriched plants may produce more fibrous litter that could serve as fuel; (3) in many regions relative humidity may fall below 65%—the threshold at which range grasses and shrubs become flammable—more frequently or for longer periods.

The factors involved in climate-driven expansion of drylands are interrelated and include declines in soil moisture, leading to higher land surface temperatures and consequently retardation of organically bound soil nitrogen as well as plant growth. The net effect is a positive feedback reinforcing a more heterogeneous pattern of water and nutrient reserves. A concomitant effect will be intensified erosion of dryland soils, which would favor the removal of fine nutrient-rich soil organic and clay particles, thus enhancing soil degradation.

VI. Transition Zones

The initial response of North American ecosystems to CO_2-induced climate change is likely to be greatest along ecosystem boundaries. The long-term response of the shift to warmer, drier climates, predicted to accompany the doubling of atmospheric CO_2 levels, would be migration of boreal and temperate forests northward. Models suggest that along the boreal–cool temperate forest boundary, tree growth will be enhanced on soils that retain adequate water and will decline on moisture-deficient soils (Pastor and Post, 1988). Thus, different tolerances of species to water stress may result in their segregation along moisture gradients in the landscape. This in turn influences rates of nitrogen and carbon cycling because species differ in growth rates, nutrient uptake, litter production, and release of nitrogen and carbon from decomposing litter. As nitrogen and water are major factors limiting forest growth, the responses of trees to drier climates and the resulting feedbacks to vegetation–soil dynamics may influence the regional and possibly global distribution of carbon and nitrogen between geospheres.

The tentative information available indicates a northward shift in the northern boundary of the boreal forest by 100–700 km and in the southern boundary by 250–900 km. Predicted biomass productivity changes include decreases from normal of 2–12% for southern locations and increases from normal of about 50% in the northern part of the forest zone (Climate Change Digest, 1989). The predictions for the southern boreal zones is that soil moisture deficits would restrict growth (Kauppi and Posch, 1988).

Results for the boreal–prairie boundary (Schindler *et al.*, 1990) indicate that over the past two decades both air and lake temperatures have risen by 2°C and snow cover has diminished. Moisture losses due to decreased snowfall were accompanied by more frequent forest fires, resulting in increased particulate fluxes into lakes, which took longer to renew themselves because of the decreased spring runoff. These findings indicate the complex and unpredictable chain of events that climatic change may set in motion.

At the desert–grassland transition, long-term studies illustrate the tight coupling between vegetation and animal dynamics. In one situation a small perturbation in material and energy flows caused by removing "keystone" rodent species was amplified and resulted in the transformation of a "desert shrub dominated" habitat into one typical of arid grasslands (Brown and Heske, 1990).

VII. Conclusions

The simulated changes in distribution of major North American ecosystems as a consequence of doubling atmospheric CO_2 levels include decreases in the extent of boreal forest ecosystems and increases in dryland areas and possibly temperate forests (Shugart *et al.*, 1986). These changes in ecosystem configurations would have an impact on, at least, regional biogeochemical cycles. The assessment of change requires both qualitative and quantitative information on trends in ecosystem patterns and processes.

References

Adams, R. M., Rosenzweig, C., Peart, R. M., Ritchie, J. T., McCarl, B. A., Geyer, J. D., Curry, J. B., Jones, J. W., Boote, K. J., and Allen, L. H. (1990). Global climate change and US agriculture. *Nature* **345**, 219–224.

Blake, D. R., and Rowland, F. S. (1988). Continuing worldwide increase in tropospheric methane, 1978–1987. *Science* **239**, 1129–1131.

Brown, J. H., and Heske, F. J. (1990). Continuing worldwide increase in tropospheric methane, 1978 to 1987. *Science* **250**, 1705–1707.

Bruenig, E. F. (1987). The forest ecosystem: tropical and boreal. *Ambio* **16**(2/3), 68–79.

Cicerone, R. J., and Oremland, R. S. (1988). Biogeochemical aspects of atmospheric methane. *Global Biogeochem. Cycles* **2**, 299–327.

Clark, J. S. (1988). Effect of climate change on fire regimes in northwestern Minnesota. *Nature* **334**, 233–235.

"Climate Change Digest: Exploring the Implications of Climatic Change for the Boreal Forest and Forestry Economics of Western Canada." (1989). Environment Canada, CCD 89-02.

D'Arrigo, R., Jacoby, G. C., and Fung, I. Y. (1987). Boreal forests and atmosphere–biosphere exchange of carbon dioxide. *Nature* **329**, 321–323.

Deighton, L. (1990). New life for old forest. *New Sci.* **13** (October), 25–29.

Delcourt, P. A., and Delcourt, H. R. (1987). "Long-Term Forest Dynamics of the Temperate Zone." Springer-Verlag, New York.

Emmanuel, W. R., Shugart, H. H., and Stevenson, M. P. (1985). Climatic change and the broad-scale distribution of terrestrial ecosystem complexes. *Climatic Change* **7**(1), 31–43.

Grier, C. C. (1975). Wildfire effects on nutrient distribution and leaching in a coniferous ecosystem. *Can. J. For. Res.* **5**, 599–607.

Harmon, M. E., Ferrell, W. K., and Franklin, J. F. (1990). Effects on carbon storage of conversion of old-growth forests to young forests. *Science* **247**, 699–702.

Houghton, R. A., and Woodwell, G. M. (1989). Global climate change. *Sci. Am.* **260**(4), 36–44.

Jones, C. G., and Shachak, M. (1990). Fertilization of the desert soil by rock-eating snails. *Nature* **346**, 839–841.

Kauppi, P., and Posch, M. (1988). A case study of the effects of CO_2-induced climatic warming on forest growth and the forest sector: productivity reactions of northern boreal forests. *In* "The Impact of Climatic Variations on Agriculture," Vol. 1, "Assessments in Cool Temperate and Cold Regions" (M. L. Parry, T. R. Carter, and N. T. Konijn, eds.), pp. 183–195. Kluwer, Dordrecht.

Koerner, R. M., and Fisher, D. A. (1990). A record of Holocene summer climate from a Canadian high-Arctic ice core. *Nature* **343**, 630–631.

Kvenholden, K. A. (1988). Methane hydrates and global climate. *Global Biogeochem. Cycles* **2**, 221–229.

Lachenbruch, A. H., and Marshall, B. V. (1986). Methane hydrate and global climate. *Science* **234**, 689–696.

Likens, G. E., Bormann, G. H., Pierce, R. S., and Reiners, W. A. (1978). Recovery of a deforested ecosystem. *Science* **199**, 492–496.

MacLean, D. A., Woodley, S. J. Weber, M. G., and Wein, R. W. (1983). Fire and nutrient cycling. *In* "The Role of Fire in Northern Circumpolar Ecosystems," pp. 111–131. Wiley, New York.

Makogon, Y. F. (1982). Perspectives for the development of gas-hydrate deposits. *In* "Proceedings 4th Canadian Permafrost Conference" (H. M. French, ed.), pp. 299–304. National Research Council of Canada, Ottawa.

Matthews, E., and Fung, F. (1987). Methane emission from natural wetlands: global distributions, and environmental characteristics of sources. *Global Biogeochem. Cycles* **1**, 61–86.

McLaren, C. (1990). Heartwood. *Equinox* **53**, 43–55.

Moore, B., and Bolin, B. (1987). The oceans, carbon dioxide and global climate change. *Oceanus* **29**(4), 9–15.

Naiman, R. J., Melillo, J. M., and Hobbie, J. E. (1986). Ecosystem alteration of boreal forest streams by beaver *(Castor canadensis)*. *Ecology* **67**, 1254–1269.

Nisbet, E. G. (1989). Some northern sources of atmospheric methane: production, history and future implications. *Can. J. Earth Sci.* **26**, 1603–1611.

Ovendene, L. (1989). Peatlands: a leaky sink in the global carbon cycle. *GEOS* **18**(3), 19–24.

Pastor, J., and Post, W. M. (1988). Responses of northern forests to CO_2-induced climate change. *Nature* **334**, 55–58.

Post, W. M., Emmanuel, W. R., Zinke, P. J., and Stangeberger, A. G. (1982). Soil carbon pools and world life zones. *Nature* **298**, 156–159.

Prentice, K. C., and Fung, F. Y. (1990). The sensitivity of terrestrial carbon storage to climate change. *Nature* **346**, 48–50.

Schindler, D. W., Beaty, K. G., Fee, E. J., Cruickshank, D. R., DeBruyn, E. R., Findlay, D. L., Linsey, G. A., Shearer, J. A., Stainton, M. R., and Turner, M. A. (1990). Effects of climatic warming on lakes of the central boreal forest. *Science* **250**, 967–970.

Schlesinger, W. H., Reynolds, J. S., Cunningham, G. L., Huenneke, L. F., Jarrel, W. M., Virginia, R. A., and Whitford, W. G. (1990). Biological feedbacks in global desertification. *Science* **247**, 1043–1048.

Shugart, H. H., Antonovsky, M. Ya., Jarvis, P. G., and Sandford, A. P. (1986). CO_2, climatic change and forest ecosystems. *In* "The Greenhouse Effect, Climate Change and Ecosystems," pp. 475–521. Wiley, New York.

Smith, D. W., and Sparling, J. H. (1966). The temperature of surface fires in jack pine barrens: the variation of temperature with time. *Can. J. Bot.* **44**, 1285–1292.

Whalen, S. C., and Reeburgh, W. S. (1990a). Methane flux transect along the trans-Alaska pipeline haul road. *Tellus* **42B**, 237 –249.

Whalen, S. C., and Reeburgh, W. S. (1990b). Consumption of atmospheric methane by tundra soils. *Nature* **346**, 160–162.

Woodwell, G. M. (1987). Forests and climate: surprises in store. *Oceanus* **29**(4), 71–75.

Woodwell, G. M., and Dykeman, W. R. (1966). Respiration of a forest measured by carbon dioxide accumulation during temperature inversions. *Science* **154**, 1031–1034.

WRI (1990–91). "World Resources 1990–1991," Table 19.1. Oxford University Press, New York.

C H A P T E R   10

Terrestrial Biogeochemical Feedbacks in Global Warming: Some Predictions for South America

EUGENIO SANHUEZA

I. Introduction

The global radiative balance of the atmosphere largely determines the earth's climate; during the present postglacial epoch, and before humans significantly perturbed the composition of the atmosphere, the absorbed and the emitted radiation were in equilibrium on the global scale. This balance depends mainly on the input of solar radiation and the abundance of greenhouse gases, clouds, and aerosols. Figure 1 is a schematic representation of the global energy balance. About 70% of the solar radiation that reaches the earth's atmosphere is absorbed by the earth's surface and atmosphere. The other 30% is reflected back to space by clouds and the earth's surface.

In accordance with its temperature, the earth emits radiation in the infrared region of the spectrum, which can be absorbed by many atmospheric gases. These radiatively active gases (greenhouse gases) reemit in the infrared and the earth's surface receives an additional 95 units of radiation. In this way, the surface is heated to a higher temperature (15°C in average) than it would be without the greenhouse gases (−18°C). Therefore the natural greenhouse effect keeps the surface warmer by about 33°C, providing the climate to support earth's living forms.

The atmospheric concentrations of greenhouse gases (i.e., CO_2, CH_4, chlorofluorocarbons, N_2O) have increased as a result of human activities. The influence on the radiation balance of a particular greenhouse gas depends on various factors:

131

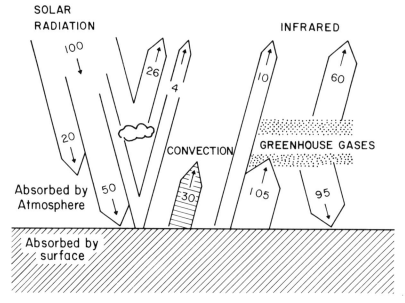

Fig. 1. Energy balance of the earth. (Adapted from Siegenthaler and Sanhueza, 1990.)

atmospheric concentration, lifetime, and ability to absorb outgoing longwave terrestrial radiation. Table I summarizes the concentrations, lifetimes, rates of increase, and radiative forcing of these gases. Excluding water vapor, CO_2 is the main gas responsible for the radiative forcing (~50%), but the contribution of the sum of all the other gases is comparable to that of CO_2.

Based on assumptions about the future growth of the economy and population and about energy policies, scenarios for future emissions of greenhouse gases have been developed by the Intergovernmental Panel on Climate Change (IPCC), and the predicted future changes in atmospheric concentrations as well as their climatic effects have been estimated by means of different general circulation models (GCMs) (Mitchell *et al.,* 1990).

As a consequence of global warming, several feedback effects may become active. For instance, the increase of atmospheric water vapor (the most important greenhouse gas) will amplify considerably the initial warming. Other effects will be decrease of the earth's albedo due to a reduction in snow- and ice-covered areas, changes in cloud cover, and changes in the biogeochemical cycles.

A major factor that regulates trace gas fluxes in terrestrial ecosystems is the physical (temperature, rainfall) and chemical (the chemical composition of the atmosphere) climate of the system. Future climate changes will modify the biogeochemical cycles of various greenhouse gases (CO_2, CH_4, N_2O), and in this chapter possible terrestrial feedback mechanisms are discussed. Because of many uncertainties in the evaluation of the biogeochemical feedbacks, they have not been incorporated in the

TABLE I
Characteristics of Greenhouse Gases

Characteristic	CO_2	CH_4	N_2O	O_3	CFCs
Preindustrial atmospheric concentration	280 ppmv	0.8 ppmv	288 ppbv	—	0
Current atmospheric concentration (1990)	353 ppmv	1.72 ppmv	310 ppbv	~20 ppbv[a]	[b]
Current annual rate of atmospheric increase (%)	0.5	0.9	0.25	0.5[a]	4
Atmospheric lifetime (years)	50–200[c]	10	150	~0.1	>65
Specific potential relative to CO_2[d]	1	32	150	2000	>14,000
Greenhouse forcing 1980–1990 (%)	50	19	4	8	19

Source: Adapted from Siegenthaler and Sanhueza (1990).
[a]Northern Hemisphere.
[b]CFC-11, 280 pptv; CFC-12, 484 pptv.
[c]The way in which CO_2 is absorbed by the oceans and biosphere is not simple and a single value cannot be given.
[d]Impact of one additional molecule released into the atmosphere at present concentrations.

GCMs to estimate future warming. It is likely that biogeochemical feedbacks would lead to more rapid climate change than is predicted by the GCMs.

II. Carbon Dioxide

Carbon dioxide is continuously exchanged (cycled) among various reservoirs: atmosphere, oceans, land biota, marine biota, and, on geological time scales, sediments and rocks (Fig. 2). As documented by ice core measurements, these fluxes must have been balanced in the past (Delmas *et al.*, 1980; Neftel *et al.*, 1982). The net inputs into the atmosphere from fossil fuel combustion and deforestation are relatively small but are large enough to modify the natural balance, and the atmospheric CO_2 concentration currently is rising at about 1.8 ppmv (0.5%) per year due to these atmospheric perturbations (Keeling *et al.*, 1989).

A. Biogeochemical Feedbacks on CO_2 Fluxes

Under nontropical conditions, photosynthesis and respiration (plants and microbes) increase with increasing temperature. However, because respiration is the more sensitive process, global warming may result in a period of net release of CO_2 to the atmosphere. The magnitude of this release is uncertain, but it is estimated that this additional global emission of CO_2 might be larger than 1 Gt of carbon per year (Watson *et al.*, 1990). Climate is also an important controller of the decomposition of both surface litter and soil organic matter (SOM), and there is concern that global warming will accelerate the process on a global scale.

At present, it is not possible to predict in a reliable way the geographic distribu-

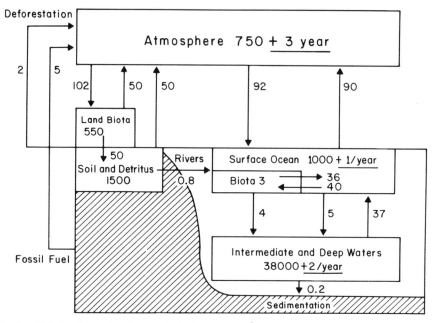

Fig. 2. Global carbon cycle. Flows are in Gt C per year (10^9 tonnes of C per year) and carbon reservoirs in Gt C. (From Watson *et al.,* 1990.)

tion of changes in soil moisture (Mitchell *et al.,* 1990), and therefore it is difficult to estimate the possible effect of these changes on carbon cycling. Within a large uncertainty, the GCMs predict for South America (with few exceptions) a general decrease in soil moisture during both winter and summer (Mitchell *et al.,* 1990). The three GCMs used in the IPCC assessment predict a large decrease in soil moisture in the northern part of the South American continent (Venezuela–Colombia) for the period June–August and in the southern part (Chile–Argentina) for the period December–February. Therefore, it is likely that in South America the distribution of vegetation will shift in response to global warming (i.e., increasing arid and semiarid zones). Because less CO_2 will be stored in drier ecosystems, these shifts will lead to an increase of the release of carbon to the atmosphere.

B. CO_2 Fertilization

Increased atmospheric CO_2 can alter plant ecosystem metabolism. Short-term experiments under controlled conditions show that elevated levels of CO_2 increase the rates of photosynthesis and growth in most plants (Strain and Cure, 1985). Also, it seems that water stress can be alleviated, at least in the short term, when plants are exposed to increased atmospheric CO_2 (Melillo *et al.,* 1990). However, at present it is not clear whether the increase in CO_2 uptake will persist for more than a few growing seasons and there is no evidence that an elevated CO_2 concentration has

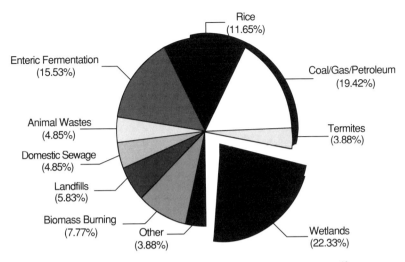

Fig. 3.   Sources of methane. The total emission is 515 Tg CH_4 per year (1 Tg = 10^{12} g). (From Watson *et al.*, 1992.)

increased net carbon storage in natural forests. If elevated levels of CO_2 increase the productivity of natural ecosystems (especially of forests which conduct about two-thirds of global photosynthesis), more carbon may be stored in woody tissue or soil organic carbon, producing a negative feedback on the atmospheric CO_2 increase. How the Amazon forest is reacting to the CO_2 increase is unknown. Field studies to answer this question are urgently needed.

III. Methane

The present atmospheric concentration of CH_4 is 1.7 ppmv, more than double its preindustrial value (Watson *et al.*, 1990). Ice core analyses indicate that atmospheric CH_4 was relatively constant during the 2000 years preceding industrialization (Etheridge *et al.*, 1988). Methane is produced mainly by a wide variety of anaerobic biogenic sources and some anthropogenic activities. Figure 3 summarizes the estimated sources of methane. The main sink is reaction with the hydroxyl (OH) radical in the atmosphere, 500–600 Tg CH_4 per year, resulting in an atmospheric lifetime of 10 ± 2 years, which is relatively short compared to that of other greenhouse gases. Removal by soil of 15–45 Tg CH_4 per year is also estimated (Watson *et al.*, 1990).

A. Biogeochemical Feedbacks on CH_4 Fluxes

Methane emissions from natural wetlands and rice paddies are sensitive to temperature and soil moisture, hence future climate changes would significantly alter the fluxes of CH_4 from these important sources. However, neither the size nor the sign of the change is known at present. Most methanogenic bacteria have temperature optima of 30–40°C; therefore, increasing the temperature would stimulate CH_4 production in most environments (Melillo et al., 1990).

The geographic distribution of freshwater natural wetlands indicates that about 1.5×10^6 km² (27% of the total) are located in South America; these are mainly swamps and floodplains in the tropical region (Aselmann and Crutzen, 1989). At present there are no estimates of how climate changes will affect this important source of atmospheric methane. However, since these systems tend to be carbon limited, it is likely that temperature change per se will not have a significant impact (Lashof, 1989).

Measurements in the Venezuelan savannah region have shown that "dry" soils are a small source of methane (Hao et al., 1988; Scharffe et al., 1990), whereas most soils in the temperate region are a sink for CH_4. It is likely that in a warmer world, with larger semiarid regions (as expected in South America), removal of CH_4 by soils will decrease, possibly leading to a positive warming feedback.

IV. Nitrous Oxide

The present-day atmospheric concentration of N_2O is 310 ppbv, about 8% greater than during the preindustrial era (Watson et al., 1990). The main sources are given in Table II; the estimates have large uncertainties. The main sink is photochemical destruction in the stratosphere, resulting in a long atmospheric lifetime of about 150 years. According to the values in Table II, it seems that either sources of N_2O are missing or the strength of some of the identified sources has been underestimated. The current annual rate of atmospheric increase of 0.25% is probably due to the growing use of nitrogen fertilizers, especially in the tropics (Prinn et al., 1990).

A. Biogeochemical Feedbacks on N_2O Fluxes

A very large percentage of global N_2O emissions is from soils (mainly from denitrifying bacteria); therefore, the potential for biogeochemical feedbacks in the N_2O flux is high. Tropical forests are the largest source of N_2O (Table II), and the Amazon forest plays an important role in the global N_2O budget. Data on soil fluxes of N_2O from South American tropical ecosystems are summarized in Table III. Results for rain forests show large variability (a factor of approximately 15), but it seems that there is no significant difference (or trend) between the dry and rainy seasons. On the other hand, in the savannah region there appears to be a significant seasonal variation in emissions, with lower emissions occurring during the dry season. The predicted decrease in soil moisture in South America, which will probably produce longer dry seasons, suggests a future decrease of the emissions of N_2O from savannah soils (negative feedback). In addition to climate changes, deforestation and land conversion in the tropics should lead to changes in the N_2O fluxes (see Section VI), making this region a priority for additional research.

At present there is no information about the N_2O fluxes in the temperate region of South America. However, larger arid and semiarid zones are predicted for this region, and it is likely that an increase of the aerobic condition of the soils will lead to decreased N_2O emissions to the atmosphere in the future.

TABLE II
Estimated Sources and Sinks of Nitrous Oxide

	Range (Tg N per year)
Sources	
A. Natural	
Oceans	1.4–2.6
Tropical soils	
Wet forests	2.2–3.7
Dry savannas	0.5–2.0
Temperate soils	
Forests	0.05–2.0
Grasslands	?
B. Anthropogenic	
Cultivated soils	0.03–3.0
Biomass burning	0.2–1.0
Stationary combustion	0.1–0.3
Mobile sources	0.2–0.6
Adipic acid prod.	0.4–0.6
Nitric acid prod.	0.1–0.3
Sinks	
Removal by soils	?
Photolysis in the stratosphere	7–13
Atmospheric increase	3–4.5

Source: Watson *et al.* (1992).

V. Tropospheric Ozone

Ozone is an important greenhouse gas, but because of its short lifetime in the troposphere (several weeks) its distribution in space and time is rather heterogeneous and therefore difficult to evaluate globally. According to Ramanathan *et al.* (1987), there is a strong relationship between the earth's surface temperature change and O_3 changes near the tropopause (i.e., 8–20 km in altitude). However, data for these altitudes are insufficient to indicate a clear trend.

Tropospheric ozone is transported down from the stratosphere and is also produced *in situ* by photooxidation of CH_4, CO, and nonmethane hydrocarbons (NMHCs) in the presence of NO_x (NO and NO_2). It is destroyed by vegetative surfaces and by reaction with the HO_2 radical.

In northern midlatitudes annual mean tropospheric concentrations seem to be a factor of 2–3 higher now than they were 100 years ago, and they have increased roughly 1% per year during the past 20 years (Watson *et al.*, 1990). In contrast, no increase or even decrease is indicated by the scarce data from the Southern Hemisphere (Janach, 1989). This difference between hemispheres is probably due to

TABLE III
N₂O Soil Fluxes from Undisturbed South American and Central America Ecosystems

Site	Season/obs.	N_2O (molecules cm^{-2} $s^{-1} \times 10^9$)
Tropical savannah climatic regions		
Venezuela, Chaguarama[a]	Dry/woodland	2.5
Venezuela, Chaguarama[a]	Dry/grassland	2.5
Venezuela, Chaguarama[a]	Simulated rainfall	10.0
Venezuela, Guri[b]	Rainy/grassland	1.7–10
Venezuela, Guri[b]	Rainy/semideciduous forest	15.0
Venezuela, Guri[c]	Dry/grassland	−2.6
Venezuela, Guri[c]	Dry/semideciduous forest	2.4
Venezuela, Calabozo[d]	Rainy/Trachipogon	~0.0
Tropical rain forest		
Brazil, Manaus[e]	Rainy (1983)	26
Brazil, Manaus[e]	Early rainy	7.8
Brazil, Manaus[e]	Rainy (1984)	18.5
Brazil, Manaus[f]	Dry (1985)	6.6
Brazil, Manaus[g]	Dry (1985)	13.8
Brazil, Manaus[h]	Dry	5.7
Brazil, Manaus[h]	Dry/sand	2.1
Brazil, Manaus[i]	1 year (1987–1988)	7.6
Brazil, Manaus[j]	Rainy/terra firme (1987)	11.4
Brazil, Manaus[j]	Rainy/sand (1987)	0.3
Brazil, Porto Alegre[h]	Dry	14.9
Brazil, Esteio[h]	Dry	9.7
Brazil, Campina[h]	Dry/sand	3.8
Ecuador, primary forest[e]	Dry (1984)	4.5
Ecuador, secondary forest[e]	Dry (1984)	4.4
Puerto Rico, Tabanoco[e]	Dry (1984)	25.3
Puerto Rico, Colorado-Palm[e]	Dry (1984)	1.7
Costa Rica, La Selva[e]	Rainy (1985)	12.8
Costa Rica, Turrialba[k]		9.9
Tropical cloud forest		
Puerto Rico, Elfin[e]	Dry (1984)	6.5
Venezuela, A. Pipe[d]	Dry (1989)	5.6

[a]Hao et al. (1988).
[b]Sanhueza et al. (1990).
[c]Donoso et al. (1990).
[d]IVIC, unpublished results.
[e]Keller et al. (1986).
[f]Livingston et al. (1988).
[g]Keller et al. (1988).
[h]Goreau and de Mello (1988).
[i]Luizao et al. (1989).
[j]Matson et al. (1990).
[k]Matson and Vitousek (1987).

different anthropogenic emissions (CO, NO_x) between both hemispheres and is related mainly to the concentration of NO_x in the atmosphere. In the Northern Hemisphere, in the presence of "high" concentrations of NO_x, the oxidation of CO and CH_4 leads to net production of ozone. In the Southern Hemisphere, at "low" concentrations of NO_x, ozone is consumed during the oxidation of CO, and in the case of methane HO and HO_2 radicals are consumed, decreasing the reactivity of the atmosphere.

A. Biogeochemical Feedbacks on O_3 Production

The photochemical production of O_3 depends on the concentrations of CO, CH_4, NMHCs and NO_x, and the effect of climate changes on the biogeochemical cycles of these compounds will affect the ozone concentrations. As discussed earlier, decomposition of soil organic matter will increase in a warmer world, and it is likely that CO emission will increase with higher soil temperatures (Conrad and Seiler, 1985; Scharffe *et al.*, 1990). Possible changes in CH_4 fluxes were discussed in a previous section. The concentration of NO_x is crucial in determining ozone formation; however, since NO_x fluxes from the surface depend on many variables (temperature, soil moisture, vegetation cover), at present, with the scarce information available, it is very difficult to estimate how NO_x fluxes will respond globally to climate and ecosystem changes.

The present trend of decreasing concentration of O_3 in the Southern Hemisphere may continue if CO (regional) and CH_4 (global) concentrations increase and the NO_x concentration remains the same. However, it is likely that in the long term, because of a future increase in anthropogenic emissions of NO_x, ozone levels will also increase in the Southern Hemisphere.

It is interesting that higher concentrations of atmospheric CO (due to larger emissions in a warmer world) will decrease the OH radical concentrations, leading to an increase in the atmospheric lifetime of CH_4 and thus to greater methane accumulation in the atmosphere.

VI. Tropical Deforestation

Natural forests hold 20 to 100 times more carbon per unit area than agricultural systems (Houghton, 1990). The carbon released globally between 1850 and 1985 because of changes in land use (primarily deforestation) has been estimated to be about 115 ± 35 Gt (Houghton and Skole, 1990). The greatest release of carbon in the nineteenth and early twentieth centuries was from the temperate zone, but during the past several decades the main contribution has come from deforestation in the tropics, with a significant increase since 1950. Estimates of the flux in 1980 range from 0.6 to 2.5 Gt C, which was practically all produced in the tropics. Tropical America accounts for about 50% of tropical deforestation (Houghton *et al.*, 1987). The current total release from tropical deforestation could be 2 to 3 Gt C per year (Houghton, 1990).

Myers (1990) proposes that there will be continued acceleration of deforestation in the tropics; "the last tracts of tropical forest could be eliminated far faster than one might expect." Several simulation studies of the climatic impact of complete deforestation of the Amazon Basin have been made at regional scales (Lean and Warrilow, 1989; Shukla *et al.,* 1990). The results show a regional reduction of about 20% in rainfall when forest is replaced by grassland. In the simulation by Shukla *et al.* (1990), the length of the dry season increased, which could prevent future forest recovery. At present it is not known how large-scale deforestation in the Amazon region would affect the climate in the rest of South America.

Deforestation produces a drastic change in the physicochemical characteristics of soil, alters the composition of flora and fauna, and transforms hydrology and microclimate (IGBP, 1990). At present, very little is known about how these changes alter production and consumption of trace gases by soil microorganisms. For instance, the impact of deforestation on the emissions of N_2O is unclear. Some studies suggest that emissions of N_2O from deforested land are enhanced (Bowden and Bormann, 1986; Luizao *et al.,* 1989), whereas others suggest that N_2 fluxes decrease when vegetation does not return (Goreau and de Melo, 1988; Robertson and Tieje, 1988; Sanhueza *et al.,* 1990).

A. Reforestation

It appears that one of the most cost-effective and technically feasible ways to counter greenhouse effects is reforestation (afforestation) (Myers, 1990). If deforestation were halted and massive reforestation took place, the net flux of CO_2 would be reversed. A complete analysis for the tropics has been made by Houghton (1990). According to Houghton, halting all deforestation would reduce emission by 2 to 3 Gt C per year and a massive reforestation program might remove from the atmosphere an average of 1.5 Gt C per year over the next 100 years; "reforestation could reduce the emission of CO_2 to the atmosphere indefinitely if wood fuels were to replace fossil fuels."

In Latin America (mainly in South America), the area available for reforestation is estimated to be 100×10^6 ha. The criteria for availability were that the land has supported forest in the past and that the land is unused for cropland or settlements at present (Houghton, 1990).

Massive reforestation would change not only the biogeochemical cycles of CO_2 but also those of the other greenhouse gases (CH_4, N_2O, ozone precursors). Since practically nothing is known about this situation, the long-term impact of these changes should be urgently evaluated.

VII. Summary

The global radiative balance of the atmosphere largely determines the earth's climate. This balance depends mainly on the input of solar radiation and the abundance of greenhouse gases, clouds, and aerosols. Future climate changes will modify the biogeochemical cycles of major greenhouse gases (CO_2, CH_4, N_2O, O_3). In general, within a large uncertainty, the general circulation models predict for South America a decrease in soil moisture. Therefore, it is likely that the distribution of vegetation will shift in response to global warming, leading to an increase in semiarid and arid zones. Positive feedbacks are predicted for CO_2 and CH_4. Less carbon would be stored in drier ecosystems, and removal of methane in drier soils would decrease. On the other hand, biogenic production of N_2O would decrease because of an increase of the aerobic conditions of soils. At present, it is difficult to predict how tropospheric ozone in the Southern Hemisphere will change with future climate changes. Deforestation of the Amazon forest is producing significant emission of CO_2; an afforestation program could reverse the net flux of CO_2 to the atmosphere.

References

Aselmann, I., and Crutzen, P. J. (1989). The global distribution of natural freshwater and rice paddies, their net primary productivity, seasonality and possible methane emissions. *J. Atmos. Chem.* **8**, 307–358.

Bowden, W. B., and Bormann, F. H. (1986). Transport and loss of nitrous oxide in soil water after forest clear-cutting. *Science* **233**, 867–869.

Conrad, R., and Seiler, W. (1985). Influence of temperature, moisture, and organic carbon on the flux of H_2 and CO between soil and atmosphere, field studies in subtropical regions. *J. Geophys. Res.* **90**, 5699–5709.

Delmas, R. J., Ascencio, J. M., and Legrand, M. (1980). Polar ice evidence that atmospheric CO_2 20,000 year B.P. was 50% of present. *Nature* **284**, 155–157.

Donoso, L. R., Santana, R., Crutzen, P. J., and Sanhueza, E. (1990). Seasonal variation of N_2O fluxes at a tropical savanna site: soil consumption of N_2O during the dry season. *7th International Symposium of CACGP,* Chamrousse, September 1990 (abstract).

Etheridge, D. M., Pearman, G. I., and de Silva, F. (1988). Atmospheric trace-gas variations as revealed by air trapped in ice core from Law Dome, Antarctica, *Ann. Glaciol.* **10**, 28–33.

Goreau, T. J., and de Mello, W. Z. (1988). Tropical deforestation: some effects on atmospheric chemistry. *Ambio* **17**, 275–281.

Hao, W. M., Scharffe, D., Crutzen, P. J., and Sanhueza, E. (1988). Production of N_2O, CH_4 and CO_2 from soils in the tropical savanna during the dry season. *J. Atmos. Chem.* **7**, 93–105.

Houghton, R. A. (1990). The future role of tropical forest in affecting the carbon dioxide concentration of the atmosphere. *Ambio* **19**, 204–209.

Houghton, R. A., and Skole, D. L. (1990). Changes in the global carbon cycle between 1700 and 1985. *In* "The Earth Transformed by Human Action" (B. L. Turner, ed.), pp. 393–408. Cambridge University Press, Cambridge, United Kingdom.

Houghton, R. A., Boone, R. D., Fruci, J. R., Hobbie, J. E., Melillo, J. M., Palm, C. A., Peterson, B. J., Shaver, G. R., Woodwell, G. M., Moore, B., Skole, D. L., and Myers, N. (1987). The flux of carbon from terrestrial ecosystems to the atmosphere in 1980 due to changes in land use: geographic distribution of the global flux. *Tellus* **39B**, 122–139.

IGBP. (1990). Tropical land-use change and trace-gas emissions. *In* "Terrestrial Biosphere Exchange with Global Atmospheric Chemistry" (P. A. Matson and D. S. Ojima, eds.), pp. 14–21. IGBP Report No. 13, Stockholm.

Janach, W. E. (1989). Surface ozone: trend details, seasonal variations and interpretation. *J. Geophys. Res.* **94**, 18289–18295.

Keeling, C. D., Bacastow, R. B., Carter, A. F., Piper, S. C., Whorf, T. P., Heimann, M., Mook, W. G., and Roeloffzen, H. (1989). A three dimensional model of atmospheric CO_2 transport based on observed winds. 1. Analysis of observational data. *In* "Aspects of Climate Variability in the Pacific and the Western Americas" (D. H. Peterson, ed.), pp. 305–363. Geophysical Monograph 55, AGU, Washington, D.C.

Keller, M. K., Kaplan, W. A., and Wofsy, S. C. (1986). Emission of N_2O CH_4, and CO_2 from tropical soils. *J. Geophys. Res.* **91,** 11791–11802.

Keller, M. K., Kaplan, W. A., Wofsy, S. C., and da Costa, J. M. (1988). Emission of N_2O from tropical forest soils: response to fertilization with NH_4^+, NO_3^- and PO_4^{3-}. *J. Geophys. Res.* **93,** 1600–1604.

Lean, J., and Warrilow, D. A. (1989). Simulation of the regional climatic impact of Amazon deforestation. *Nature* **342,** 411–413.

Lashof, D. A. (1989). The dinamic greenhouse: feedback processes that influence future concentrations of trace gases and climate changes. *Climatic Change* **14,** 213–242.

Livingstone, G. P., Vitousek, P. M., and Matson, P. A. (1988). Nitrous oxide flux and nitrogen transformation across a landscape gradient in Amazonia. *J. Geophys. Res.* **93,** 1593–1599.

Luizao, F. P., Matson, P., Livingston, G., Luizo, R., and Vitousek, P. (1989). Nitrous oxide flux following tropical land clearing. *Global Biogeochem. Cycles* **3,** 281–285.

Matson, P., and Vitousek, P. (1987). Cross-system comparations of soil nitrogen transformation and nitrous oxide flux in tropical forest ecosystems. *Global Biogeochem. Cycles* **1,** 163–170.

Matson, P. A., Vitousek, P. M., Livingston, G. P., and Swanberg, N. A. (1990). Sources of variations in nitrous oxide flux from Amazonian ecosystems. *J. Geophys. Res.* **95,** 16789–16798.

Melillo, J. M., Callaghan, T. V., Woodward, F. I., Salati, E., and Sinha, S. K. (1990). Effects on ecosystems. *In* "Climate Change, the IPCC Scientific Assessment" (T. J. Houghton, G. J. Jenkins, and J. J. Ephroums, eds.), pp. 282–310. Cambridge Univ. Press, Cambridge.

Mitchell, J. F. B., Manabe, S., Meleshko, V., and Tokioka, T. (1990). Equilibrium climate change—and its implications for the future. *In* "Climate Change, the IPCC Scientific Assessment" (J. T. Houghton, G. J. Jenkins, and J. J. Ephroums, eds.), pp. 131–172. Cambridge Univ. Press, Cambridge.

Myers, N. (1990). Tropical forest. *In* "Global Warming, the Greenpeace Report" (J. Leggett, ed.), pp. 372–399. Oxford Univ. Press, Oxford.

Neftel, A., Oeschger, H., Schwander, J., Stauffer, B., and Zumbrunn, R. (1982). Ice core measurements give atmospheric CO_2 content during the past 40,000 years. *Nature* **295,** 222–223.

Prinn, R. D., Cunnold, D., Rasmussen, R., Simmonds, P., Alyea, F., Crawford, A., Fraser, P., and Rosen, R. (1990). Atmospheric trends and emissions of nitrous oxide deducted from ten years of ALE-GAGE data. *J. Geophys. Res.* **95,** 18369–18385.

Ramanathan, V., Callis, L., Cess, R., Hansen, J., Isaksen, I., Kuhn, W., Lacis, A., Luther, F., Mahlman, J., Reck, R., and Schlesinger, M. (1987). Climate chemical interactions and effects on changing atmospheric trace gases. *Rev. Geophys.* **25,** 1441–1482.

Robertson, G. P., and Tieje, J. M. (1988). Deforestation alters denitrification in a lowland tropical rain forest. *Nature* **336,** 756–759.

Sanhueza, E., Hao, W. M., Scharffe, D., Donoso, L., and Crutzen, P. J. (1990). N_2O and NO emissions from soils of the northern part of the Guayana shield, Venezuela. *J. Geophys. Res.* **95,** 22481–22488.

Scharffe, D., Hao, W. M., Donoso, L., Crutzen, P. J., and Sanhueza, E. (1990). Soil fluxes and atmospheric concentrations of CO and CH_4 in the northern part of the Guayana shield, Venezuela. *J. Geophys. Res.* **95,** 22475–22480.

Shukla, J., Nobre, C., and Sellers, P. (1990). Amazon deforestation and climate change. *Science* **247,** 1322–1325.

Siegenthaler, V., and Sanhueza, E. (1990). Greenhouse gases and other climate forcing agents. *In* "Climate Change: Science, Impacts and Policy" (J. Jager and H. L. Ferguson, eds.), pp. 47–58. Cambridge Univ. Press, Cambridge.

Strain, B. R., and Cure, J. D., eds. (1985). Direct effect on increasing carbon dioxide on vegetation. *DOE/ER-0238,* U.S. Department of Energy, Washington, D.C.

Watson, R. T., Rodhe, H., Oeschger, H., and Siegenthaler, V. (1990). Greenhouse gases and aerosols. *In* "Climate Change, the IPCC Scientific Assessment" (J. T. Houghton, G. J. Jenkins, and J. J. Ephroums, eds.), pp. 1–40. Cambridge Univ. Press, Cambridge.

Watson, R. T., Meira Filho, L. G., Sanhueza, E., and Janetos, A. (1992). Greenhouse gases: Sources and sinks. *In* "Climate Change 1992. The Supplementary Report to the IPCC Scientific Assessment" (J. T. Houghton, B. A. Callander, and S. K. Varney, eds.), pp. 25–46, Cambridge Univ. Press, Cambridge.

C H A P T E R 11

North–South Comparisons: Biogeochemistry

BARBARA I. KRONBERG   EUGENIO SANHUEZA

Chapters 9 and 10 emphasize the strong coupling between climate and biogeochemical processes at the ecosystem level. Climate changes accompanying a doubling of atmospheric carbon dioxide (CO_2) would result in modifications to biogeochemical (including hydrologic) cycles. The contrasts in modified biogeochemical patterns between continents would be governed principally by the different distributions of major ecosystems with respect to latitude. Major South American ecosystems are situated in tropical and subtropical regions in which changes in temperature and precipitation are predicted to be much less than in major North American ecosystems located at high temperate latitudes. In the short term, the responses from both continents would be reflected in alterations of biogeochemical feedback patterns to the atmosphere and these could be significant at regional and possibly global levels.

There are large (~100 Gt/year) continual material exchanges between the atmosphere and terrestrial ecosystems associated with the photosynthetic fixation of atmospheric CO_2 into plant matter and CO_2 release to the atmosphere accompanying plant respiration. Because respiration rates are more sensitive to temperature than those for photosynthesis, CO_2 contributions to the atmosphere from plant respiration could increase substantially with global warming. This perturbation of atmospheric carbon fluxes would be more significant for northern boreal and temperate forest regions, for which predicted average temperature increases are two- to threefold above globally averaged values. The simulated long-term response of these northern seasonal forests is a migration northward by hundreds of kilometers. Climate-driven trends could be reinforced by currently high deforestation rates (millions of hectares per year) in both boreal and tropical biomes. If deforestation in South American forests were halted and reforestation were effective in achieving net biomass production, the South American continent could possibly operate as a net CO_2 sink.

Work by Townsend *et al.* (1992) indicates that in the short term (decadal scale) greater amounts of carbon could be released from tropical than boreal and temperate lands because of higher rates of soil respiration in tropical ecosystems, which appear highly sensitive to small changes in temperature. This suggestion is corroborated by the high correlation found between measurements of CO_2 anomalies at Mauna Loa between 1958 and 1987 and temperature averaged over equatorial regions (Marston *et al.*, 1991).

The extensive high northern latitude systems (tundra–wetland–forest) contribute approximately 13% of global methane (CH_4) emissions to the atmosphere. These contributions account for about 60% of methane emitted from all natural wetlands (115 Tg CH_4/year). Methane emissions from northern ecosystems could increase as warmer climates lower permafrost boundaries in soils. Limited studies indicate that methane is consumed in nonwaterlogged tundra soils, and this net negative feedback process would be reinforced if warmer climates result in extensive lowering of tundra water tables. Vast amounts of gas hydrates stored in frozen ground at high northern latitudes will be the largest potential source of CH_4 emissions to the atmosphere if the simulated increases (10–15°C) in Arctic air temperatures are realized. Tropical wetlands tend to be carbon limited, and carbon budgets in these terrains may not be affected by climate change. Diminished rainfall could lead to a reduction in tropical wetland area and possibly smaller CH_4 contributions to the atmosphere from these regions. Decreases in rainfall would be enhanced by deforestation of tropical forests, which participate in local and regional hydrologic cycles by recycling significant amounts of rain.

In South America the ratio of dryland to forest is greater than in North America. In simulated ecosystem responses to doubling atmospheric CO_2 concentrations, North American grasslands and deserts are predicted to expand. In South America disruption of the Amazonian hydrologic cycle could affect drought-prone areas that receive air masses leaving Amazonia. In both continents the natural biogeochemical cycles in drylands are extensively perturbed by human activities. The naturally low water and carbon reserves in these systems make the biogeochemical cycles associated with them vulnerable to disruption. There is evidence for strong linkages in these systems between abiotic and biotic processes. This close coupling of processes appears to promote the amplification in habitats of small perturbations in flows of energy and materials.

It is likely that global warming will affect the budgets of other atmospheric components (N_2O, NO_x, CO, O_2). Reliable assessments of responses at regional scales are precluded by the paucity of information on biogeochemical cycles of C, N, S, and O in major ecosystems. More detailed understanding of biogeochemical interactions and more reliable regional predictions from general circulation models are required to deal with the uncertainties in possible future responses to climate changes in terrestrial ecosystems of North and South America. For example, current GCMs do not reliably predict the geographic distribution of future changes in soil moisture, a key variable for assessing regional responses.

References

Marston, J. B., Oppenheimer, M., Fujita, R. M., and Gaffin, S. R. (1991). Carbon dioxide and temperature. *Nature* **349,** 573–574.

Townsend, A. R., Vitousek, P. M., and Holland, E. A. (1992). Tropical soils could dominate the short-term carbon cycle feedbacks to increase global temperatures. *Climatic Change* **22,** 293–303.

Part V: Intertidal

C H A P T E R   12

Possible Ecological Responses to Global Climate Change: Nearshore Benthic Biota of Northeastern Pacific Coastal Ecosystems

JANE LUBCHENCO   SERGIO A. NAVARRETE
BRIAN N. TISSOT   JUAN CARLOS CASTILLA

I. Introduction

Anticipating ecological responses to global climate changes is a formidable challenge. The complexities of the mechanisms driving both global climate changes and biotic responses to these changes render precise predictions of biotic responses impossible at present. Such predictions would require a comprehensive understanding of the multiple and complex processes that determine patterns of distribution and abundance of organisms. These processes range from direct effects of abiotic factors on individual organisms to indirect and complex interactions within and between the biotic and abiotic elements. Moreover, these processes operate across multiple scales: spatial scales ranging from microscopic to global, temporal scales ranging from nanoseconds to centuries, and levels of biological organization from molecules and cells to individual organisms up through populations, communities, ecosystems, and landscapes to the biosphere.

A. Useful Approaches

Despite the inherent difficulties in making precise predictions of biotic responses to global changes, it is possible and useful to anticipate likely classes of biotic responses, to identify species or groups of species that may be at particular risk, to define the range of possible responses, and to seek mechanistic understanding of the processes driving responses. Aid in accomplishing these tasks comes from three main sources: (1) records of past changes in distribution or abundance of species, (2) studies of causes of present-day patterns of distribution and abundance, and (3) studies of biotic responses to present-day, large-scale perturbations of the abiotic environment. Each of these approaches should incorporate empirical and theoretical information. Each offers different kinds of insight.

The first approach, a paleontological one, can describe biotic changes that are correlated with past large-scale climate alterations. This approach may be descriptive (simply delineating the observed changes in a particular place at a particular time) or comparative (looking for similarities and differences between observed changes in different places or at different times). Paleoecological information has the potential advantage of encompassing large spatial and long temporal scales but the disadvantage that the actual mechanisms of biotic responses are unknown. Although past changes in biota may be observed to be correlated with climate alterations, conclusions that these alterations caused the observed biotic changes are unwarranted. We do not know, for example, whether an extinction or change in a species' range was a direct response to an alteration in a single abiotic factor (such as elevated air temperature), a direct response to changes in multiple abiotic factors (warmer temperature coupled with decreased soil moisture), a response to changes in one or more other species (a competitor, predator, or mutualist, for example) that were affected by the abiotic changes, or a combination of these factors. Correlations thus indicate possible responses but offer limited predictive insight.

The second approach involves studies of factors causing present-day patterns of distribution and abundance of biota and is within the domain of evolutionary, physiologic, behavioral, population, community, and ecosystem ecology and biogeography. Descriptive, comparative, and experimental approaches in these areas provide complementary information (Lubchenco and Real, 1991). *Descriptions* of biogeographic ranges, for example, indicate correlations between a species' distribution and abiotic factors (such as water temperature or ocean currents). *Comparisons* of patterns or correlations (e.g., those existing on different continents) may refine hypotheses about possible causes of the patterns. *Experiments* testing the hypotheses generated from these comparison provide insight into mechanisms determining the patterns (Lubchenco and Real, 1991). *Comparative experiments* (i.e., comparable experiments in geographically distinct areas) allow comparison of mechanisms in different communities and therefore a broader spatial and temporal scale of insight (Menge 1991).

There are thus important differences between the first (paleo) and the second (present-day) approaches. The temporal distinction between them is obvious. Less apparent may be the difference in the kinds of understanding that each affords. Paleo studies have the potential to "sample" a greater range of conditions (e.g., climates) and may thus provide insight into a broader range of possible biotic responses. Present-day studies, on the other hand, have the unique potential for testing causality. The opportunity to manipulate components of the biotic and abiotic environments experimentally allows identification of the factors causing (not just correlated with) the observed responses. This ability to alter environmental conditions is required for an understanding of mechanisms driving responses and sets apart studies of present-day patterns from those of the past.

Although experiments can provide powerful insight into causes of local patterns of distribution and abundance and can suggest general mechanisms, extrapolations from the experiments to specific conditions that are outside the range of the experiments are limited. Specifically, because field experiments can normally be performed only over relatively small spatial and short temporal scales, the validity of extrapolations to larger or longer scales is uncertain. Similarly, extrapolations to more extreme abiotic conditions or to different rates of change in the environment (e.g., to warmer temperatures, more variance in temperature, or different rates of change in temperature) are tentative. Thus, although the experimental approach provides the most direct and unequivocal evidence for understanding the mechanisms of interaction between abiotic and biotic factors in determining local patterns, it is limited in providing specific insight into responses to changes outside the range of the conditions created by the experimenter. Predictions about the consequences of large-scale or large-magnitude alterations in temperature, for example, are outside the domain of the specific understanding provided by this approach. A third, thus far underutilized approach may provide additional insight.

The third approach, studies of biotic responses to present-day, large-scale perturbations of the abiotic environment, affords the opportunity to investigate mechanisms driving biotic responses to conditions outside the range of those produced by local, investigator-initiated experiments. Examples of such perturbations include anomalies like the 1982–1983 El Niño–Southern Oscillation phenomenon. If studies of anomalies incorporate field experiments (the ideal situation) or if they build on an understanding of mechanisms known from previous field experiments, opportunities exist for unparalleled insight into mechanisms of biotic responses to these unusual conditions. This combined approach affords the opportunity to extend the power of the experimental approach and to provide a bridge between the paleoecological and present-day–experimental approaches. If large-scale perturbations are anticipated, recognized as they are occurring, or occur fortuitously during ongoing descriptive and experimental studies, they can provide unique insight. If comparable anomalies occur in different biogeographic regions, comparative analyses of biotic responses to anomalies would be particularly useful.

B. Biotic Responses to Anomalous Events: A Case Study

This chapter integrates the second and third approaches to gain insight into possible alterations in the biota in response to climate change. It focuses on changes correlated with recent, anomalous events in light of experimental studies, specifically for shallow-water, benthic biota of the northeastern Pacific Ocean. In particular, studies documenting effects of the 1982–1983 El Niño–Southern Oscillation (ENSO) phenomenon provide information about responses of entire communities to perturbations with some affinities to those predicted under global warming scenarios.

In the following, we provide a brief overview of some relevant oceanographic and ecological characteristics of the northeastern Pacific nearshore ecosystem, summarize predicted relevant alterations in the physical environment under global warming scenarios, suggest possible biotic responses to these changes, present lessons from the relevant biotic responses to the 1982–1983 ENSO phenomenon, and discuss areas for productive research efforts.

II. Ecological, Oceanographic, and Biogeographic Background

Rocky intertidal and shallow subtidal communities along the coasts of California, Oregon, and Washington are dominated by seaweed and invertebrate assemblages. These benthic assemblages are comparable in many respects to those along the western coasts of Peru and Chile, as described in Castilla *et al.* (Chapter 13). These North American communities occur at the edge of, and are strongly affected by, the dynamics of the California Current system (Bernal, 1981; Chelton *et al.*, 1982; Tegner and Dayton, 1987; Peterson *et al.*, Chapter 2). Upwelling of cold, nutrient-rich waters is an important factor influencing the characteristics of these coastal communities (Barber and Smith, 1981).

Local community dynamics in these assemblages are fairly well understood. Over the past 30 or so years, a plethora of experimental studies along the coasts of California, Oregon, and Washington has resulted in a particularly rich understanding of community dynamics. These findings have been amply described (Connell, 1972, 1975; Dayton, 1971, 1975, 1985; Dayton and Tegner, 1984a,b; Dayton *et al.*, 1984; Denny, 1988; Dethier, 1984; Dethier and Duggins, 1984; Gaines and Lubchenco, 1982; Lubchenco and Gaines, 1981; Menge and Farrell, 1989; Menge and Lubchenco, 1981; Menge and Sutherland, 1987; Paine, 1980; Sousa, 1984). Of particular interest here are studies indicating variation in the types or strengths of biological interactions along latitudinal gradients (Dethier and Duggins, 1988; Fawcett, 1984; Frank, 1975; Gaines and Lubchenco, 1982) and studies indicating the links between onshore and nearshore events (Dayton and Tegner, 1984a,b; Gaines and Roughgarden, 1985, 1987; Roughgarden *et al.*, 1988; Tegner and Dayton, 1987).

Most benthic species inhabit specific latitudinal regions along this coastline. The coincidence of many species' boundaries results in the designation of biogeographic "provinces" (Fig. 1C). Provinces are not rigid units: many species "cross" provincial lines, and alternative provincial division are used by different biogeographers. Nonetheless, there is a clear replacement of species along latitudinal gradients and the provincial designation in Fig. 1C provides a useful framework in which to discuss these general biogeographic trends.

There are obvious correlations between the provincial boundaries and a number of interrelated abiotic factors, including latitude, sea surface temperature, and major

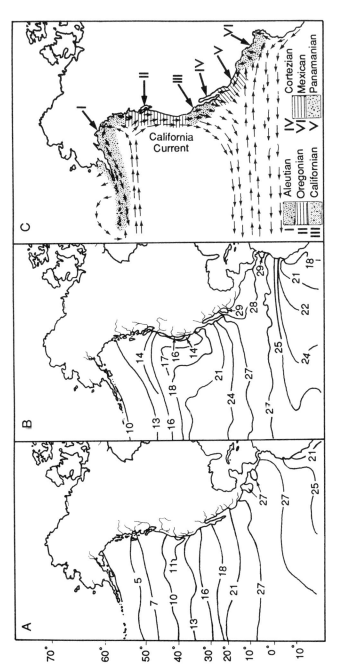

Fig. 1. Sea surface temperatures, currents, and biogeographic provinces of the northeastern Pacific Ocean. (A) Isotherms for average sea surface temperatures (SST, °C) in February (annual minima). (B) Isotherms for SST in August (annual maxima). (C) Major surface currents and main biogeographic provinces. [Parts (A) and (B) modified from Sea-Surface Temperature Chart, U.S. Hydrographic Office, Washington, D.C., 1976. Part (C) modified from Seapy and Littler (1980) and Pickard and Emery (1982).]

ocean surface currents (Fig. 1A and B). Although temperature per se has often been assumed to determine species' boundaries, there is no clear evidence that it does so directly. Present and past correlations between biogeographic patterns and oceanic currents or water temperatures (Fig. 1C) are just correlations and do not necessarily indicate a causal relationship.

III. Predicted Biotic Responses to Changes in Physical Parameters

The distribution and abundance of organisms are determined by multifarious interactions among abiotic and biotic factors. The abiotic environment for nearshore biota consists of a complex array of factors, including air and water temperature, chemistry, and movement; salinity; nutrients; sedimentation; and quality and quantity of light. The effects of these factors on the organisms may depend not only on the mean values of the factors but also on their extreme values, variances, spatial and temporal distribution, or rates of change. Because organisms create habitat and microhabitat for one another and provide food for, eat, protect, poison, remove, decompose, compete for resources with, or otherwise affect other organisms, biotic interactions often exert strong effects on the distribution and abundance of organisms. The abiotic environment may thus impinge on individual organisms both directly and indirectly. The indirect effects devolve from abiotic factors affecting one organism that in turn affects another. Thus changes in the abiotic environment are expected to result in complex alterations in biotic patterns, via both direct and indirect effects. In the following we consider direct impacts of specific changes in certain abiotic factors.

A. Predicted Changes in Physical Parameters

Some of the changes in abiotic factors that are likely to affect nearshore marine organisms may be inferred from the predictions of the Intergovernmental Panel on Climate Change (Houghton et al., 1990). The IPCC "business-as-usual scenario," that is, an increase in greenhouse gas concentrations equivalent to a doubling of the preindustrial levels of CO_2 by the middle of the next century, predicts (1) an increase in global mean temperature during the next century of about 0.3°C (0.2–0.5°C) per decade and (2) an increase in global mean sea level rise of about 6 cm (3–10 cm) per decade over the next century. The first prediction would result in increases in both sea surface temperatures (Trenberth, 1990; Chapter 3, this volume) and air temperatures. As discussed by Peterson et al. (Chapter 2), increases in the concentration of CO_2 are not expected to have much direct effect on marine biota.

Both the rates at which the changes occur and the spatial patterns of the changes are immediately relevant to biota. Changes in the global mean temperature or average global sea level may have little direct relevance to organisms in a particular place; the local changes are the meaningful measures. Some spatial patterns, such as latitudinal variation in the rates of change in temperature, are predicted (Stouffer et al., 1989;

Trenberth, Chapter 3; Bernal, 1991). Other spatial patterns are likely but more difficult to predict.

Other abiotic changes will undoubtedly occur, but their likelihood, rate, and magnitude of change are highly uncertain. For example, alterations in ocean circulation or air–land–sea dynamics may result in alterations of upwelling patterns (Bakun, 1990), ocean currents, frequency or intensity of storms, frequency or intensity of anomalous events such as ENSO phenomena, or rainfall and thus runoff and salinity patterns. Any of these changes has the possibility of dramatically altering the local nearshore environment in a critical fashion (see also Castilla *et al.*, Chapter 13). The complexity of the earth system is such, however, that predictions of these vitally important parameters are impossible at present. Nonetheless, because the consequences of some of these changes are so great, they must be borne in mind.

B. Possible Biotic Responses

Species may respond to changes in the abiotic environment by shifting in abundance or distribution (without evolving), by becoming extinct, or by evolving (with or without a concomitant shift in abundance or distribution; Holt, 1990). The responses that occur are a complex function of the rates and magnitudes of the environmental changes, environmental characteristics affecting species' responses (such as oceanic currents as vehicles for dispersal), and species' characteristics (such as phenotypic plasticity, dispersal ability, population size, generation time, reproductive output, and genetic variation). For example, marine invertebrates with long larval phases should have greater potential for dispersing to new, more favorable environments than those with direct development and no larval phase. Species with obligate interactions with other species, such as many hermatypic corals and their symbiotic zooxanthellae, might have more limited ability to adapt to changes in water temperature (since both species must be able to respond) than would species without close, obligate interactions.

In the following sections we consider possible responses to changes in single abiotic factors. We make simple predictions of possible biotic responses as a heuristic device to help formulate and delineate the problems. These predictions should not be taken literally. They serve as a useful starting point for discussion and examination of the lessons from biotic responses to anomalous events (see Section IV).

1. Possible Responses to Changes in Sea Surface Temperatures. Because adverse effects of extreme temperatures on marine organisms are well known and correlations exist between biogeographic patterns of nearshore marine taxa and water temperature (Fig. 1C), it is tempting to conclude that species' ranges are determined directly by water temperature and thus to predict that organisms will respond directly to changes in sea surface temperature by simple alterations in biogeographic distributions. Because each species should respond somewhat differently to altered temperatures, its distribution might be expected to shift according to the magnitude of the change in temperature. The simplest and most direct prediction is thus:

PREDICTION 1: Species will migrate poleward to stay within current water temperature ranges.

This prediction is simplistic, but it serves as a useful starting point. Many factors could prevent this migration or could affect migration rates. Unavailability of suitable habitat could present formidable barriers to migration. For example, a species requiring a stable substratum and having short-lived dispersal stages might be unable to migrate across long stretches of sandy beaches. Lack of appropriate dispersal vehicles could also preclude or slow down migration. For example, currents flowing toward the equator (such as the California Current) would transport larvae in the "wrong" direction. Thus the accuracy of the prediction would depend on the match between the life history characteristics and behavior of the species in question and environmental specifics such as habitat availability and dispersal vehicles.

Unless a species with a mismatch could adapt to the new temperature regime, its biogeographic range would shrink or, if the range was sufficiently constricted, the species would become extinct. Range reductions and extinctions would be particularly likely for polar species, which would lack higher latitudes into which to expand.

2. Possible Responses to Changes in Air Temperature. Increased global temperatures would mean not only increased sea surface temperatures but also increased air temperatures. The former would be expected to affect both intertidal and subtidal organisms; the latter would affect only intertidal organisms. The exact consequences for intertidal plants and animals would depend in part on the amount and rates of the change in air versus water temperature. Some species might be able to withstand warmer water but not warmer air, or vice versa, or they might be stressed only by the cooccurrence of both. In the absence of evolutionary changes resulting in individuals with greater tolerance to higher air temperatures, species would eventually migrate poleward to sites with air temperatures comparable to those presently occurring. Thus,

PREDICTION 2: Species will migrate to higher latitudes to stay within the current range of air temperatures.

The most obvious direct consequence of higher air temperatures would be an increase in desiccation of intertidal organisms. Because the upper limits of many intertidal plants and animals appear to be determined by desiccation (see reviews in Connell, 1972; Lubchenco, 1980; but see Cubit, 1984), a direct, immediate consequence of increased air temperatures could be a lowering of the upper limits of many species. Thus, bathymetric shifts may occur. If this downward shift in vertical distribution on the shore occurred for large numbers of intertidal organisms, the result could be a general compression of the intertidal region. Thus, we predict that over ecological time, and in the absence of changes in sea level, the vertical extent of intertidal communities might decrease. The impact of this intertidal compression on biological diversity or on the dynamics of the rest of the nearshore ecosystem is unknown. This leads to:

PREDICTION 3: The bathymetric range of intertidal biota will be compressed because of increased desiccation.

3. Responses to Sea Level Rise. Potential consequences of sea level rise have already received considerable attention because of their possible impact on human coastal structures, water quality, and water supply. Coastal flooding, shoreline erosion, intrusion of salt water into fresh water, and alteration of sediment transport have been predicted (Smith and Tirpak, 1989). Estuaries are expected to be particularly vulnerable. Rocky shore biota, on the other hand, should be more resistant to direct negative impacts of sea level rise. We predict that rocky intertidal biota will simply migrate or disperse to higher shore levels as sea level rises. Negative impacts will occur if no higher shore is available for colonization. Hence:

PREDICTION 4: Rocky intertidal biota will shift upward on the shore as sea level rises and thus maintain current bathymetric distributions relative to sea level.

Altered sea level may also affect the intensity of wave action and thus may have strong indirect negative or positive consequences for nearshore biota. These consequences will vary as a function of local topography. Increases in sea level during the 1982–1983 ENSO phenomenon were reported to result in greatly intensified wave action on shores of southern California (Tegner and Dayton, 1987).

4. Responses to Other Possible Changes. Alterations in ocean–atmosphere–land interactions could drastically affect ocean circulation (Broecker, 1987), the frequency and intensity of anomalous events such as the ENSO phenomenon, up welling patterns (Bakun, 1990), or regional storm and precipitation patterns, but uncertainties about the likelihood of these events render predictions about biotic responses unwarranted. Nonetheless, these changes could dramatically affect oceanic biota. Significant shifts in upwelling regimes, for example, could have major impacts on productivity with resultant changes in ecosystem dynamics and community structure (Menge, 1992).

5. Responses to Multiple Environmental Changes—Abiotic and Biotic, Direct and Indirect. Each of the predicted alterations is considered separately in the previous section. Moreover, only direct responses of the biota to these single-factor abiotic changes are treated. In reality, multiple abiotic changes would occur (increase in sea surface temperature and air temperature and sea level rise, for example); in addition, these alterations would set in motion various biotic responses, which in turn would affect other biota. The overall responses of an ecosystem to such complex phenomena are impossible to predict. The next section considers an alternative approach, specifically an examination of the biotic responses to a recent anomalous event that involved multiple environmental changes.

IV. Lessons from a Recent Anomalous Event: The 1982–1983 El Niño–Southern Oscillation

Knowledge of biotic responses to ENSOs may provide useful insight into possible short-term responses of biota to global climate change. Although both ENSO and global change involve widespread alterations in sea surface temperature and sea level and multiple other perturbations, there are important differences between them. Three critical differences are (1) the *rate* of change in the physical and chemical parameters and therefore the opportunities for biota to acclimate or evolve adaptations to the new environment, (2) the *duration* of the change, and (3) the *past occurrence* of the change during the recent evolutionary history of the species involved. Because these differences are undoubtedly important, studies of biotic responses to ENSO events should be used primarily for heuristic purposes, that is, to gain insight into how large-scale environmental perturbations might influence broad biotic patterns.

A. Abiotic Alterations

A major ENSO phenomenon such as that occurring during 1982–1983, although centered in the southern Pacific Ocean and atmosphere, results through global tele-connections in dramatic, worldwide changes in both oceanic and atmospheric events (Cane and Zebiak, 1985; Rasmusson and Wallace, 1983; Trenberth, 1990). These events can translate into a dramatically altered nearshore environment. For example, during the 1982–1983 El Niño, the nearshore environment along the eastern edge of the Pacific Ocean was characterized by increased water temperature, higher sea level, lower nutrients, decreased salinity, and increased intensity of storms and rough seas (Glynn, 1988, 1989; McGowan, 1985; Dayton and Tegner, 1989; Barber and Chavez, 1983; Barber *et al.*, 1985). The magnitude of these changes varied latitudinally, with the most intense changes in sea surface temperature and sea level occurring in the tropics.

B. Biotic Responses—Tropical

Dramatic biotic changes were documented for many tropical nearshore ecosystems (Glynn, 1988). These responses include extinctions (two species of hydrocorals, *Millepora*), mass mortalities (corals, algae, crustaceans, mollusks, echinoderms, ascidians, seabirds, fishes, pinnipeds, and marine iguanas), and species invasions (algae, crabs, a brachiopod, a polychaete, a barnacle, a stomatopod, a shrimp, a gastropod, and bivalves; see Glynn's extensive and excellent review, 1988).

Coral reef bleaching (loss of symbiotic zooxanthellae from coral polyps) was particularly widespread and resulted in coral mortality of 50 to 98% over extensive areas of Costa Rica, Panama, Colombia, and Ecuador (Glynn, 1988). Because corals create the coral reef habitat, their decimation results in substantial alteration of the rest of the community. Dramatic changes in species composition, richness, and diversity of reefs were reported from different locales (Glynn, 1988).

Many of the initial perturbations set in motion a network of other biotic alterations (Glynn, 1988). For example, coral mortality triggered the emigration or death of a number of obligate crustacean symbionts of corals (Glynn *et al.*, 1985; Glynn and D'Croz, 1990). These crabs and shrimp normally deter predation by the corallivore *Acanthaster*, the crown of thorns seastar. Elimination of these guardians from protective coral barriers allowed *Acanthaster* access to previously inaccessible corals, which resulted in even greater coral mortality, much of which was not directly attributable to the initial warming event (Glynn, 1985). Additional reverberating effects were reported. Massive corals suffering partial mortality were colonized by algae. Algal patches were subsequently invaded by herbivorous sea urchins and damselfish, both of which caused further coral mortality. Amplification of initial effects via trophic interactions was observed in pelagic nearshore ecosystems as well. Most of the effects of ENSO on vertebrates (Peruvian anchovy, sardine, guano birds) are related to depletion of their food supply and subsequent starvation (Barber and Chavez, 1983, 1986; Barber and Kogelschatz, 1989; Glynn, 1988).

The changes just described include both direct responses to alterations in physical and chemical parameters and responses mediated through other species. Coral reef bleaching and coral mortality were probably due to direct effects of higher water temperatures at some sites and a drop in sea level at others; reproductive failure of Christmas Island seabirds appeared to be in response to extensive flooding of nesting sites (Schreiber and Schreiber, 1984); and disappearance of a number of species of intertidal red and green algae in the Galápagos was probably a direct response to decreased salinity and higher sea surface temperatures (Laurie, 1984, 1985). These responses trigger other alterations through trophic, competitive, mutualistic, or structural interactions. For example, reductions of 30–55% of the marine iguana population in the Galápagos were probably a response to the disappearance of most of the red algae on which they normally graze (Laurie, 1984, 1985); loss of crustacean symbionts, invasion of algae and grazers, and subsequent secondary coral mortality were indirect responses to primary coral mortality, decimation of higher trophic levels reflects fundamental perturbations at lower levels. The end result is a complex combination of all of these effects.

C. Biotic Responses—Extratropical

The 1982–1983 ENSO was sufficiently strong that the equatorial oceanographic anomalies extended northward as far as Alaska (Glynn, 1988). Because many of these anomalies extended well into 1984 along the North American coast, this ENSO is often referred to as the 1982–1984 ENSO by North American workers. Profound effects occurred throughout the California Current system. Peterson *et al.* (Chapter 2, this volume) review many of the consequences of the ENSO for pelagic communities. The following sections treat effects on nearshore, benthic communities.

Biological effects of this ENSO in Northern Hemisphere nearshore habitats ranged from strong to undetectable (Tegner and Dayton, 1987; Peterson *et al.*, Chapter 2; Wooster and Fluharty, 1985; Paine, 1986; Glynn, 1988). Some of the most dramatic responses occurred in subtropical kelp forests in California, where beds of the giant kelp *Macrocystis pyrifera* experienced up to 90% declines in canopy cover

(Glynn, 1988; Tegner and Dayton, 1987). Other strong responses include mass mortality of intertidal black abalone populations throughout southern California (Davis *et al.,* 1992; Tissot, 1988, 1991).

A few studies found negligible effects of the 1982–1984 ENSO, especially in intertidal communities. Two studies in California found no substantial changes in intertidal algae (Gunnill, 1985; Murray and Horn, 1989). One study in Washington (Paine, 1986) found significant perturbations in algal communities, but it was pointed out that other factors (coincidence of winter low tides with freezing conditions) provide a more likely explanation for the observed changes. Thus, with the exception of the black abalone, few biotic responses were recorded for intertidal communities.

The perturbations of southern California giant kelp forests were particularly well documented and the causal mechanisms relatively well understood, primarily because long-term studies of these communities were in progress (Tegner and Dayton, 1987). Abiotic perturbations of these systems included sea levels about 3 standard deviations above normal, sea surface temperatures up to 4°C above normal, large-scale depression of the thermocline, nutrient depletion, and the most severe storm season in many decades involving more storms, longer wave periods, and bigger waves (six events were recorded in which waves exceeded 6 m in 1982–1983 versus 18 such events from 1900 to 1984; Seymour *et al.,* 1989; Glynn, 1988; Tegner and Dayton, 1987). Kelp forest responses to this and other ENSO events vary considerably within and between ENSO events (Glynn, 1988; Tegner and Dayton, 1987).

Decimation of Point Loma *Macrocystis* forests in southern California was caused primarily by the combination of nutrient depletion, warm temperatures, and winter storms (Dayton and Tegner, 1984b; Dayton, 1985; Tegner and Dayton, 1987; Dayton and Tegner, 1989; Seymour *et al.,* 1989). *Macrocystis* can withstand temperatures up to 25°C when nutrients are abundant—for example, when they are added experimentally (North and Zimmermann, 1984; Dean and Jacobsen, 1986); however, nitrate usually becomes limiting above 15°C (Jackson, 1977; Gerard, 1982). Storms, which occurred more frequently and were more intense during the 1982–1984 ENSO (in response to shifts in the jet stream), caused extensive damage to kelp forests, far exceeding that due to warm waters and nutrient depletion. In Santa Barbara, which is more protected from winter storms, kelp beds were less damaged by storms (Ebeling *et al.,* 1985).

Significant losses of giant kelps have serious consequences for the entire ecosystem because *Macrocystis* is such an integral component of its community. It provides food and habitat for kelp forest denizens as well as providing food for adjacent communities. Within kelp forests, kelps provide food directly to herbivores and detritivores and, by leaking substantial amounts of photosynthate, also provide dissolved organic material to the community. The three-dimensional structure of kelps forms a habitat for a diverse assemblage of fishes, marine mammals, and numerous invertebrates (Foster and Schiel, 1985). Kelps also enrich adjacent communities. In one study, 70% of kelp biomass ended up as drift algae, which was transported to intertidal, beach wrack, and submarine canyon communities (Gerard, 1976). Giant kelps also provide food, reduce wave impact, and provide recreational sites for humans. In California, 100,000 to 170,000 wet tons yielding in excess of $35 million per year are harvested annually (Foster and Schiel, 1985).

In addition to direct effects of loss of habitat and food within kelp forests, many indirect effects were reported. For example, there were significant declines of species in adjacent communities that feed on drift algae originating in kelp forests. The mass mortality of black abalone occurring in 1986–1988 throughout southern California emphasizes the importance of this trophic link between subtidal and intertidal communities and illustrates a number of points. The decimation of black abalone populations on Santa Cruz Island (Fig. 2) mirrors that occurring elsewhere in southern California (Tissot, 1991). This mortality appears to be due a combination of three factors: starvation, abnormally high water temperatures, and potential infection of a coccidial parasite (Tissot, 1991; Steinbeck *et al.*, 1992; Davis *et al.*, 1992). No single fact or pair of these factors results in mortality: the coincidence of all three is thought to be necessary. Mortality was density dependent and temperature dependent and occurred primarily during periods of low food availability in the fall. The population declines observed in 1986–1988 appear to be a response to an accumulation of these stresses during both the 1982–1984 and 1986–1987 ENSO events. Davis *et al.* (1992) speculate that prior to the ENSO events, black abalone populations were unstable because of abnormally high abalone densities. The high abundances reflect absence of natural predators such as sea otters, lobsters, crabs, and sheepshead fish, all of which have been overharvested by humans.

A second delayed, indirect effect of kelp forest devastation emphasizes the potential importance of the reverberation of effects throughout the community after initial responses. As *Macrocystis* in the Point Loma kelp bed began to recover in 1985, it was infested with an outbreak of grazing amphipods that completely denuded portions of the forest (Tegner and Dayton, 1987). Canopy cover was reduced by almost 60% during the season when it should have been expanding. These amphipods

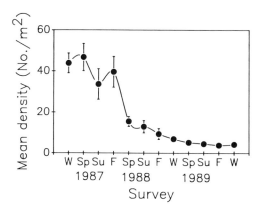

Fig. 2.   Mean density (±1 SE) of black abalone, *Haliotis cracherodii*, on permanent intertidal transects on the west end of Santa Cruz Island, California. (After Tissot, 1991)

are normal inhabitants of kelp forests but are usually kept in check by their predators, in particular kelp surf perch. The fish declined sharply or disappeared entirely from kelp beds during earlier kelp canopy loss and were apparently unable to reestablish as quickly as the amphipods. Initial canopy loss thus resulted in an outbreak of a herbivore that delayed recovery of the canopy.

These indirect effects may extend to adjacent communities as well. Gaines and Roughgarden (1987) report that juvenile rockfish, which inhabit kelp canopies, normally feed extensively on barnacle larvae and other meroplankton and may significantly deplete the supply of larvae recruiting to nearby rocky shores. Destruction of the kelp forest habitat resulted in depletion of the fish populations (Bodkin *et al.*, 1987). The subsequent banner recruitment year for barnacles on adjacent shores has been attributed to a significant decline in fish predation on the larvae (Gaines and Roughgarden, 1987).

Thus the effects of the 1982–1984 ENSO on extratropical communities were variable, with little *direct* impact observed for intertidal systems and profound effects reported for certain giant kelp forests. Lower-latitude kelp forests sustained significantly greater impacts than those at higher latitudes. Storms were responsible for significant amounts of damage to kelp beds, with significant canopy loss reported in some locales. Temperature and nutrient anomalies contributed to the devastation. The disappearance of kelps set in motion numerous, complex other responses, with immediate consequences for kelp forest communities but also delayed consequences for populations in adjacent communities (mass mortality of black abalone, increased barnacle recruitment) as well as recovery of kelp forests (amphipod outbreak).

V. Discussion and Conclusions

The complexities of likely biotic responses to global change are sufficiently overwhelming that some level of simplification is required. The challenge is to identify the appropriate level of simplification. Because there are good correlations between the distribution of species and physical factors such as air temperature and soil moisture for terrestrial vegetation or sea surface temperature for coastal biota (Fig. 1C), it is tempting to simplify the likely responses to the level of single species and their direct responses to these abiotic factors. The simpleminded predictions made in Section IIIB are very general examples of this approach. More sophisticated versions would involve models to predict changes in species' distributions given knowledge about the individual species' tolerances or requirements, for example, particular combinations of air temperature and soil moisture or sea surface temperature. One could, for example, use expected temperatures at different latitudes from a general circulation model (e.g., Stouffer *et al.*, 1989) to predict biogeographic changes in species distributions. Use of models such as these does not imply any assumption that other factors are unimportant; rather, these models assume that other factors are either less important than the effects of the primary independent variable (e.g., temperature) or that other factors act in concert with the primary independent variable, so that reasonable predictions may be made without the added detail.

How useful and how accurate are these simple predictions? Would they, for example, have predicted the consequences of temperature change and sea level rise for nearshore biota during and after the 1982–1983 ENSO? The earlier summary of biotic responses to this ENSO suggests that a simplification limited to information about temperature changes, sea level rise, and species' direct responses to those changes would probably be grossly insufficient to predict the ensuing changes. It would be insufficient for two reasons. First, some abiotic changes (such as winter storms in California kelp forests) may be more important than—or at least amplify substantially—the direct effects of temperature and the associated nutrient depletion of the water (e.g., Dayton and Tegner, 1984a; Tegner and Dayton, 1987). Second, the primary responses of the biota to the altered abiotic environment catalyze a plethora of secondary responses, as alterations in the distribution or abundance of some species reverberate throughout the community via trophic or other biotic interactions. The magnitude of these secondary effects is such that they cannot be ignored.

Some of these secondary effects are obvious. The multiple direct and indirect, immediate and delayed consequences of the decimation of *Macrocystis* in some California kelp beds emphasize the critical role that certain key species play in creating habitat and food for an entire ecosystem. Corals play the same role in coral reef ecosystems; mussels and seaweeds do the same in littoral communities. Therefore, improved predictions of the consequences of environmental changes for these systems must focus on factors affecting the species creating the habitat and forming the base of the food web and on effects of changes in these key species on the rest of the ecosystem.

Other species may play critical roles for other reasons. Keystone predators (Paine, 1966) or keystone species (National Research Council, 1986) that are known to have strong effects on community patterns and processes are obvious examples. Certain other species may be critical at larger scales, for example, in biogeochemical or hydrologic cycles.

However, some of the secondary effects are not so obvious. Based on a consideration of species' tolerances for thermal changes and on trophic or habitat interactions, one would not have predicted, for example, the mass mortality of black abalone, a species that is quite tolerant of increased water temperatures but not of the combination of starvation, thermal stress, and disease (Tissot, 1991) or the demise of corals that were apparently tolerant of warm waters but not of *Acanthaster* predation resulting from the demise of less tolerant corals and their associated mutualists (Glynn, 1985).

Thus predictions of the consequences of alterations in the physical environment must include these kinds of biological interactions. Habitat creation and trophic interactions stand out as interactions that should be included. Others, involving cumulative effects, diffuse interactions, indirect effects, or mutualistic interactions, may be equally important but more difficult to identify or incorporate. The black abalone case underscores the importance of cumulative or synergistic effects. The mortality of massive corals underscores the importance of mutualistic and indirect interactions.

Additional realism is thus required for better and useful predictions. Can this realism be achieved without undue and overwhelming complexity? Although some biotic responses, in particular those associated with invasions and diseases, may be impossible to predict, we believe that a better understanding of likely responses to unprecedented conditions is possible but will require new approaches. These approaches must include (1) a refined understanding of combined effects of biotic and abiotic interactions, especially as they occur along gradients (Menge and Sutherland, 1987; Lubchenco, 1986); (2) understanding of the transfer of information across temporal and spatial scales and levels of ecological organization (Levin, 1992); (3) incorporation of biologically important phenomena such as cumulative effects, thresholds, nonlinearities, heterogeneity, variability, and patchiness; (4) a focus on key species that are likely to have unduly strong effects on the community, ecosystem, or even the entire earth system; and (5) a focus on species at particular risk.

Certain species may be at greater risk than others as environmental parameters change. The risk of extinction should be greater for species with limited distributions, species with small populations, or species already close to their physiologic tolerance limits. The extinction of two species of hydrocorals in the Bay of Panama (Glynn 1984) is an example and underscores the possibility of such events. Other species with increased likelihood of extinctions include those with long-lived individuals, low recruitment, limited dispersal, or limited mobility. Particular attention should be paid to species with these characteristics.

Given the complexities of the system and the spatial and temporal variability inherent in ecological systems, there will always be surprises. Diseases and invasions are probably phenomena that will prove highly difficult to predict. Hence workers would be well advised to acknowledge the limitations inherent in making complete predictions for these systems.

In summary, the difficult task of predicting responses of the earth system to unprecedented changes will benefit from the synthesis of contributions from diverse sources. The coupling of small-scale, process-oriented ecological experiments with analyses of responses to large-scale environmental anomalies holds much potential for increasing our understanding of the functioning of the geobiosphere. Although our present level of understanding is insufficient to make precise predictions of possible biotic responses to global change for North American nearshore ecosystems, improved understanding will result from innovative approaches, especially those which cross scales and which couple local and regional process-oriented studies with large-scale environmental anomalies.

Acknowledgments

We gratefully acknowledge the vision and outstanding leadership of H. A. Mooney in promoting interdisciplinary, international efforts to understand global change patterns and processes and the helpful comments and ecological insight of Bruce A. Menge.

References

Bakun, A. (1990). Global climate change and intensification of coastal ocean upwelling. *Science* **247**, 198–201.

Barber, R. T., and Chavez, F. R. (1983). Biological consequences of El Niño. *Science* **222**, 1203–1210.

Barber, R. T., and Chavez, F. R. (1986). Ocean variability in relation to living resources during the 1982–1983 El Niño. *Nature* **319**, 279–275.

Barber, R. T., and Kogelschatz, J. E. (1989). Nutrients and productivity during the 1982/1983 El Niño. *In* "Global Ecological Consequences of the 1982–1983 El Niño–Southern Oscillation" (P. Glynn, ed.), pp. 21–53. Elsevier, Amsterdam.

Barber, R. T., and Smith, R. L. (1981). Coastal upwelling ecosystems. *In* "Analysis of Marine Ecosystems" (A. R. Longhurst, ed.), pp. 31–68. Academic Press, London.

Barber, R. T., Chavez, F. R., and Kogelschatz, J. E. (1985). Biological effects of El Niño. *In* "Ciencia, tecnología y agresión ambiental: El fenomeno El Niño," pp. 399–425. Consejo Nacional de Ciencia y Tecnología (CONCYTEC), Lima, Peru.

Bernal, P. A. (1981). A review of the low frequency response of a pelagic ecosystem in the California Current. *Calif. Coop. Ocean. Fish. Invest. Rep.* **22**, 49–64.

Bernal, P. A. (1991). Consequences of global change for oceans. A review. *Climate Change* **18**, 339–359.

Bodkin, J. L., Van Blaricom, G. R., and Jameson, J. J. (1987). Mortalities of kelp forest fishes associated with large oceanic waves off central California [USA] 1982–1983. *Environ. Biol. Fishes* **18**, 73–76.

Broecker, W. S. (1987). Unpleasant surprises in the greenhouse. *Nature* **328**, 123–126.

Cane, M. A., and Zebiak, S. E. (1985). A theory for El Niño and the Southern Oscillation. *Science* **228**, 1085–1087.

Chelton, D. B., Bernal, P. A., and McGowan, J. A. (1982). Large-scale interannual physical and biological interactions in the California Current. *J. Mar. Res.* **40**, 1095–1125.

Connell, J. H. (1972). Community interactions on marine rocky intertidal shores. *Annu. Rev. Ecol. Syst.* **3**, 169–192.

Connell, J. H. (1975). Some mechanisms producing structure in natural communities: a model and evidence from field experiments. *In* "Ecology and Evolution of Communities" (M. L. Cody and J. M. Diamond, eds.), pp. 460–490. Belknap Press of Harvard University Press, Cambridge, Massachusetts.

Cubit, J. D. (1984). Herbivory and the seasonal abundance of algae on a high intertidal rocky shore. *Ecology* **65**, 1904–1917.

Davis, G. E., Richards, D. V., Haaker, P. L., and Parker, D. O. (1992). Abalone population declines and fishery management in southern California. *In* "Abalone of the World: Biology, Fisheries, and Culture" (S. A. Shepherd, M. J. Tegner, and S. A. Guzman Del Proo, eds.), pp. 237–252. *Proceedings of the First International Symposium Abalone,* Fishing News Books, Blackwell Scientific Publications, Oxford.

Dayton, P. K. (1971). Competition, disturbance, and community organizations: the provision and subsequent utilization of space in a rocky intertidal community. *Ecol. Monogr.* **41**, 351–389.

Dayton, P. K. (1975). Experimental evaluation of ecological dominance in a rocky intertidal algal community. *Ecol. Monogr.* **45**, 137–159.

Dayton, P. K. (1985). Ecology of kelp communities. *Annu. Rev. Ecol. Syst.* **16**, 215–245.

Dayton, P. K., and Tegner, M. J. (1984a). The importance of scale in community ecology: a kelp forest example with terrestrial analogs. *In* "A New Ecology: Novel Approaches to Interactive Systems" (P. W. Price, C. N. Slobodchikoff, and W. S. Gaud, eds.), pp. 457–481. Wiley, New York.

Dayton, P. K., and Tegner, M. J. (1984b). Catastrophic storms, El Niño, and patch stability in a Southern California kelp community. *Science* **224**, 283–285.

Dayton, P. K., and Tegner, M. J. (1989). Bottoms beneath troubled waters: benthic impacts of the 1982–1984 El Niño in the temperate zone. *In* "Global Ecological Consequences of the 1982–1983 El Niño–Southern Oscillation" (P. Glynn, ed.), pp. 433–472. Elsevier, Amsterdam.

Dayton, P. K., Currie, V., Gerrodette, T., Keller, B. D., Rosenthal, R., and ven Tresca, D. (1984). Patch dynamics and stability of some California kelp communities. *Ecol. Monogr.* **54**, 253–289.

Dean, T. A., and Jacobsen, F. R. (1986). Nutrient-limited growth of juvenile kelp, *Macrocystis pyrifera,* during the 1982–1984 "El Niño" in southern California [USA]. *Mar. Biol.* **90**, 597–502.

Denny, M. W. (1988). "Biology and the Mechanics of the Wave-Swept Environment." Princeton University Press, Princeton, New Jersey.

Dethier, M. N. (1984). Disturbance and recovery in intertidal pools: maintenance of mosaic patterns. *Ecol. Monogr.* **54**, 99–118.

Dethier, M. N., and Duggins, D. O. (1984). An indirect commensalism between marine herbivores and the importance of competitive hierarchies. *Am. Natur.* **124**, 205–219.

Dethier, M. N., and Duggins, D. O. (1988). Variation in strong interactions in the intertidal zone along a geographical gradient: a Washington–Alaska comparison. *Mar. Ecol. Prog. Ser.* **50**, 97–105.

Ebeling, A. W., Laur, D. R., and Rowley, R. J. (1985). Severe storm disturbances and reversal of community structure in a southern California kelp forest. *Mar. Biol.* **84**, 287–294.

Fawcett, M. H. (1984). Local and latitudinal variation in predation on an herbivorous marine snail. *Ecology* **65**, 1214–1230.

Foster, M. S., and Schiel, D. R. (1985). The ecology of giant kelp forests in California: a community profile. *U.S. Fish Wildl. Serv. Biol. Rep.* **85**(7.2).

Frank, P. W. (1975). Latitudinal variation in the life history features of the black turban snail *Tegula funebralis* (Prosobranchia: Trochidae). *Mar. Biol.* **31**, 181–192.

Gaines, S. D., and Lubchenco, J. (1982). A unified approach to marine plant–herbivore interactions. II. Biogeography. *Annu. Rev. Ecol. Syst.* **13**, 111–138.

Gaines, S. D., and Roughgarden, J. (1985). Larval settlement rate: a leading determinant of structure in an ecological community of the marine intertidal zone. *Proc. Natl. Acad. Sci. USA* **82**, 3707–3711.

Gaines, S. D., and Roughgarden, J. (1987). Fish in offshore kelp forests affect recruitment to intertidal barnacle populations. *Science* **235**, 479–481.

Gerard, V. A. (1976). Some aspects of material dynamics and energy flow in a kelp forest in Monterey Bay, California. Ph.D. dissertation, Univ. of California Santa Cruz.

Gerard, V. A. (1982). Growth and utilization of internal nitrogen reserves by the giant kelp *Macrocystis pyrifera* in a low-nitrogen environment. *Mar. Biol.* **66**, 27–35.

Glynn, P. W. (1984). Widespread coral mortality and the 1982/83 El Niño warming event. *Environ. Conserv.* **11**, 133–146.

Glynn, P. W. (1985). El Niño-associated disturbance to coral reefs and post disturbance mortality by *Acanthaster. Mar. Ecol. Prog. Ser.* **26**, 295–300.

Glynn, P. W. (1988). El Niño–Southern Oscillation 1982–1983: nearshore population, community and ecosystem responses. *Annu. Rev. Ecol. Syst.* **19**, 309–345.

Glynn, P. W., ed. (1989). "Global Ecological Consequences of the 1982–1983 El Niño–Southern Oscillation." Elsevier, New York.

Glynn, P. W., and D'Croz, L. (1990). Experimental evidence for high temperature stress as the cause of El Niño-coincident coral mortality. *Coral Reefs* **8**, 181–191.

Gunnill, F. C. (1985). Population fluctuations of seven macroalgae in southern California during 1981–1983 including effects of severe storms and an El Niño. *J. Exp. Mar. Biol. Ecol.* **85**, 149–164.

Holt, R. D. (1990). The microevolutionary consequences of climate change. *Trends Ecol. Evol.* **5**, 311–315.

Houghton, J. T., Jenkins, G. J., and Ephraums, J. J., eds. (1990). "Climate Change: The IPCC Scientific Assessment." Cambridge Univ. Press, New York.

Jackson, G. A. (1977). Nutrients and production of the giant kelp, *Macrocystis pyrifera,* off southern California. *Limnol. Oceanogr.* **22**, 979–995.

Laurie, W. A. (1984). El Niño causa extragos nunca vistos en la población de iguanas marinas. *Biol. Erfen* **11**, 15–18.

Laurie, W. A. (1985). The effects of the 1982–83 El Niño on marine iguanas. *In* "El Niño en Las Islas Galápagos: El Evento de 1982–1983" (G. Robinson and E. M. del Pino, eds.), pp. 199–209. Fundación Charles Darwin para Las Islas Galápagos, Quito, Equador.

Levin, S. A. (1992). The problem of pattern and scale in ecology. The Robert H. MacArthur Award Lecture. *Ecology* **73**(6), 1943–1967.

Lubchenco, J. (1980). Algal zonation in the New England rocky intertidal community: an experimental analysis. *Ecology* **61**, 333–344.

Lubchenco, J. (1986). Relative importance of competition and predation: early colonization by seaweeds in New England. *In* "Community Ecology" (J. Diamond and T. J. Case, eds.), pp. 537–555. Harper & Row, New York.

Lubchenco, J., and Gaines, S. D. (1981). A unified approach to marine plant–herbivore interactions. I. Populations and communities. *Annu. Rev. Ecol. Syst.* **12**, 405–437.

Lubchenco, J., and Real, L. A. (1991). Manipulative experiments as tests of ecological theory. *In* "Foundations of Ecology" (L. A. Real and J. H. Brown, eds.), pp. 715–733. Univ. of Chicago Press, Chicago.

McGowan, J. A. (1985). El Niño 1983 on the Southern California Bight. *In* "El Niño North: Niño Effects in the Eastern Subarctic Pacific Ocean" (W. S. Wooster and D. L. Fluharty, eds.), pp. 185–187. Univ. of Washington Press, Seattle.

Menge, B. A. (1991). Generalizing from experiments: is predation strong or weak in the New England rocky intertidal? *Oecologia* **88**, 1–8.

Menge, B. A. (1992). Community regulation: under what conditions are bottom-up factors important on rocky shores? *Ecology* **73**(3), 755–765.

Menge, B. A., and Farrell, T. M. (1989). Community structure and interaction webs in shallow marine hard-bottom communities: tests of an environmental stress model. *Adv. Ecol. Res.* **19**, 189–262.

Menge, B. A., and Lubchenco, J. (1981). Community organization in temperate and tropical rocky intertidal habitats: prey refuges in relation to consumer pressure gradients. *Ecol. Monogr.* **51**, 429–450.

Menge, B. A., and Sutherland, J. S. (1987). Community regulation: variation in disturbance, competition, and predation in relation to environmental stress and recruitment. *Am. Natur.* **130**, 730–757.

Murray, S. N., and Horn, M. H. (1989). Variations in standing stocks of central California macrophytes from a rocky intertidal habitat before and during the 1982–1983 El Niño. *Mar. Ecol. Prog. Ser.* **58**, 113–122.

National Research Council. (1986). "Ecological Knowledge and Environmental Problem-Solving: Concepts and Case Studies." National Academy Press, Washington, D.C.

North, W. J. and Zimmerman, B. C. (1984). Influences of macronutrients and water temperatures on summertime survival of *Macrocystis* canopies. *Hydrobiologia* **116/117**, 419–424.

Paine, R. T. (1966). Food web complexity and species diversity. *Am. Natur.* **100**, 65–75.

Paine, R. T. (1980). Food webs: linkage, interaction strength and community infrastructure. The Third Tansley Lecture. *J. Anim. Ecol.* **49**, 667–685.

Paine, R. T. (1986). Benthic community–water column coupling during the 1982–1983 El Niño. Are community changes at high latitudes attributable to cause or coincidence? *Limnol. Oceanogr.* **31**, 351–360.

Pickard, G. L., and Emery, W. J. (1982). "Descriptive Physical Oceanography: An Introduction" (4th ed.). Pergamon Press, Oxford, U.K.

Rasmusson, E. M., and Wallace, J. M. (1983). Meteorological aspects of the El Niño/Southern Oscillation. *Science* **222**, 1195–1202.

Roughgarden, J., Gaines, S. D., and Possingham, H. (1988). Recruitment dynamics in complex life cycles. *Science* **241**, 1460–1466.

Seapy, R. R., and Littler, M. M. (1980). Biogeography of rocky intertidal macroinvertebrates of the southern California Islands. *In* "The California Channel Islands: Proceedings of a Multidisciplinary Symposium" (D. M. Power, ed.), pp. 307–324. Santa Barbara Museum of Natural History, Santa Barbara.

Seymour, R. J., Tegner, M. J., Dayton, P. K., and Parnell, P. E. (1989). Storm wave induced mortality of giant kelp, *Macrocystis pyrifera*, in Southern California. *Estuarine Coastal Shelf Sci.* **28**, 277–292.

Schreiber, R. W., and Schreiber, E. A. (1984). Central Pacific seabirds and the El Niño Southern Oscillation: 1982 to 1983 perspectives. *Science* **225**, 713–716.

Smith, J. B., and Tirpak, D. A., eds. (1989). The potential effects of global climate change on the United

States. Appendix B—Sea level rise. Office of Policy, Planning and Evaluation, U.S. Environmental Protection Agency, Washington, D.C.

Sousa, W. P. (1984). The role of disturbance in natural communities. *Annu. Rev. Ecol. Syst.* **15,** 353–391.

Steinbeck, J. R., Groff, J. M., Friedman, C. S., McDowell, T., and Hedrick, R. P. (1992). Investigations into a mortality among populations of the California black abalone, *Haliotis cracherodii,* on the central coast of California, USA. *In* "Abalone of the World: Biology, Fisheries, and Culture." (S. A. Shepherd, M. J. Tegner, and S. A. Guzman Del Proo, eds.), pp. 203–213. *Proceedings of the First International Symposium on Abalone,* Fishing News Books, Blackwell Scientific Publications, Oxford.

Stouffer, R. J., Menabe, S., and Bryan, K. (1989). Interhemisphere asymmetry in climate response to a gradual increase of atmospheric CO_2. *Nature* **342,** 660–662.

Tegner, M. J., and Dayton, P. K. (1987). El Niño effects on southern California kelp forest communities. *Adv. Ecol. Res.* **17,** 243–279.

Tissot, B. N. (1988). Mass mortality of black abalone in southern California. *Am. Zool.* **27,** 69A.

Tissot, B. N. (1991). Geographic variation and mass mortality in the black abalone: The roles of development and ecology. Ph.D. dissertation, Department of Zoology, Oregon State Univ., Corvallis.

Trenberth, K. E. (1990). General characteristics of El Niño–Southern Oscillation. *In* "ENSO Teleconnections Linking Worldwide Climate Anomalies: Scientific Basis and Societal Impact" (M. Glantsy, R. Katz, and N. Nicholls, eds.), pp. 13–42. Harvard Univ. Press, Cambridge, Massachusetts.

Wooster, W. S., and Fluharty, D. L., eds. (1985). "El Niño North." Washington Sea Grant Program, University of Washington, Seattle.

CHAPTER 13

Southeastern Pacific Coastal Environments: Main Features, Large-Scale Perturbations, and Global Climate Change

JUAN CARLOS CASTILLA   SERGIO A. NAVARRETE
JANE LUBCHENCO

I. Introduction

Although the southeastern Pacific (SP) coastal and oceanic realms of Peru and Chile are in some aspects similar to their Northern Pacific counterparts, they also have some special characteristics that make them unique (Castilla, 1979; Santelices, 1989). Among these characteristics are their upwelled waters and marine resources that rank among the richest in the world (Pauly and Tsukayama, 1987) and the particularly strong influence of large-scale perturbations, such as the El Niño–Southern Oscillation (ENSO) and coseismic coastal uplifts or subsidences due to earthquakes (Lomnitz, 1970; Kelleher, 1972; Cane, 1983; Nishenko, 1985; Bernal and Ahumada, 1985; Comte et al., 1986; Castilla, 1988; Enfield, 1989; Castilla and Oliva, 1990). Knowledge about the impact of such anomalies on biological systems is still poor, but some of their major consequences have been documented in the literature both for SP oceanic ecosystems (Barber and Chávez, 1983) and for coastal, intertidal, and subtidal ecosystems [see Arntz et al., 1985; special volume of Investigación Pesquera (Chile), 1985; and Castilla and Oliva, 1990]. Another important characteristic of the coastal SP ecosystems is their key role in the establishment and development of ancient cultures along the coasts of Peru and Chile (Parsons, 1970; Llagostera, 1979a,b) and in the well-being of the present inhabitants as a direct source of food (Durán *et al.*, 1987) and economic resources for the Pacific rim countries (Bustamante and Castilla, 1987).

In this chapter we first present a review of the main geomorphic, climatic, oceanographic, biogeographic, and ecological features of the SP realm, with an emphasis on Chilean coastal environments. We have arbitrarily defined coastal environments as those encompassing the waters within about 3 miles off the shoreline and with a maximum depth of about 150 m. These environments sustain four types of well-defined fishery activities: (1) sport, (2) manual extraction during low tides, (3) free "hooka artisanal" diving (diver receives air from air compressor on small boat), and (4) small-scale artisanal fishery (boats ≤15 tons) (see Durán *et al.*, 1987; Bustamante and Castilla, 1987) (Fig. 1). Within this realm, also arbitrarily, we distinguish three major ecosystems: (1) intertidal, ranging from the extreme high water spring (EHWS) to the extreme low water spring (ELWS); (2) nearshore subtidal, ranging roughly from the ELWS to depths of about 30 m, which can be reached by hooka divers; and (3) farshore subtidal, ranging from about 30 to 150 m in depth and in which most of the artisanal (nondiving) fishery occurs (Fig. 1). Offshore waters outside this realm will be referred to as the interior ocean. Second, we analyze the biological effects of regional large-scale anomalies by analyzing examples of species, populations, or communities that have been monitored for long periods of time. Third, we project these examples to the years 2050–2100 under scenarios set by the Intergovernmental Panel on Climate Change (IPCC), Working Group II (1990). The goal of this exercise is to focus on SP examples in which large-scale changes similar to those predicted by global climate change models (GCCMs) have occurred at least temporarily in the past decades—namely rise in sea water level and temperature.

Finally, we incorporate human influences and activities into the resulting scenarios and address questions such as: How can we prepare to adapt to this kind of change? How should we promote more effective interdisciplinary investigation and what can be done to provide incentives for regional, interregional, and interhemispheric cooperative research?

II. Main Geomorphic and Climatic Features with Emphasis on the Chilean Coastline

Most of the SP rim occurs along the subduction of the oceanic Nazca plate under the continental South American plate. Earthquakes of large magnitude are frequent, producing coastal uplifts or subsidences (Kelleher, 1972; Castilla and Oliva, 1990). Deep oceanic trenches and a narrow continental shelf are characteristic features of the area. The Andes Mountains run close and parallel to the coast all along this Pacific

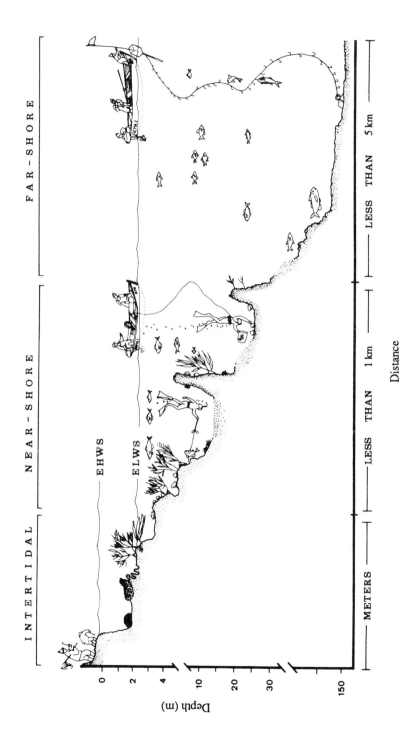

Fig. 1. Sketch of intertidal nearshore and farshore environments in the southeastern Pacific. EHWS, extreme high water spring; ELWS, extreme low water spring. Depths and distances are approximate.

rim, creating special climatic conditions. According to Viviani (1979) and Paskoff (1989), four distinct sections can be identified along the Chilean coast:

1. The *arid coast,* extending from central Peru to 27°S in Chile, is characterized by a northward-trending cliff that is several hundred meters high in some areas (for example, between Arica, near the Peruvian border, and Iquique: 18°30′–20°30′S) (see also Araya-Vergara, 1976). This section shows an almost straight coastal rim with very few sheltered bays and no high-waterflow rivers reaching the ocean. The land climate is defined as arid (di Castri and Hajek, 1976) or hyperarid (Atacama Desert; Paskoff, 1989), but littoral zones and coastal waters are not correspondingly warm because of the coastal upwelling. Surface waters in this section show isotherms nearly parallel to the coastline, ranging from about 22 to 24°C depending on the season of the year and prevailing winds (Robles *et al.,* 1974; Viviani, 1979). Upwelling processes weaken during warm seasons of the year or warm periods such as the ENSO.

2. The *semiarid coast,* which extends from about 27 to 32°S, is characterized by well-developed Plio-Quaternary marine terraces (mostly granitic) extending from the shoreline to about 150–200 m offshore. This section also has an almost straight coastline with very few sheltered bays or offshore islands and few rivers reaching the sea (Limarí, Choapa, Aconcagua). The climate is mediterranean, with rainfall concentrated during the winter season and increasing with latitude (Viviani, 1979). There are seasonal trends in seawater temperature, but in general they are 2–4°C lower than in the arid section (Robles *et al.,* 1974; Viviani, 1979). Furthermore, numerous upwelling areas exist along the coast (e.g., Coquimbo and Valparaíso).

3. *Central coast,* which extends between approximately 32 and 42°S. This section is rougher than the previous ones in terms of wave impact and is interrupted by extensive beaches and dune fields. There are many high-waterflow rivers (e.g., Bío Bío, Cautín, Tolten, Calle-Calle) reaching the sea, and the oceanic temperate climate (di Castri and Hajek, 1976) favors runoff erosion (and hence sedimentation), particularly in winter. This coastal section has a mediterranean climate, with rainfall occurring mostly during winter and periods of drought during summer. Sheltered bays are common. South of 39°S there are many "ria"-type estuaries (e.g., Maullín, approximately 41°40′S), which result from submergence by Holocene transgressions. Chiloé Island, approximately 42–43°50′S, is the main coastal geomorphologic feature of this section of the country.

4. *Fjords coast.* This section extends from about 43°30′ to 56°S (Cape Horn) and is one of the best examples of deep and rugged fjord coasts in the world (Pickard, 1971). Its highly indented shoreline, the numerous islands, and the relief of 500–2000 m around inlets make it comparable to the fjord coast between southern British Columbia and Alaska in the northern Pacific. In southern Chile the fjords penetrate inland through the glacially dissected Andes Mountains, and numerous short and low-waterflow rivers can be found between 52 and 56°S. The climate is oceanic type with rainfall ranging between 2000 and 3000 mm/year and homogeneously distributed throughout the year. Air temperature is also rather homogeneous throughout the year, showing some increases during the summer.

III. Main Oceanographic Features with Emphasis on the Chilean Sector

A. Oceanography of the Area

Since the early 1960s several oceanographic cruises, particularly off the coast of northern Chile, have gathered information to elucidate the main geostrophic current fields (Brandhorsts, 1971; Sievers and Silva, 1975; Silva and Sievers, 1981). Several circulation patterns have been proposed by different authors. Bernal and Ahumada (1985) summarized the main oceanographic features of the Chilean sector. The main cell of circulation is connected to the South Pacific anticyclone gyre and driven by the West Wind Drift, reaching the South American continent near latitude 40°S (or 45°S, Johnson *et al.,* 1980). At those latitudes, branching into two main systems of currents is observed. The southern branch is known as the *Cape Horn system,* and the northern branch comprises what we will call the *Humboldt* or *Chile-Peru Current system* (Fig. 2A and B). These two oceanic branches are particularly evident in the interior ocean but not very clear in coastal waters (Fig. 2B). Within the first 150 miles off the northern coast of Chile and within the Humboldt Current system, Silva and Fonseca (1983) identified three main flows that exhibit seasonal variation. During summers, a permanent northward current, part of the Humboldt Current, is clearly observed in the near oceanic area, while farther west the southward Peru–Chile Undercurrent that reaches superficial levels is evident (Fig. 2C). Closer to the coast, another northward flow is found, but there is no evidence of its connection with the Humboldt Current. Finally, Fig. 2D shows a summarized interpretation of Bernal and Ahumada (1985) for the main subsuperficial and superficial water masses found in the SP realm during a "normal" summer. In general, little is known about the circulation patterns in truly coastal waters of the SP (see Fig. 1). Much emphasis has been placed on the oceanography of the interior ocean and too little on that of the near shore or far shore.

One of the main oceanographic features of SP coastal waters is the upwelling of subsurface waters into surface layers (Svedrup, 1938; Wyrtki, 1963; Smith, 1968), which creates anomalous low temperatures and high productivities. In the SP there are numerous well-identified areas of upwelling (Valparaíso, Coquimbo, Iquique, southern and central Peru) that together sustain one of the richest pelagic fisheries of the world.

B. The ENSO Events

The main oceanographic anomaly affecting the SP realm is the El Niño event and its atmospheric counterpart, the Southern Oscillation (ENSO). One of the prevailing interpretations is that ENSO events occur as an internal cycle of positive and negative feedbacks within the coupled ocean–atmosphere climate system of the tropical Pacific (see review by Enfield, 1989). Baroclinic equatorial Kelvin waves are generated and propagate eastward toward South America, depressing the thermocline and raising the sea level. The net result of ENSO events is an increase in surface

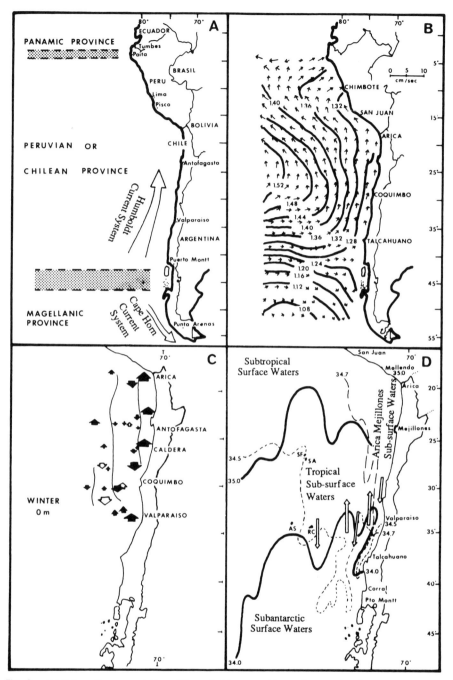

Fig. 2. (A) Schematic representation of the main biogeographic provinces and surface current systems in the South Pacific. (B) Main vectors of surface currents obtained over long periods of observations (after Bernal and Ahumada, 1985). (C) Geostrophic velocity composition at surface waters between Arica and Valparaíso during the summer (after Silva and Fonseca, 1983). (D) Main surface and subsurface water masses in the South Pacific. (Parts B, C, and D adapted from Bernal and Ahumada, 1985.)

seawater temperature of about 3–5°C and an increase in sea level of up to 20 cm from Peru to Chile (Enfield, 1989; Glynn, 1988a).

IV. Biogeography: Past and Present

A. The Past Environment

Some paleontologic and paleoclimatologic studies enable a general understanding of the Tertiary and Quaternary coastal environments in the SP. Paleoclimatic and glaciation evidence shows major climatic changes during the late Tertiary for the coast of Chile (Caviedes, 1972, 1975). Pliocene coastal fossils of mollusks and palm trees found in central and southern Chile suggest that at that time there was about a 30° southward displacement of warm coastal waters compared to the present situation. According to Caviedes, the late Tertiary western coast of South America, down to about 37°S, was warm and humid, much like the present-day western coast of the equator. On the other hand, Herm and Paskoff (1966) and Herm (1969), analyzing Pliocene and Pleistocene coastal fauna (mainly mollusks), concluded that at the Miocene–Pliocene boundary (about 10 million years ago) a morphologic and regional constellation in the southern part of South America led to the formation of the Humboldt Current. This major oceanographic change contributed to the alteration of the coastal fauna of the SP. Large numbers of "Atlantic" genera disappeared and the fauna received a more "Pacific" character. Then, at the beginning of the Pleistocene (about 1 million years ago) a second major event took place, a worldwide climatic change, and once again the Chilean marine coastal fauna was strongly modified. There was a further reduction of warm-water genera (e.g., *Anadara, Anomia*), a reduction in the number of species of several genera (e.g., the scallops *Chlamys* and mussels *Chorus*), and a substitution of dominant species (e.g., venerid mollusks). Several species that appeared in the Pleistocene as accessory forms (e.g., *Mesodesma donacium* or "macha" and *Mulinia* sp.) became dominant in their habitats. Several studies suggest the existence of a new major event during the Recent, which can be considered as the third critical phenomenon affecting the coastal fauna of the SP (Burnett, 1980; De Vries and Schader, 1981; Rollins *et al.,* 1986). This last major event contributed to the transformation of this fauna into its present form, and it has been interpreted by Richardson (1981) as a "reorganization" of the Pacific Current system and the birth of El Niño, which appeared for the first time during a short interval of 500 years around 5000 years b.p. (see also Rollins *et al.,* 1986). The evidence suggests that during the Recent, between about 11,000 and 5000 years b.p., the warm Panamic province extended some 500 km farther south from its actual limit at 5°S (Camus, 1990). So, the present-day arid coastal climate in north and central Peru is probably post-5000 years b.p.

Thus, the marine fauna along the Pacific realm has suffered at least three major reorganizations since the Pleistocene, with dramatic changes in the limits of distribution of species and boundaries of the main biogeographic provinces. One of these major changes occurred as recently as 10,000 years ago. The exact causes of these faunistic rearrangements are somewhat unclear, but they seem to be related to major changes in ocean currents, global climatic changes, and the birth of oceanographic "anomalies" such as El Niño.

B. Present Biogeography

At present, the SP area is within the so-called Temperate Pacific Realm, south of the 20°S minimum water isotherm found around the Galápagos Island in Ecuador. Its northern biogeographic border with the Panamic Province (Fig. 2A) is set around Paita or Tumbes (4–5°S) in Peru. Zoogeographically, the most accepted view is the division of this realm into three provinces: (1) Peru–Chile: Paita to Valparaíso; (2) central Chile, Valparaíso to Chiloé Island; and (3) Magellanic, south of Chiloé (Balech, 1954; Knox, 1960; Dell, 1971; see review by Castilla, 1976). However, two provinces have also been postulated: Peruvian or Chilean, Paita to Chiloé Island and Magellanic, south of Chiloé Island (Fig. 2A).

On the other hand, Santelices (1980) described three phytogeographic regions: Tumbes (4–5°S), Transitional (5–53°S), and Fuegia (53°S to the tip of South America). Despite this confusion, it appears that some of the most critical biogeographic boundaries are located around Paita and Tumbes in Peru and the Chiloé Island in Chile. Not surprisingly, these two areas are also key areas in the main water circulation of the SP (Fig. 2).

V. Global Change Scenarios

Although there is some controversy about the reality and the causes of global climate changes (e.g., Ellsaesser, 1989; Landsberg, 1989), the IPCC Working Group II (1990) proposed different likely scenarios. These scenarios (summarized in Table I) are based on an effective doubling of CO_2 in the atmosphere over preindustrial levels between 1990 and 2025–2050 and an increase of mean global temperature in the

TABLE I
Global Change Scenarios According to IPCC Working Group II, 1990[a]

	Years		
	2025–2050	2050–2100	
CO_2	Global temperature	Sea level rise	Sea surface temperature
Doubling of present	Mean 1.5–4.5°C[b]	0.3–1 m 0.2–0.65 m[c]	0.2–2.5°C

[a]Intergovernmental Panel on Climatic Change, Working Group II, June 1990.
[b]Unequal global distribution, smallest increase of half global mean in tropical regions and about twice the global mean in polar regions.
[c]"Business as usual" scenario (Scenario A, IPCC Working Group I Report) for years 2030 and 2100, respectively.

range of 1.5 to 4.5°C with an unequal global distribution, namely half of the global mean in the tropical regions and twice the global mean in the polar regions. Furthermore, the scenarios include a sea level rise of about 0.3 to 0.5 m by 2050 and approximately 1 m by 2100, coupled with a 0.2 to 2.5°C rise of temperature of the ocean surface layer. Unfortunately, different theoretical models produce different scenarios (e.g., Bryan et al., 1982; Bryan and Spelman, 1985; Stouffer et al., 1989; for review see Bernal, 1991). Results of Stouffer et al. (1989) are particularly interesting; they show that warming will be particularly slow in the Antarctic circumpolar ocean because of the larger fraction of the globe covered by ocean waters in the Southern Hemisphere and because of the strong surface westerlies (particularly between 45 and 65°S). Their simulations of mean surface air temperature predict time lags of about 30 to 35 years between latitude 45°S and 45°N for a 2°C increase (slower increase of air temperature in the Southern Hemisphere). Furthermore, the pattern of air temperature rise decreases with increasing latitude in the Southern Hemisphere, in sharp contrast with the Northern Hemisphere (Stouffer et al., 1989).

These models suggest very complex scenarios for the surface air temperature and emphasize interhemispheric asymmetries, in opposition to previous models predicting rather symmetric responses (e.g., Spelman and Manabe, 1984). The manner in which these interhemispheric asymmetric responses will influence ocean circulation patterns, local or regional currents, or global oceanic phenomena is a highly controversial matter. Bakun (1990) proposed that because greenhouse warming will lead to less global temperature contrast between tropical and polar regions, the basin-scale atmospheric and oceanic circulation will lead to a decrease in the intensity of the Humboldt Current in the SP and probably to an increase in ENSO events. Furthermore, since heating of the earth will increase and greenhouse gas could inhibit nighttime cooling, onshore–offshore atmospheric pressure gradients (and winds) could be intensified, accelerating coastal upwelling circulation and increasing primary production in the SP.

It is difficult to make certain predictions about any of the colateral effects associated with global warming or to indicate how these changes will affect specific regions of the world oceans (e.g., Bernal, 1991). What is important, and rather alarming, is that global warming could at least potentially modify the major large-scale structural processes of the SP ecosystems, namely current patterns, frequency and intensity of ENSO events, and intensity of upwellings.

The most important aspect of global change scenarios is the predicted rates of change. According to these predictions, changes in the near future will occur so fast that no parallel can be found in the paleoarcheological record (Davis, 1989). For instance, for temperate latitudes even conservative models predict changes in air temperature of 0.5 to 1.0°C per decade (Bernal, 1991). These rates are 5 to 10 times greater than those observed in the past century or between pre-Holocene glacial intervals. Similarly, the estimated rate of sea level rise (about 10–15 cm over the last century; Milliman 1989, Jacobson 1990) is faster than that recorded over the past 5000 years (1–10 cm per century).

VI. Large-Scale Perturbations

A. ENSO and Direct Effects on Species Abundance

Although numerous studies have been carried out on the coastal effects of ENSO, it is difficult to assess the real impact on SP temperate ecosystems because there is a general lack of long-term records covering years without ENSO events (but see papers referenced in Arntz *et al.,* 1985, and Glynn, 1988a). Moreover, there are at least two problems in comparing ENSO effects with the predicted global climate scenarios. First, it is usually difficult to single out causal variables (e.g., temperature) from the complex set of multiple interacting variables accompanying ENSO (Glynn, 1988a; Lubchenco *et al.,* Chapter 12, this volume). Second, the rate at which environmental changes occur during ENSO events is much faster (days, weeks, or months) than the rates predicted by global climate scenarios.

Since we have no information for South America about ecological effects of temperature increases produced by nuclear plant discharges (unlike North America, Lubchenco *et al.,* Chapter 12, this volume), ENSO events provide the closest test example of large-scale temperature increases in coastal waters. Effects of ENSOs might help us foresee the effects of global climate change, but the problems mentioned earlier should be borne in mind when drawing conclusions from them.

Abnormally high temperatures during the 1982–1983 ENSO seem to have been the main cause of mortality of invertebrates and algae in Peru and northern Chile. For instance, dramatic mortalities (> 80%) in subtidal banks of mussels *(Aulacomya)* and of prominent space occupiers (cirriped barnacles and mytilid mussels) and molluscan grazers in the intertidal zone of Peru (Tarazona *et al.,* 1985; Arntz, 1986), as well as macroalgae in northern Chile (Stanley, 1984; Fuenzalida, 1985), have been attributed to temperature increases during ENSO. The direct effects of sea level rise itself are less clear. One of the few accounts is that of Hays (1986), who noted that nesting sites of the Humboldt penguin were adversely affected by sea level rise, rough seas, and flooding. In general, sea level rise during ENSO events does not seem to have extensive direct effects on temperate SP ecosystems; the more dramatic direct effects of ENSO on these ecosystems can be attributed to the sudden increase in sea surface temperature, but undoubtedly nutrient depletion played an important role too (see Glynn, 1988a). Some of the most important and long-lasting ecological changes observed during ENSO were produced directly by the changes in physical conditions. For instance, invasion of intertidal and shallow subtidal communities by novel species after reductions in the populations of dominant competitors or controlling predators have had profound and prolonged (several years) effects on tropical and temperate ecosystems (e.g., Tarazona *et al.,* 1985; Glynn, 1988b). Similarly, distributional changes, generally involving range extensions of tropical or subtropical species into higher latitudes, produced important community changes and could potentially lead to biogeographic rearrangements. Glynn (1988a) discusses in detail examples that illustrate the importance of biological interactions in mediating the direct environmental changes during ENSO events (see also comprehensive reviews by Glynn, 1989, and Arntz *et al.,* 1985).

An example of ENSO effects on the abundance of two species in Peru and Chile is presented in Figs. 3 and 4. This example is particularly interesting because of the economic value of the species and the existence of long-term programs monitoring their fishery and oceanographic parameters in the area. Both species of bivalve mollusks, the mussel *Aulacomya ater* ("cholga" or "cholgua") and the scallop *Argopecten purpuratus* ("concha abanico" or "ostión del norte"), are harvested exclusively by artisanal fleets (small scale, relatively unequipped, low cost, and numerous people involved). Although the fishery of these species has shown a persistent increasing trend for the past 20–30 years in both Chile (Bustamante and Castilla, 1987) and Peru (IMARPE, unpublished information), their landings have fluctuated markedly. The landing of mussels in Peru between 1956 and 1972 increased steadily from about 2000 to 15,000 tons per year (Fig. 3). In 1972–1973 and 1977 the fishery experienced the first two large landing decreases, and another drastic decline was observed in 1984–1985. On the other hand, scallop landings showed complementary increases in the same years or with a time lag of 1 to 2 years (Fig. 3). Particularly spectacular were landings of scallops in Peru during 1985 (Pisco, Peru; Mendo *et al.*, 1988), reaching nearly 48,000 tons. Between 1984 and 1989 a steady increase in mussel landings was observed, coupled with a substantial decrease in landings of scallops. Overexploitation of scallops has been proposed as the reason for this drastic drop (Samamé and Valdivieso, 1986; Valdivieso *et al.*, 1988).

As shown in Fig. 3, the major fluctuations in landings of mussels and scallops occur during post-ENSO events (at least between 1972 and 1989). This seems particularly evident when extraordinary ENSO events, such as the 1982–1983 one, are accompanied by extreme surface water anomalies of about +6° to +8°C (Arntz and Valdivia, 1985; Brainard and McLain, 1987). A similar pattern in the landings of these species was observed in Antofagasta, northern Chile (Fig 4). Although total landings of mussels are smaller than in Peru, the main fluctuations also occurred during post-ENSO events. No fishery of scallops was recorded before 1980 in Antofagasta, but following the 1982–1983 ENSO event the landings increased dramatically. Note also that temperature anomalies at Antofagasta are smaller (less than +2°C) than those reported for central and northern Peru (compare Figs. 3 and 4).

Do the complementary fluctuations in scallop and mussel landings reflect ecological interactions between these species, or are they the result of independent and opposite specific responses to similar oceanographic anomalies? To gain insight into these questions it is important to notice that *Aulacomya ater* is a southern cold species (Cancino and Becerra, 1978; distributed from Callao, Peru, to southernmost South America), whereas *Argopecten purpuratus* is a rather northern tropical species (distributed from Paita, Peru, about 5°S, to Coquimbo, Chile, about 30°S; Valdivieso *et al.*, 1988). Hence, the central coast of Peru (Callao-Pisco) represents roughly the northern limit of distribution for *Aulacomya*, but it is just the center of the distribution of *Argopecten*. During the 1982–1983 ENSO *Aulacomya* suffered large mortalities and its growth rate was dramatically decreased (Soenens, 1985), while *Argopecten* exhibited (1) an elevated gonad index, (2) reduced time for metamorphosis, (3) reduced mortality, and (4) increased natural collection of larvae (Wolff, 1985; Mendo

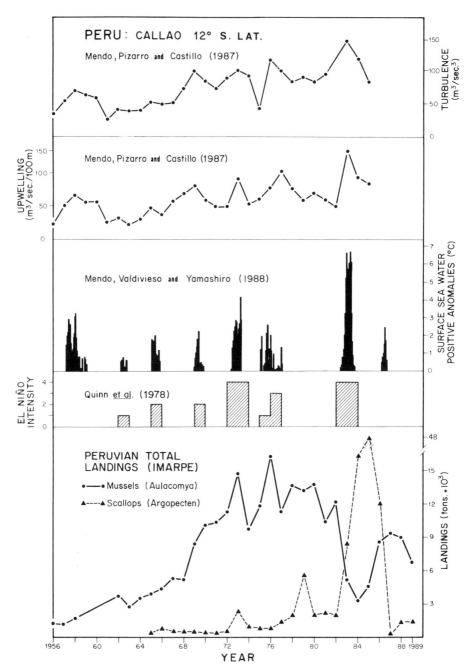

Fig. 3. Total landings of the mussel *Aulacomya ater* ("cholga") and the scallop *Argopecten purpuratus* in Peru (Instituto del Mar del Peru) and oceanographic indexes for Callao, Peru, between 1956 and 1989.

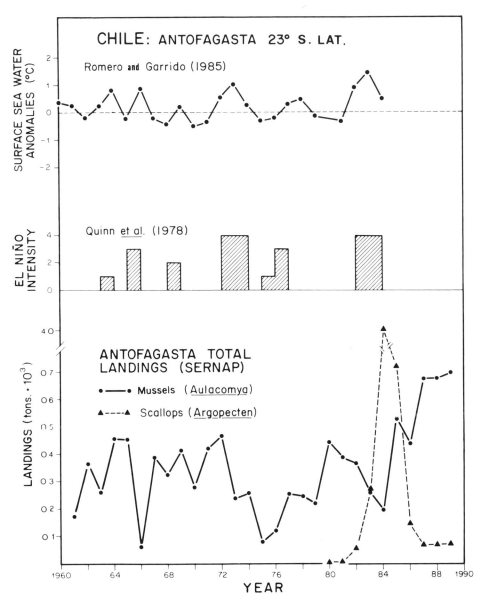

Fig. 4.   Total landings of the mussel *Aulacomya ater* and the scallop *Argopecten purpuratus* in Antofagasta, Chile (Servicio Nacional de Pesca) and oceanographic indexes for Antofagasta, Chile, between 1960 and 1989.

et al., 1988; Illanes *et al.*, 1985). On the other hand, these species usually occupy different kinds of habitats (DiSalvo *et al.*, 1984; Cancino and Becerra, 1978) and hence it is difficult to imagine direct interference between them (although competition for food during ENSO events cannot be ruled out). Thus, an oceanographic anomaly involving a combination of environmental factors at a single geographic point had opposite effect on two economically important species with different biogeographic origins.

B. Coseismic Coastal Subsidence and Uplifts Due to Earthquakes

The SP coast is periodically subjected to intense seismic activity. The coast of central Chile, for instance, has been affected by large earthquakes at a remarkably regular period of 83 ± 9 years (mean ± SD) (Comte *et al.*, 1986), and many of them have produced important uplifts and subsidences of coastal rocks. These vertical uplifts translate into sea level changes for coastal benthic organisms (Plafker and Savage, 1970; Castilla and Oliva, 1990; see also Lomnitz, 1970). The magnitude of these changes can be remarkable. For instance, the May 1960 earthquake in southern Chile produced uplifts and subsidences of up to 5 m along more than 700 m of coast (Table II); an earthquake in central Chile in March 1985 produced an uplift of about 40–60 cm on sectors of the coast, which represents up to one-third of the tidal range in those areas (Castilla, 1988; Castilla and Oliva 1990).

Understanding the effects of coseismic subsidences on coastal ecosystems might provide insight into the potential effects of sea level rise under the global change scenarios. Changes due to earthquakes, however, occur much more rapidly (virtually instantaneously) and often exceed the magnitude of those predicted for the next 100 years (compare Tables I and II). Unfortunately, these natural large-scale "sea level rises" have not been properly studied in Chile and we have very little and mostly anecdotal ecological information. One of the few biological effects actually recorded was that on the macroalga *Gracilaria*—economically the most important macroalga in Chile—in the estuary of the river Maullín (Santelices and Ugarte, 1987). Before the 1960 earthquake 5000 ha of the estuary was populated by *Gracilaria*. After the earthquake the population expanded about 1200 ha (25%) by colonizing suitable new estuarine habitats generated by the subsidence. Santelices and Ugarte (1987) described a time lag of several years between the earthquake and the establishment of the new algal beds, probably corresponding to the time required to change a substratum suited for cultivation of land crops to one suited for colonization and propagation of a marine crop (see also Davis, 1989). This kind of expansion of appropriate habitats for some estuarine species might be a common event between latitudes 38 and 42°S (Santelices and Ugarte, 1987).

We have no information on the biological effects of subsidences on intertidal or shallow subtidal areas on the open coast. However, two detailed studies documented the effects of coseismic uplifts on intertidal communities (Castilla, 1988; Castilla and Oliva, 1990). In this case, long-term monitoring programs allowed us to identify and differentiate the effects of the coseismic event from natural population fluctuations. The uplift eventually produced mortality of the structurally and competitive dominant

TABLE II

Vertical Displacement Relative to Sea Level as a Result of Coseismic Uplift and Subsidence at 20 Localities along about 800 km of Coast after the May 1960 Chilean Earthquake

Locality	Latitude (S)	Longitude (W)	Subsidence (−m)	Uplift (+m)
Isla Mocha[a]	38°23.3′	73°53.4′		0.9–1.0
Pto. Saavedra	38°43.6′	73°25.7′	1.20–1.60	
Toltén	39°13.1′	73°12.8′	2.0	
Queule	39°26.0′	73°12.4′	1.60	
Valdivia	39°47.4′	73°12.5′	2.70	
Corral	39°51.7′	73°25.4′	1.80–2.10	
Pucatrihue	40°30.4′	73°49.2′	1.30	
Puerto Montt	41°31.2′	73°02.7′	0	0
Maullín	41°37.0′	73°37.0′	1.50–1.60	
Calbuco	41°45.2′	73°07.7′	0.40–0.60	
Ancud	41°51.5′	73°57.7′	1.80	
Quemchi	42°08.5′	73°28.2′	1.50	
Cucao	42°37.7′	74°07.0′	1.00	
Chaitén[a]	42°55.0′	72°43.5′		1.00
Quellón	43°07.5′	73°37.7′	1.20	
Isla Guafo[a]	43°34.0′	74°50.0′		2.8–3.6
Isla Ipun	44°38.5′	74°52.2′	0	0
Isla Stokes	44°39.7′	74°54.4′	0	0
I.Guamblín[a]	44°55.0′	75°03.0′		5.5
Archipiélago Chonos	~45°30.0′		1.00–2.00	

Source: Plafker and Savage (1970).

[a]Localities east of the submarine fault which showed coseismic uplift.

macroalga *Lessonia nigrescens* and its associated fauna (see Santelices, 1990), inducing several community rearrangements in the low intertidal zone (Castilla, 1988). After 5 years the community is returning to its preearthquake structure (Castilla and Oliva, 1990).

VII. Conclusions

A. General Predictions and Recommendations

As indicated by Davis (1989), global climate change can be understood only if we coordinate research on the different levels of ecological organization. In this chapter we have focused on some aspects of coastal environments of the temperate SP realm and selected specific examples that could help us interpret the effects of regional or global anomalies on SP species, communities, or ecosystems. Some very general conclusions about the potential effect of global climate change and a few recommendations can be drawn from this exercise.

1. Effects of Sea Level Rise. The information about the effects of sea level rise during ENSO events and coseismic subsidences suggests that sea level rise itself will not have extensive influence on temperate intertidal and shallow subtidal ecosystems of open coasts. The dramatic effects of coastal uplifts on low intertidal communities reported by Castilla (1988) and Castilla and Oliva (1990) seemed to be caused by the speed with which these sea level changes occurred, exposing sessile animals to unusual desiccation rates. At the rate at which sea level changes occur under the global change scenarios, the time for appreciable sea level rises might well exceed the life span of most intertidal organisms (although we have very little accurate information about life spans of invertebrates) and by far exceeds recruitment rates of most of them. This might allow continuous recolonizations and changes in the vertical limits of distribution of organisms in the intertidal zone. Moreover, unlike corals in tropical systems (e.g., see Glynn and D'Croz, 1990; Glynn, 1991), temperate species do not seem to be extremely sensitive to changes in depth, as their extensive bahtymetric distributions suggest (Brattström and Johansen, 1983; Santelices, 1989).

On the other hand, sea level rises in estuaries produce direct habitat modifications, in general expanding estuarine conditions inland but also modifying local circulation patterns. As shown earlier, this habitat modification might have strong consequences. Thus, the area potentially most sensitive to sea level rise would be the central coast of Chile (see Section I). Knowledge about abundant estuaries in these area is very preliminary; estuarine fauna have just begun to be explored and their ecology is for the most part completely unknown. Intensive basic exploratory and ecological studies in this area are urgent, and establishment of permanent monitoring of physical conditions as well as biological systems is necessary.

2. Effects of Surface Temperature. Increases in sea surface temperature during ENSO events have had extensive and dramatic effects in tropical and temperate biological systems of the SP. Most of the direct mortality produced by temperature rises (e.g., mortality of algae and invertebrates) might be attributable to the rate at which these changes occurred. Unlike some tropical species (e.g., Glynn and D'Croz, 1990; Jokiel and Coles, 1990), temperate invertebrates have wide ranges of tolerance to temperatures when time is allowed for acclimatization (Southward, 1958; Jabloski *et al.*, 1985). However, temperature can have important sublethal effects on the performance of organisms and hence, rather indirectly, can ultimately determine the limits of distribution of species. An important complication is that temperature can affect coexisting species in different and even opposite ways (see Section VIA). Laborious individual studies of physiologic responses to temperature might help to identify particularly sensitive species but do not help much when making predictions for real situations because of the intricate biological interactions among species in natural communities (see Lubchenco *et al.*, Chapter 12, this volume; Glynn, 1988a). Thus, the most pervasive and dramatic effects of rises in temperature due to global climate change might be produced not directly by temperature changes (e.g., mortality of species) but by the combination of physical conditions and biological interactions. For instance, successful invasion of novel species might substantially modify the network of species interactions that maintain a particular community structure and landscape. Expansions or contractions in the distribution ranges of keystone species (in the sense of Paine, 1966) or structurally dominant species (e.g., mussels) will have profound consequences in the entire community.

Experimental studies conducted mainly in central Chile provide a good starting basis to identify species that play key roles in their communities (e.g., Castilla and Paine, 1987; Castilla and Durán, 1985; Santelices *et al.*, 1981). For instance, we know that a starfish species *(Heliaster helianthus)* plays a key role in some Chilean communities (Paine *et al.,*1985), that structurally dominant mussels beds serve as nursery habitats for a great number of invertebrate species (Cancino and Santelices, 1981; Castilla *et al.,* 1989; Navarrete and Castilla, 1990a), and that recruitment of these mussels requires very special conditions (Navarrete and Castilla, 1990b). However, we know little about the ecology of these species at different latitudes. Particularly attractive study areas for ecological studies and establishment of permanent monitoring programs are Paita in Peru and Chiloé Island in Chile, where the boundaries of the main biogeographic provinces are located. In these areas the limits of distribution of several species are found, and ecological interactions might be very different from those observed at other latitudes (e.g., S. A. Navarrete and J. C. Castilla, in preparation; see also Dethier and Duggins, 1988 for a Northern Hemisphere example).

3. Collateral Effects of Global Climate Change. Some global circulation models predict important changes in the current patterns, intensity of upwellings, and frequency of oceanographic anomalies such as ENSO (see Section V). These processes are the main large-scale structural forces in SP ecosystems (see Sections IV and VI), and drastic changes in them would have profound, extensive, and unpredictable biological consequences.

B. The Uniqueness of SP Coastal Environments

An array of unique climatic, oceanographic, and ecological characteristics makes the SP temperate realm of the world one of the critical geographic areas to be studied in view of future global climate changes. Moreover, some global circulation models predict that the SP will be affected differently from the northern Pacific, which reinforces the need for comprehensive studies in this area of the world.

The strong influence and periodicity of ENSO events and coseismic coastal uplifts and subsidences offer unparalleled opportunities to study biological effects of large-scale perturbations somewhat like those predicted by global change scenarios. Because of lack of basic ecological information and permanent monitoring programs in the past, there has not been appropriate study of the biological effects of these climate perturbations. Comparative studies and the increased ecological knowledge gained in the last decade should facilitate the establishment of permanent monitoring programs in particularly sensitive areas.

C. SP Coastal Environments and Their Significance to Humans

Since ancient times coastal environments of Peru and Chile have represented a rich source of food for humans and many cultures have flourished along the over 6000 km of coastline bordering these two countries (Parsons, 1970; Llagostera, 1979a,b).

Bustamante and Castilla (1987) reviewed the artisanal fishery of shellfish resources along the Chilean territory and showed that more than 60 different species of coastal invertebrates are fished. Over 55,000 shellfish artisanal workers are directly engaged in the activity, which generates for the country well over $100 million U.S. per year. A number of these commercial species are endemic to the SP and their fisheries present clear indications of overexploitation. Similarly, there is much evidence that intertidal and shallow subtidal ecosystems are experiencing heavy fishing pressures along the SP (e.g., Durán *et al.*, 1987). In many cases the overexploitation of unique species of invertebrates and macroalgae is the result of a combination of fishing pressure by intertidal collectors, free divers, commercial hooka divers, and artisanal fishermen (Oliva and Castilla, 1990; Castilla, 1990).

How global climate change will affect the abundances of these resource species and the human populations (and countries) that depend on them is hard to envision. The effects do not necessarily need to be negative (e.g., Santelices and Ugarte, 1987; and see Bardach 1989 for a general analysis); however, we are ignorant about the final positive–negative balance.

In this chapter we selected some examples and singled out attractive scientific possibilities for biological studies in the SP. This contribution and that of Lubchenco *et al.* (Chapter 12) provide a basis for suggesting recommendations and outlining specific strategies to deal with global climate change in coastal environments (e.g., Navarrete *et al.*, Chapter 14, this volume).

References

Araya-Vergara, J. F. (1976). Reconocimiento de tipos e individuos geomorfológicos regionales en la costa de Chile. *Inf. Geogr. (Chile)* **23**, 9–30.

Arntz, W. E. (1986). The two faces of El Niño 1982–1983. *Meersforschung* **31**, 1–46.

Arntz, W. E., and Valdivia, J. (1985). *In* "El Niño, su impacto en la fauna marina." *Bol. Inst. Mar. Perú* (special volume), pp. 5–10.

Arntz, W. E., Landa, A., and Tarazona, J. (1985). Visión integral del problema "El Niño": Introducción. *In* "El Niño, su impacto en la fauna marina." *Bol. Inst. Mar. Perú* (special volume), pp. 5–10.

Bakun, A. (1990). Global climate change and intensification of coastal ocean upwelling. *Science* **247**, 198–201.

Balech, E. (1954). División zoogeográfica del litoral Sudamericano. *Rev. Biol. Mar. Valparaíso* **4**, 184–195.

Barber, R. T., and Chávez, F. P. (1983). Biological consequences of El Niño. *Science* **222**, 1203–1210.

Bardach, J. E. (1989). Global warming and the coastal zone. *Climatic Change* **15**, 117–150.

Bernal, P. A. (1991). Consequences of global change for oceans. A review. *Climate Change* **18**, 339–359.

Bernal, P., and Ahumada, R. (1985). Ambiente Oceánico. *In* "Medio Ambiente en Chile" (F. Soler, ed.), pp. 55–105. Ediciones Universidad Católica de Chile, Santiago.

Brainard, R. E., and McLain, D. R. (1987). Seasonal and interannual subsurface temperature variability off Peru, 1952 to 1954. *In* "The Peruvian Anchoveta and Its Upwelling Ecosystem: Three Decades of Change." ICLARM Studies and Reviews, Manila, Vol. 15, pp. 14–45.

Brandhorsts, W. (1971). Condiciones oceanográficas estivales frente a la costa de Chile. *Rev. Biol. Mar. Valparaíso* **14**, 45–84.

Brattström, H., and Johanssen, A. (1983). Ecological and regional zoogeography of the marine benthic fauna of Chile. *Sarsia* **68**, 289–337.

Bryan, K., and Spelman, M. J. (1985). The ocean's response to a CO_2-induced warming. *J. Geophys. Res.* **90,** 11679–11688.

Bryan, K., Komro, F. G., Manabe, S., and Spelman, M. J. (1982). Transient climate response to increasing atmospheric carbon dioxide. *Science* **215,** 56–58.

Burnett, W. (1980). Apatite-glauconite associations off Peru and Chile: paleo-oceanographic implications. *J. Geol. Soc. London* **137,** 757–764.

Bustamante, R. H., and Castilla, J. C. (1987). The shellfishery in Chile: an analysis of 26 years of landings (1960–1985). *Biología Pesquera (Chile)* **16,** 79–97.

Camus, P. (1990). Procesos regionales y patrones fitogeográficos. *Rev. Chil. Hist. Nat.* **63,** 11–17.

Cancino, J. C., and Becerra, R. (1978). Antecedentes sobre la biología y tecnología del cultivo de *Aulacomya ater* (Molina, 1782) (Mollusca: Mytilidae). *Biol. Pesq. Chile* **10,** 27–45.

Cancino, J. C., and Santelices, B. (1981). Importancia ecológica de los discos adhesivos de *Lessonia nigrescens* Bory (phaephyta) en Chile central. *Rev. Chil. Hist. Nat.* **57,** 23–33.

Cane, M. A. (1983). Oceanographic events during El Niño. *Science* **222,** 1189–1194.

Castilla, J. C. (1976). Parques y reservas marítimas: Creación, probables localizaciones y criterios básicos. *Medio Ambiente (Chile)* **2,** 70–80.

Castilla, J. C. (1979). Características bióticas del Pacífico sur oriental con especial referencia al sector Chileno. *Rev. Com. Perm. Pacífico Sur* **10,** 167–182.

Castilla, J. C. (1988). Earthquake-caused coastal uplift and its effects on rocky intertidal kelp communities. *Science* **242,** 440–443.

Castilla, J. C. (1990). El erizo chileno *Loxechinus albus:* Importancia pesquera, historia de vida, cultivo en laboratorio y repoblación natural. *In* "Cultivo de Moluscos en América Latina" (A. Hernández, ed.), pp. 83–98. Red Regional de Entidades y Centros de Acuicultura de América Latina, CIID-Canadá, Bogotá, Colombia.

Castilla, J. C., and Durán, L. R. (1985). Human exclusion from the rocky intertidal zone of central Chile: the effects on *Concholepas concholepas* (Gastropoda). *Oikos* **45,** 391–399.

Castilla, J. C., and Oliva, D. (1990). Ecological consequences of coseismic uplift on the intertidal kelp belts of *Lessonia nigrescens* in central Chile. *Estuarine Coastal Shelf Sci.* **31,** 45–56.

Castilla, J. C., and Paine, R. T. (1987). Predation and community organization on eastern Pacific, temperate zone, rocky intertidal shores. *Rev. Chil. Hist. Nat.* **60,** 131–151.

Castilla, J. C., Luxoro, C., and Navarrete, S. A. (1989). Galleries of the crabs *Acanthocyclus* under intertidal mussel beds: their effects on the use of primary substratum. *Rev. Chil. Hist. Nat.* **62,** 199–204.

Caviedes, C. (1972). On the paleoclimatology of the Chilean littoral. *Iowa Geogr.* **29,** 8–14.

Caviedes, C. (1975). Quaternary glaciations in the Andes of north-central Chile. *J. Glaciol.* **14,** 155–170.

Comte, D., Eisemberg, S. K., Lorca, E., Pardo, M., Ponce, L., Saragoni, R., Singh, S. K., and Suárez, G. (1986). The 1985 central Chile earthquake: a repeat of previous great earthquakes in the region? *Science* **233,** 449–453.

Davis, M. B. (1989). Insights from paleoecology on global change. *Bull. Ecol. Soc. Am.* **70,** 222–228.

Dethier, M. N., and Duggins, D. O. (1988). Variation in strong interactions in the intertidal zone along a geographical gradient: a Washington–Alaska comparison. *Mar. Ecol. Progr. Ser.* **50,** 97–105.

De Vries, T., and Shrader, H. (1981). Variation of upwelling/oceanic conditions during the latest Pleistocene through Halocene off the central Peruvian coast: a diatom record. *Mar. Micropaleontol.* **6,** 157–164.

Dell, R. K. (1971). The marine mollusca of the Royal Society Expedition to southern Chile, 1958–1959. *Rec. Dom. Mus. Wellington* **7,** 155–233.

di Castri, F., and Hajek, E. (1976). "Bioclimatología de Chile." Vicerectoría Académica, Universidad Católica de Chile, Santiago, Chile.

DiSalvo, L. H., Alarcón, E., Martínez, E., and Uribe, E. (1984). Progress in mass culture of *Clamys (Argopecten) purpurata* Lamarck (1819) with notes on its natural history. *Rev. Chil. Hist. Nat.* **57,** 35–45.

Durán, R. L., Castilla, J. C., and Oliva, D. (1987). Intensity of human predation on rocky shores at Las Cruces in central Chile. *Environ. Cons.* **14,** 143–149.

Ellsaesser, H. W. (1989). Response to W. W. Kellog's paper. *In* "Global Climate Change" (S. F. Singer, ed.), pp. 67–80. Paragon House, New York.

Enfield, D. B. (1989). El Niño past and present. *Rev. Geophys.* **27**, 159–187.

Fuenzalida, F. R. (1985). Aspectos oceanográficos y meteorológicos de El Niño 1982–83 en la zona costera de Iquique. *Invest. Pesq. (Chile)* **32**, 47–52.

Glynn, P. W. (1988a). El Niño–Southern Oscillation 1982–83: nearshore population, community, and ecosystem responses. *Annu. Rev. Ecol. Syst.* **19**, 309–345.

Glynn, P. W. (1988b). El Niño warming, coral mortality and reef framework destruction by echinoid bioexosion in the eastern Pacific. *Galaxea* **7**, 129–160.

Glynn, P. W., ed. (1989). "Global Ecological Consequences of the 1982–83 El Niño–Southern Oscillation." Elsevier, New York.

Glynn, P. W. (1991). Coral reef bleaching in the 1980s and possible connections with global warming. *Trends Ecol. Evol.* **6**, 175–179.

Glynn, P. W., and D'Croz, L. (1990). Experimental evidence for high temperature stress as the cause of El Niño-coincident coral mortality. *Coral Reefs* **8**, 181–191.

Hays, C. (1986). Effects of the 1982–1983 El Niño on Humboldt penguin [*Spheniscus humboldti*] colonies in Peru. *Biol. Coserv.* **36**, 169–180.

Herm, D. (1969). Marines pliozän und pleistozä in Nord- und Mittel-Chile unter besonderer Berücksichtigund der Entwicklung der mollusken-faunen. *Zitteliana* **2**, 1–158.

Herm, D., and Paskoff, R. (1966). Note preliminaire sur le Tertiaire superiour du Chili centre-nord. *Bull. Soc. Geol. Fr.* **7**, 760–765.

Illanes, J. E., Akabo hi, S., and Uribe, E. (1985). Efectos de la temperatura en la reproducción del ostión del norte *Clamys (Argopecten) purpuratus* en la Bahía de Tongoy durante el fenómeno El Niño 1982–1983. *Invest. Pesq. (Chile)* **32**, 167–173.

Intergovernmental Panel on Climate Change. (1990). Policymaker's Summary of the Potential Impacts of Climate Changes. Report from Group II, pp. 1–46.

Investigación Pesquera, Chile. (1985). Special Number, **32**, 1–25.

Jablonski, D., Flessa, K. W., and Valentine, J. W. (1985). Biogeography and paleobiology. *Paleobiology* **11**, 75–90.

Jacobson, J. L. (1990). Holding back the sea. *In* "State of the World, 1990" (L. R. Brown *et al.*, eds.), pp. 79–97. Norton, New York.

Johnson, D. R., Fonseca, T., and Sievers, H. (1980). Upwelling in the Humboldt Coastal Current near Valparaíso, Chile. *J. Mar. Res.* **38**, 1–16.

Jokiel, P. L., and Coles, S. L. (1990). Response of Hawaiian and other Indo-Pacific reef corals to elevated temperature. *Coral Reefs* **4**, 153–154.

Kelleher, J. A. (1972). Rupture zones of large South American earthquakes and some predictions. *J. Geophys. Res.* **77**, 2087–2103.

Knox, G. A. (1960). Littoral ecology and biogeography of the southern oceans. *Proc. R. Soc. London Ser. B.* 1952, 577–624.

Lansdsberg, H. E. (1989). Where do we stand with the CO_2 greenhouse effect problem? *In* "Global Climate Change" (S. F. Singer, ed.), pp. 87–89. Paragon House, New York.

Llagostera, A. (1979a). Tres dimensiones en la conquista prehistórica del mar: un aporte para para el estudio de las formaciones pescadoras de la costa sur andina. *In* "Actas del Octavo Congreso Nacional de Arqueología, Valdivia, Chile," pp. 217–245. Sociedad Chilena de Arqueología, Santiago.

Llagostera, A. (1979b). 9700 years of maritime subsistence on the Pacific: an analysis by means of bioindicators in the north of Chile. *Am. Antiquity* **44**, 309–324.

Lomnitz, C. (1970). Major earthquakes and tsunamis in Chile during the period 1935–1955. *Geol. Rundsch.* **59**, 138–960.

Mendo, J., Pizarro, L., and Castillo, S. (1987). Monthly turbulence and Ekman transport indexes 1953 to 1985, based on local wind records from Trujillo and Callao, Perú. *In* "The Peruvian Anchoveta and Its Upwelling Ecosystem: Three Decades of Change." *ICLARM Studies and Reviews,* Manila, Vol. 15, pp. 75–88.

Mendo, J., Valdivieso, V., and Yamashiro, C. (1988). Cambios en la densidad, numero y biomasa de la población de concha de abanico *Argopecten purpuratus* en la Bahía Independencia (Pisco, Peru) durante 1984–1987. *In* "Recursos y dinámica del ecosistema de afloramiento peruano," *Bol. Inst. Mar. Perú* (special volume), pp. 153–162.

Milliman, J. D. (1989). Sea levels: past, present, and future. *Oceanus* **32,** 40–42.

Navarrete, S. A., and Castilla, J. C. (1990a). Resource partitioning between intertidal predatory crabs: interference and refuge utilization. *J. Exp. Mar. Biol. Ecol.* **143,** 101–129.

Navarrete, S. A., and Castilla, J. C. (1990b). Barnacle walls as mediators of intertidal mussel recruitment: effects of patch size on the utilization of space. *Mar. Ecol. Prog. Ser.* **68,** 113–119.

Nishenko, P. S. (1985). Seismic potential for large and great interplate earthquakes along the Chilean and southern Peruvian margins of South America: a quantitative reappraisal. *J. Geophys. Res.* **90,** 3589–3615.

Oliva, D., and Castilla, J. C. (1990). Repoblación natural: el caso del loco *Concholepas concholepas* (Gastropoda: Muricidae) en America Latina. *In* "Cultivo de Moluscos en América Latina" (A. Hernández, ed.), pp. 273–295. Red Regional de Entidades y Centros de Acuicultura de América Latina, CIID-Canadá, Bogotá, Colombia.

Paine, R. T. (1966). Food web complexity and species diversity. *Am. Nat.* **100,** 65–75.

Paine, R. T., Castilla, J. C., and Cancino, J. (1985). Perturbation and recovery patterns of starfish-dominated intertidal assemblages in Chile, New Zealand, and Washington State. *Am. Nat.* **125,** 769–691.

Parsons, M. H. (1970). Preceramic subsistence on the Peruvian coast. *Am. Antiquity* **35,** 292–304.

Paskoff, R. (1989). Zonality and main geomorphic features of the Chilean coast. *Essener. Geogr. Arbeiten.* **18,** 237–267.

Pauly, D., and Tsukayama, I. (1987). "The Peruvian Anchoveta and Its Upwelling Ecosystem: Three Decades of Change." *ICLARM Studies and Reviews,* Manila, Vol. 15.

Pickard, G. L. (1971). Some physical oceanographic features of inlets of Chile. *J. Fish. Res. Bd. Can.* **28,** 1077–1106.

Plafker, G., and Savage, J. C. (1970). Mechanism of the Chilean earthquakes of May 21 and 22, 1960. *Geol. Soc. Am. Bull.* **81,** 1001–1030.

Quinn, W. H., Zopf, D. O., Short, K. S., and Kuo Yang, R. T. (1978). Historical trends and statistics of the Southern Oscillation, El Niño, and Indonesian droughts. *Fish. Bull.* **76,** 663–678.

Richardson, J. B. (1981). Modelling the development of sedentary maritime economics on the coast of Peru: a preliminary statement. *Ann. Carnegie Mus.* **50,** 139–150.

Robles, F., Alarcón, E., and Ulloa, A. (1974). Water masses in the northern Chilean zone and their variation in a cold period (1967) and warm periods (1969, 1971–1973) *In* "Reunión de Trabajo sobre el Fenómeno de El Niño," pp. 1–68. COI, Guayaquil.

Rollins, H. B., Rhichardson, J. B., and Sandweiss, D. H. (1986). The birth of El Niño: geoarcheological evidences and implications. *Geoarcheology* **1,** 3–15.

Romero, H., and Garrido, A. M. (1985). Influencias genéticas del fenómeno El Niño sobre los patrones climáticos de Chile. *Invest. Pesq.* (Chile) **32,** 19–35.

Samamé, M., and Valdivieso, V. (1986). Informe del seguimiento de la extracción de conchas de abanico en Bahia Independencia-Pisco (16 Diciembre 1985–13 Enero 1986) *Informe Interno, Instituto del Mar, Perú,* Callao, Perú, pp. 1–9.

Santelices, B. (1980). Phytogeographic characterization of the temperate coast of Pacific South America. *Phycologia* **19,** 1–12.

Santelices, B. (1989). "Algas Marinas de Chile." Ediciones Universidad Católica de Chile, Santiago.

Santelices, B. (1990). Patterns of organizations of intertidal and shallow subtidal vegetation in wave exposed habitats of central Chile. *Hydrobiologia* **192,** 35–57.

Santelices, B., and Ugarte, R. (1987). Production of *Gracilaria:* problems and perspectives. *Hydrobiologia* **151/152,** 295–299.

Santelices, B., Montalva, S., and Oliger, P. (1981). Competitive algal community organization in exposed intertidal habitats from central Chile. *Mar. Ecol. Progr. Ser.* **6,** 267–276.

Sievers, H., and Silva, N. (1975). Masas de agua y circulación en el Océano Pacifico Sud Oriental, latitudes

18°S–33°S. *Cienc. y Tec. del Mar, CONA* **1,** 7–67.

Silva, N., and Fonseca, T. (1983). Geostrophic component of the ocean flow off northern Chile. *In* "Recursos Marinos del Pacífico" (P. Arana, ed.), pp. 59–70. Escuela de Ciencias del Mar, Universidad Católica de Valparaíso, Chile.

Silva, N., and Sievers, H. (1981). Masas de agua y circulación en la región de la zona costera de la Corriente de Humboldt latitudes 18°S–33°S (operación oceanográphica MarChile X-ERFEN I). *Cienc. y Tec. del Mar, CONA* **5,** 5–50.

Smith, R. L. (1968). Upwelling. Dept. of Oceanography, Oregon State Univ., Corvallis, 7 (345).

Soenens, P. (1985). Estudios preliminares sobre el efecto del fenómeno El Niño 1982–1983 en comunidades de *Aulacomya ater*. *In* "El Niño" su impacto en la fauna marina. *Bol. Inst. Mar. Perú* (special volume), pp. 51–53.

Southward, A. J. (1958). Note on the temperature tolerances of some intertidal animals in relation to environmental temperatures and geographic distribution. *J. Mar. Biol. Assoc. U.K.* **37,** 49–66.

Spelman, A. J., and Manabe, S. (1984). Influence of oceanic heat transport upon the sensitivity of a model climate. *J. Geophys. Res.* **89,** 571–586.

Stanely, S. M. (1984). Marine mass extinctions: a dominant role for temperature. *In:* "Extinctions" (M. H. Nitecki, ed.), pp. 60–117. Univ. of Chicago Press, Chicago.

Stouffer, R. J., Manabe, S., and Bryan, K. (1989). Interhemispheric asymmetry in climate response to a gradual increase of atmospheric CO_2. *Science* **342,** 660–662.

Sverdrup, H. V. (1938). *J. Mar. Res.* **1,** 155–164.

Tarazona, J., Paredes, C., Romero, L., Blaskovich, V., Guzmán, S., and Sánchez, S. (1985). Características de la vida planctónica y colonización de los organismos bentónicos epilíticos durante el fenomeno "El Niño." *Bol. Inst. Mar. Perú,* (special volume), pp. 41–49.

Valdivieso, V., Yamashiro, C., Samamé, M., and Mendez, M. (1988). Informaciones sobre el recurso bentónico conchas de abanico *(Argopecten purpuratus),* Pisco, Peru. *Informe UNESCO Ciencias del Mar* **47,** 149–159.

Viviani, C. A. (1979). Ecogeografía del litoral chileno. *Studies on Neotropical Fauna and Environment* **14,** 65–123.

Wolff, M. (1985). Fishcherei, Oecologie und Populationsdynamik der Pilgermuschel *Argopecten purpuratus* (L) im Fischereigebiet von Pisco (Perú) unter dem einfluss des El Niño 1982–1983. Tesis Karl-Albrecht-Univertität, Kiel, Bundesrepublik Deutschland.

Wyrtki, K. (1963). The horizontal and vertical field of motion in the Peru Current. *Bull. Scripps Inst. Oceanogr.* **8,** 313–346.

CHAPTER 14

Pacific Ocean Coastal Ecosystems and Global Climate Change

SERGIO A. NAVARRETE   JANE LUBCHENCO
JUAN CARLOS CASTILLA

There is clear evidence that the earth's environment has suffered profound global alterations due to human activities in the past few decades (IGBP, 1986, 1990). The threat of rapid and historically unprecedented global climate change, probably as a result of these activities, is now an increasing reality. In the midst of these discouraging presages, ecologists are being challenged to provide a basis for understanding the consequences of these changes for the earth's biological systems (Davis, 1989; Graham and Grimm, 1990; Lubchenco *et al.*, 1991).

To face the enormous task of predicting the effects of such a large-scale phenomenon on the great variety of biological systems, ecologists are using different approaches (Levin, 1992; Menge, 1991; Mooney *et al.*, 1991; Rosswall *et al.*, 1988). Certainly, none of these approaches will provide the kind of definite and precise answers that scientists and policy makers would like. Both the inaccuracy of the predictions of the magnitude and geographic distribution of climate changes and the inherently complex dynamics of ecological systems make such a task immense. Chapters 12 and 13 in this volume illustrate the usefulness and limitations of some of these approaches for marine coastal ecosystems and provide a basis for comparison with other kinds of ecosystems and for the formulation of general recommendations.

A number of important changes in the physical environment of the ocean have been predicted as direct or indirect consequences of the sea–atmosphere interactions in a transformed climate. However, most of these predictions are highly uncertain and under intense debate (e.g., Bryan and Spelman, 1985; Stouffer *et al.*, 1989; Broecker,

1992). The most certain physical changes predicted to occur in the oceans in the next century are an increase in the sea surface temperature and a rise in sea level. To begin to understand the effects that these two changes might have on biological systems, it is helpful to consider information about natural large-scale anomalies (e.g., El Niño–Southern Oscillation events) (Chapters 12 and 13).

Sea level rise observed during the 1982–1983 El Niño–Southern Oscillation (ENSO) event in the eastern Pacific did not appear to have extensive direct effects on marine ecosystems of temperate areas. On the other hand, the changes in community structure observed during coseismic coastal uplifts and subsidences along the coast of Chile are better understood as responses to the rate of change (i.e., instantaneous) rather than to the change itself (Castilla and Oliva, 1990; Castilla *et al.,* Chapter 13, this volume). Thus, considering the short generation time, wide bathymetric distributions, and great dispersal abilities of marine invertebrates, the rise in sea level predicted by the global climate change scenarios is not likely to have profound direct effects on temperate coastal communities. The situation might be different in estuaries, as suggested by Castilla *et al.* (Chapter 13, this volume), where changes in sea level might cause important expansions or reductions of the estuarine habitat itself. Sea level rise in tropical systems might have important negative effects on coral reef species, the structurally most important components of these communities. Because coral species are in general very sensitive to changes in sea level, it is possible that some of them will not be able to grow fast enough to keep up with the predicted rate of sea level rise (Glynn, 1991).

The 2 to 5°C increase in sea surface temperature during ENSO events has had extensive and dramatic effects on coastal communities of low and middle latitudes. Although effects of ENSO have been reported in both hemispheres, both the physical changes and their biological consequences were greater at lower latitudes (see reviews by Arntz *et al.,* 1985; Glynn, 1988, 1989; Chapters 12 and 13, this volume). However, the lack of long-term studies, especially in South America, makes it difficult to differentiate unequivocally "natural" variation from ENSO-caused changes. Some of the best information in this sense concerns commercially important species, which may already be under the stress of overexploitation (Bustamante and Castilla, 1987; Chapter 13, this volume). On the other hand, the generally scarce knowledge of the processes and mechanisms regulating communities and populations precludes determining the real proximate causes of biological change. Many of the most pervasive and long-lasting changes associated with ENSO events seem to have been indirectly rather than directly caused by temperature anomalies (Glynn, 1988; Chapter 12, this volume). For instance, with a better understanding of the processes and mechanisms regulating the population of black abalones in California, North America, the importance of indirect effects has been revealed (Tissot, 1991; Chapter 12, this volume).

Increases in seawater temperatures during ENSO events have had dramatic effects on tropical ecosystems. Massive bleaching and subsequent mortality of coral reef species have been confidently attributed to temperature increases of about 2–4°C during ENSO years (Glynn, 1988; Glynn and D'Croz, 1990). The effects have been particularly devastating for coral reef communities of the Galápagos Islands (Glynn, 1991). ENSO-related temperature anomalies probably caused the extinction of two or three coral species off the coast of Panama (Glynn and deWeerdt, 1991).

ENSO events provide a unique opportunity to study the effects of increases in sea surface temperature over large spatial scales, particularly for the coast of South America. If we are going to take advantage of this, we must gather information about communities and populations in "normal" and "abnormal" years, as well as determine the processes and mechanisms regulating these systems.

Studies of kelp forests in Southern California and coral reefs in Panama and Galápagos Islands provide clear examples of the rich information that can be obtained with long-term studies and knowledge of the processes involved in community regulation (see discussions in Lubchenco et al., Chapter 12, this volume). With knowledge of the normal variability in these communities and the mechanisms producing community patterns, responses of the system to the ENSO perturbation could be understood. For example, it was clear that coastal communities did not respond as tight units shifting from a temperate mode to a more tropical one. However, the observed changes were not completely independent responses of individual species, because changes in some species were closely coupled to those in other species.

Some general recommendations about how ecological research could provide a better basis for predictions in the presence of global climate change can be made. Determination of acclimatized tolerance limits to high temperatures might be considered a first step in research, which can help to identify species with low tolerances and thus those expected to be particularly susceptible to temperature shifts. This information is especially important for species that play key roles in the structure of communities (e.g., predators; Paine, 1966; Lubchenco, 1978). Low-tolerance species should be the focus of field experimentation to determine their association with other species and to identify subsets of tightly associated species ("modules" in the sense of Paine, 1980). Information about the change of ecologically important interactions along environmental (e.g., Lubchenco, 1986) and latitudinal (e.g., Dethier and Duggins, 1988) gradients will shed light on the dependence of these interactions on the physical environment.

We believe that long-term monitoring programs on coastal environments of the eastern Pacific could produce valuable information if directed to the appropriate ecosystems and the critical species that characterize them. The structure and functioning (dynamics) of such ecosystems must be the goal of the program. Simple, yet time-persistent measurements of the phenology of the variation, both in climate and physical oceanographic parameters and in key aspects of marine ecosystems, have proved to be useful (McGowan, 1990). Incorporation of spatial structure at different scales into the monitoring will substantially enhance our capability to detect and interpret long-term trends.

A series of observatories along the eastern Pacific coastline is urgently needed to provide a common data base and comparable approaches of descriptive and experimental studies. These observations should include baseline physical, chemical, and biological data as well as support experimental studies aimed at elucidating the processes determining the patterns. This strategy will allow us to (1) build a long-term data base to better understand the critical factors regulating coastal communities, thus addressing important theoretical and yet relevant problems; (2) compare the natural dynamics of these systems against changes produced by large-scale perturbations (e.g., ENSO); and (3) provide an unambiguous signal of any future climate change having a significant impact on coastal marine ecosystems. Existing marine laboratories could serve as a starting point for these observations. Some laboratories, such as the Estación Costera de Investigaciones Marinas of the Pontificia Universidad Católica de Chile at Las Cruces, central Chile, already monitor populations and community structure on a routine basis (Castilla, 1990). Cooperative research between marine scientists covering temperate areas of the coasts of Peru–Chile and California–Oregon–Washington could be extremely valuable and enhance our understanding of both basic ecological problems and how ecological systems respond to long-term global climate change.

References

Arntz, W. E., Landa, A., and Tarazona, J. (1985). Visión integral del problema "El Niño": Introducción. *In* "El Niño, su impacto en la fauna marina." *Bol. Inst. Mar. Perú* (special volume), pp. 5–10.

Broecker, W. S. (1992). Global warming on trial. *Natural Hist.* (April), 6–14.

Bryan, K., and Spelman, M. J. (1985). The ocean's response to a CO_2-induced warming. *J. Geophys. Res.* **90**, 11679–11688.

Bustamante, R. H., and Castilla, J. C. (1987). The shellfishery in Chile: an analysis of 26 years of landings (1960–1985). *Biol. Pesq. (Chile)* **16**, 79–97.

Castilla, J. C. (1990). El erizo chileno *Loxechinus albus:* Importancia pesquera, historia de vida, cultivo en laboratorio y repoblación natural. *In* "Cultivo de Moluscos en América Latina" (A. Hernández, ed.), pp. 83–98. Red Regional de Entidades y Centros de Acuicultura de América Latina, CIID-Canadá, Bogotá, Colombia.

Castilla, J. C., and Oliva, D. (1990). Ecological consequences of coseismic uplift on the intertidal kelp belts of *Lessonia nigrescens* in central Chile. *Estuarine Coastal Shelf Sci.* **31**, 45–56.

Davis, M. B. (1989). Insights from paleoecology on global change. *Bull. Ecol. Soc. Am.* **70**, 222–228.

Dethier, M. N., and Duggins, D. O. (1988). Variation in strong interactions in the intertidal zone along a geographical gradient: a Washington–Alaska comparison. *Mar. Ecol. Prog. Ser.* **50**, 97–105.

Glynn, P. W. (1988). El Niño–Southern Oscillation 1982–83: nearshore population, community, and ecosystem responses. *Annu. Rev. Ecol. Syst.* **19**, 309–345.

Glynn, P. W., ed. (1989). Global Ecological Consequences of the 1982–83 El Niño–Southern Oscillation." Elsevier, New York.

Glynn, P. W. (1991). Coral reef bleaching in the 1980s and possible connections with global warming. *Trends Ecol. Evol.* **6**, 175–179.

Glynn, P. W., and deWeerdt, W. H. (1991). Elimination of two reef-building hydrocorals following the 1982–1983 El Niño warming event. *Science* **253**, 69–71.

Glynn, P. W., and D'Croz, L. (1990). Experimental evidence for high temperature stress as the cause of El Niño-coincident coral mortality. *Coral Reefs* **8**, 181–191.

Graham, R. W., and Grimm, E. C. (1990). Effects of global climate change on the patterns of terrestrial biological communities. *Trends Ecol. Evol.* **5**, 289–292.

IGBP (1986). The International Geosphere–Biosphere Program: a study of global change. IGBP Report Number 1, Stockholm, Sweden.

IGBP (1990). The International Geosphere–Biosphere Program: a study of global change (IGBP), the initial core projects. IGBP Report Number 12, Stockholm, Sweden.

Levin, S. A. (1992). The problem of pattern and scale in ecology. The Robert H. MacArthur Award Lecture. *Ecology* **73**(6), 1943–1967.

Lubchenco, J. (1978). Plant species diversity in a marine intertidal community: importance of herbivore food preference and algal competitive abilities. *Am. Nat.* **112**, 23–39.

Lubchenco, J. (1986). Relative importance of competition and predation: early colonization by seaweeds in New England. *In* "Community Ecology" (J. Diamond and T. J. Case, eds.), pp. 537–555. Harper & Row, New York.

Lubchenco, J., Olson, A. M., Brubaker, L. B., Carpenter, S. R. Holland, M. M., Hubbell, S. P., Levin, S. A., MacMahon, J. A., Matson, P. A., Melillo, J. M., Mooney, H. A., Peterson, C. H., Pulliam, H. R., Real, L. A., Regal, P. J., and Risser, P. G. (1991). The Sustainable Biosphere Initiative: an ecological research agenda. *Ecology* **72**(2), 371–412.

McGowan, J. A. (1990). Climate and change in oceanic ecosystems: the value of time-series data. *Trends Ecol. Evol.* **5**, 293–299.

Menge, B. A. (1991). Generalizing from experiments: is predation strong or weak in the New England rocky intertidal? *Oecologia* **88**, 1–8.

Mooney, H. A., Medina, E., Schindler, D. W., Schulze, E.-D., and Walker, B. H., eds. (1991). "Ecosystem Experiments," SCOPE 45. Wiley, Chichester, U.K.

Paine, R. T. (1966). Food web complexity and species diversity. *Am. Nat.* **100**, 65–75.

Paine, R. T. (1980). Food webs: linkage, interaction strength and community infrastructure. The Third Tansley Lecture. *J. Ani. Ecol.* **49**, 667–685.

Rosswall, T., Woodmansee, R. G., and Risser, P. G., eds. (1988). "Scales and Global Change," SCOPE 35. Wiley, Chichester, U.K.

Stouffer, R. J., Manabe, S., and Bryan, K. (1989). Interhemisphere asymmetry in climate response to a gradual increase of atmospheric CO_2. *Nature* **342**, 660–662.

Tissot, B. N. (1991). Geographic variation and mass mortality in the black abalone: the roles of development and ecology. Ph.D. dissertation, Dept. of Zoology, Oregon State Univ., Corvallis.

Part VI: Plants

C H A P T E R 15

Full and Late Glacial Paleoenvironmental Scenarios for the West Coast of Southern South America

CAROLINA VILLAGRÁN   JUAN J. ARMESTO

I. Introduction

Paleoclimatic comparisons between the west coasts of North America and South America are difficult at the present time, partly because of the incomplete data base and also because of the natural asymmetries and asynchronies of paleoclimatic events across hemispheres (Salinger, 1981). Asymmetries result mainly from the contrasting distribution of ocean and land masses in both hemispheres (Pittock, 1978; Wells, 1986). In addition, global climatic anomalies, such as the El Niño–Southern Oscillation, which currently affect the Pacific coast of America (Aceituno *et al.*, Chapter 4, this volume), could have had different effects in the past (Martin *et al.*, 1992; Mörner, 1992).

In this chapter, we discuss the vegetational history of southern South America in full Glacial and late Glacial times (18,000–10,000 years B.P.). The discussion is based mainly on the palynologic evidence available from various sites ranging from subtropical to subantarctic latitudes (32–55°S) in South America (Fig. 1). We will attempt to relate the vegetational changes observed during this period to the corresponding paleoclimates in the hope that these hypothetical scenarios can provide the groundwork for future comparative studies.

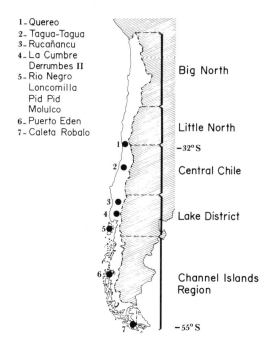

1 - Quereo
2 - Tagua-Tagua
3 - Rucañancu
4 - La Cumbre
   Derrumbes II
5 - Rio Negro
   Loncomilla
   Pid Pid
   Molulco
6 - Puerto Eden
7 - Caleta Robalo

Big North

Little North

−32°S

Central Chile

Lake District

Channel Islands
Region

−55°S

Fig. 1. Sites of fossil pollen samples discussed in this work.

II. Full Glacial Vegetation and Climate

A. Paleobiological Evidence

According to palynological studies in Isla de Chiloé (42–43°S), hygrophilous plant formations expanded down the mountain slopes and northward along the Coastal Range during the late Pleistocene in southern Chile. North Patagonian–subantarctic forests developed on Chiloé Island between 45,000 and 33,000 years B.P., probably an interstadial (Villagrán, 1985, and unpublished manuscript), at the sites Pid-Pid and Molulco (Fig. 1). This vegetation was composed mainly of *Nothofagus* woodland, associated with trees such as *Podocarpus, Lomatia Drimys, Fitzroya/Pilgerodendron,* Myrtaceae. Herbs were mainly Gramineae. The full Glacial sections (Villagrán, 1988b, 1991b; Heusser, 1990a) reveal the presence of a Magellanic moorland mosaic with wetlands, cushion bogs of *Astelia* and *Dacrydium,* and *Nothofagus* woodland. This type of landscape suggests that cooler and wetter but ice-free conditions prevailed in the lowland sites of Chiloé during the last glacial maximum. Between 12,000 and 10,000 years B.P., Magellanic moorlands persisted in Chiloé Island only on the mountaintops of the Coastal Range (Villagrán, 1991c). Today, the center of distribution of Magellanic moorland vegetation is located about six degrees farther south, along the west coast of Tierra del Fuego, where the maximal annual rainfall and the highest records of wind velocity have been recorded.

Farther north along the Chilean coast, between 37 and 42°S, late Glacial pollen sequences (Heusser, 1966, 1974, 1981) indicate the dominance of *Nothofagus dombeyi* type and substantial amounts of nonarboreal pollen (Gramineae, Compositae, and Cyperaceae). Evidence from sites east of the Andes show predominantly nonarboreal pollen during the same period (Markgraf, 1983, 1989). Both Markgraf and Heusser have interpreted these vegetational patterns as evidence of a seasonally drier and overall cooler climate for the late Pleistocene at midlatitudes.

Based on the pollen sequence from the sediments of the now dry basin of Lake Tagua-Tagua (34°30'S, 200 m altitude), Heusser (1983, 1990b) postulated that precipitation was more abundant in central Chile in the late Pleistocene than it is today, as indicated by the northward displacement of the distributional ranges of *Nothofagus dombeyi* and *N. obliqua* types, *Podocarpus saligna,* and *Prumnopitys andina.* These elements were then found 5° north of their present range. In association with this vegetation, a rich extinct megafauna, including mastodons, megatheria, giant ground sloth, native horses, Camelidae, and deer, has been documented (Moreno and Marshall, 1992). The abundance of Gramineae pollen in most glacial profiles (see later) indicates that prairies and open woodlands predominated in Chile during this period, providing habitat and food resources for the extinct megafauna.

Phytogeographic evidence also suggests wetter climatic conditions along the coast of central Chile during the Pleistocene. Patches of deciduous *Nothofagus* forests and of evergreen rain forests dominated by *Aextoxicon punctatum* occur today on the summits (about 600 m elevation) of coastal hills and in some ravines in central Chile (Villagrán, 1990), near the coastal towns of Zapallar, El Tabo, and Pichilemu (32–34°S). These patches may represent remnants of a continuous band of rain forest found along the coast of central Chile during the Pleistocene.

In contrast, north of 27°S, floristic evidence indicates that little if any climatic change occurred during the glacial age in the inland Atacama Desert and in the Andean slopes (Villagrán *et al.*, 1983; Kalin *et al.*, 1988). Specific affinity between the floras of the Andean Altiplano (18–26°S) and the Andes of central Chile (30–33°S) is presently very low (up to 6.7%), despite the continuous connection through the Andes. We infer from this low floristic similarity that the extreme aridity that prevails today in north-central Chile reached the Andean highlands during the Quaternary, obstructing the exchange of flora between the northern Altiplano and central Chile. These floristic differences are not the result of recent climatic shifts, because taxa from the mesic, northern and southern ends of the gradient do not show a discontinuity in their distribution along the Andes.

B. Full Glacial Climate

What was the paleoclimatic scenario along the coast of southern South America during the last glacial maximum? Climap's (1981) reconstruction of sea surface temperatures (SSTs) at the time of the last glacial maximum, about 18,000 years B.P., shows an increased proportion of the Southern Hemisphere ocean covered by polar water. Accordingly, subpolar marine assemblages shifted equatorward, to approximately 30°S. Transitional marine assemblages were thus compressed beneath the subtropical gyres, and there was a steepened thermal gradient. As a consequence, surface water circulation in the ice age was probably more energetic than today and surface currents were intensified, causing increased upwelling along the Pacific coast of South America.

Does this glacial paleoscenario fit with the palynologic data from mediterranean and temperate regions in Chile? Both the widespread distribution of Magellanic moorlands in Chiloé Island and the presence of *Nothofagus* parkland in the lake district and central Chile, extending 5° to the north of their present range during glacial times, have been interpreted as evidence of increased rainfall. These wetter conditions were probably the result of a northward displacement of the westerlies wind belt (Caviedes 1972, 1990; Heusser 1984a, 1989a; Villagrán, 1988b, 1990) that causes much of the winter precipitation in south-central Chile.

Data from oceanographic studies support the postulated northward displacement of the westerlies during glacial times. A deep-sea core (Groot and Groot, 1966), taken 175 km off the coast of Valparaíso (34°30'S–74°19'W) in the eastern Pacific ocean, shows five repeated cycles of cool, moist climate, possibly associated with glacial advances, alternated with warm, dry intervals, presumably interglacials. Peaks of *Podocarpus* pollen were considered indicators of glacial ages with cool, moist conditions. These peaks coincided with a higher carbonate content of samples, probably because of increased ocean upwelling. These results strongly support the view that during glacial times storm activity was more intense and displaced equatorward along the Pacific coast of South America. Stronger westerlies during the full glacial have also been suggested by Thiede (1979) to explain the patterns of deposition of windblown quartz particles off the east coast of Australia in the southwestern Pacific.

Geologic evidence documenting shifts in the position of modern and Pleistocene snowlines in the semiarid region of north-central Chile (Hastenrath, 1971) provides additional evidence for the intensification and northward displacement of the westerly wind flow during full glacial times. Pleistocenic snowlines were lower on the western than on the eastern flank of the Andes south of 28°S, marking the northern limit of the westerly wind flow at that time. Farther north the snowline depression was stronger on the eastern side, suggesting an easterly moisture source (Fox and Strecker, 1991).

Although the evidence examined so far supports the proposed northward displacement of the westerlies wind belt, fossil pollen assemblages from subtropical and midlatitudes introduce some questions (Markgraf, 1989). Why are Valdivian rain forest and sclerophyllous taxa absent from the pollen profiles in the central depression? Instead, the landscape was dominated by Gramineae, Compositae, and *Nothofagus* and conifer parkland. In both Chilean and Argentinean sites, Cyperaceae and other marshy and aquatic taxa are also well represented. Seasonal reconstructions of the SST for the last glacial maximum (Climap, 1981) show that the glacial winters in the Southern Hemisphere were characterized by increased equatorward flow of cool waters to just above 30°S. The thermal equator, and possibly the Intertropical

Convergence, should have been displaced to the north. In contrast, during the glacial summers of the Southern Hemisphere, the thermal equator was considerably farther south of its present position, which is between 0 and 20°S, and nearer the eastern Pacific (see Climap, 1988). It seems likely that the notable contrast between glacial summer and winter temperatures in the South Pacific caused a pronounced seasonality in south-central Chile during the glacial period. This marked seasonality is considered unprecedented today, except perhaps in the most continental sites, favoring the dominance of deciduous *Nothofagus* and conifer species rather than Valdivian rain forest species, which thrive in more equitable climates.

A simpler explanation of the drier vegetation in the central depression and east of the Andes, north of 35°S, could be the reinforcement of the rainshadow effect of the Coastal and Andean Ranges, which in central Chile rise up to twice their maximum elevation in the Lake District. Local climatic differences between coastal and inland sites should have been important in supporting different forest types. Valdivian rain forest taxa may have persisted in coastal refuges. Studies of fossil pollen assemblages from coastal locations, currently unavailable, are needed to test this prediction.

III. Environmental Changes during the Late Glacial

A. Palynological Evidence

All the fossil pollen assemblages known from the Lake District (39–43°S) in south-central Chile indicate that tree establishment leading rapidly to a closed forest began synchronously at around 13,000 years B.P. To explain this rapid process of forest expansion, we must postulate that glacial forest refugia were located nearby the glaciated areas, probably in the coastal ranges of the Lake District (Villagrán, 1985, 1988a,b; Heusser, 1966, 1982, 1984b). Initially, colonizing trees belonged mainly to the cold–temperate, north Patagonian forest (*Nothofagus* and conifers). The early presence of these hygrophilous taxa in the pollen profiles reveals that conditions were very rainy. Regional differences in the floristic composition of late Glacial forests have been inferred from maps of isopollen lines (Villagrán, 1991a). These maps show that *Podocarpus nubigena, Fitzroya/Pilgerodendron,* and *N. dombeyi* type first dominate the montane, oceanic sites of the Lake District. In lowland sites of the Lake District, the more thermophilic Myrtaceae taxa were the first colonizers. In the wetlands closer to the lakes, the successional series began with aquatic and swamp taxa, suggesting gradual filling of lakes and ponds after ice melting (Villagrán, 1988b, 1991b). On the summits of the coastal ranges there was a simultaneous development of Magellanic moorland. Pollen spectra show an uninterrupted sequence of development of peatlands, ranging from minerotrophic, with peaks of *Astelia pumila,* to ombrotrophic characterized by successive peaks of *Donatia fascicularis* and *Gaimardia australis* (Villagrán, 1991c).

The beginning of the Holocene in the Lake District was marked by rapid expansion of the more thermophilic Valdivian rain forest elements in all the pollen profiles. *Eucryphia/Caldcluvia* and *Weinmannia trichosperma* were the most relevant taxa. These species are known to be more resistant to seasonal drought than the north Patagonian species that prevailed in the late glacial (Villagrán, 1991b). The Andean profiles show that deciduous *Nothofagus procera/obliqua* dominated the mountain slopes at the beginning of the Holocene (Fig. 2). Since 3000 years B.P. these species began to be replaced by north Patagonian taxa in the Andes (Villagrán, 1990).

In the semiarid Norte Chico, the pollen record (Villagrán and Varela, 1990) and paleontologic evidence (Núñez *et al.*, 1983) from Quebrada Quereo, near Los Vilos (32°S), also support the hypothesis that rainfall was more abundant in the semiarid coastal desert during the late Glacial. Similar islands of relict vegetation from a more humid period are found between 400 and 700 m on the coastal mountains of Chañaral and Taltal (25–26°S) in the southern Atacama Desert (Johnston, 1929; Rundel *et al.*, 1991). At Quebrada Quereo (32°S), wet conditions prior to 10,000 B.P. are shown by the abundant pollen of swamp and wetland plants, such as Cyperaceae and *Myriophyllum*, with only traces of arboreal pollen (Villagrán and Varela, 1990). A trend toward radically drier conditions began in the early Holocene, as implied by the substantial decrease of both arboreal and aquatic taxa and a general decrease in the diversity of shrubland indicators. The fossil pollen assemblage from Lake Tagua-Tagua (Heusser, 1983, 1990b) shows, at the same time, a sudden change from woodland indicators to strong dominance by *Chenopodiaceae/Amaranthaceae*, indicators of warmer and drier conditions.

When we compare fossil pollen assemblages from mediterranean and temperate latitudes of Chile with sites located south of 50°S, we notice that paleoclimatic changes have been asynchronous and unequal over the latitudinal range analyzed (Figs. 2–4). Pollen records from Tierra del Fuego (about 55°S) provide evidence for cold and dry conditions prevailing between 13,000 and 10,000 B.P., when sites located at lower latitudes were more humid. The vegetation of Tierra del Fuego during that period was dominated by *Empetrum* and Gramineae; *Nothofagus* forests developed after 10,000 B.P. (Fig. 3), presumably as a response to increased precipitation (Heusser and Rabassa, 1987; Heusser, 1989b,c; Rabassa and Clapperton, 1990). At Puerto Edén (about 50°S), nonarboreal taxa dominate the pollen profile before 13,000 B.P., whereas pollen of *Nothofagus* increased between 13,000 and 10,000 B.P. (Fig. 3). It was only from 10,000 B.P. on that Magellanic moorland indicators became important (Fig. 4), suggesting wetter conditions (Ashworth *et al.*, 1991).

The changes in the distribution of Magellanic moorlands and the latitudinal advances and retreats of the *Nothofagus* parkland are the best indicators of changing patterns in precipitation in Chile during the Glacial–Holocene transition (Figs. 2–4). At the glacial maximum, moorlands were conspicuously present in lowland sites at 42°S (Fig. 4). From 13,000 B.P. onward, Magellanic moorlands became restricted to the mountaintops of the Coastal Range in Chiloé Island. Finally, in the last 10,000 years moorland taxa colonized the Chilean Channels Region, at 50°S, where this plant formation is a major component of the present vegetation. Accordingly, the modern climate of austral Chile was established in the early Holocene.

Nothofagus dombeyi-type and Gramineae (Fig. 3) were the dominant taxa of

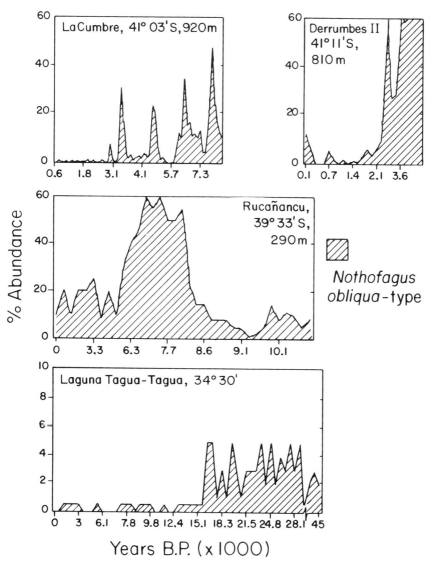

Fig. 2. Importance of *Nothofagus obliqua* type in various fossil pollen assemblages in south-central Chile.

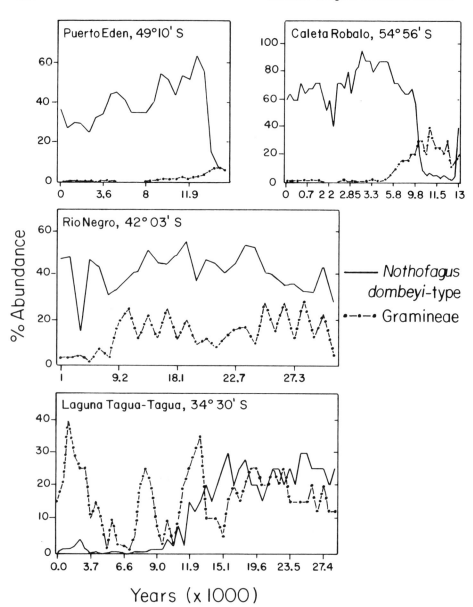

Fig. 3. Importance of *Nothofagus dombeyi* type and Gramineae in various fossil pollen assemblages from subtropical to austral latitudes in Chile.

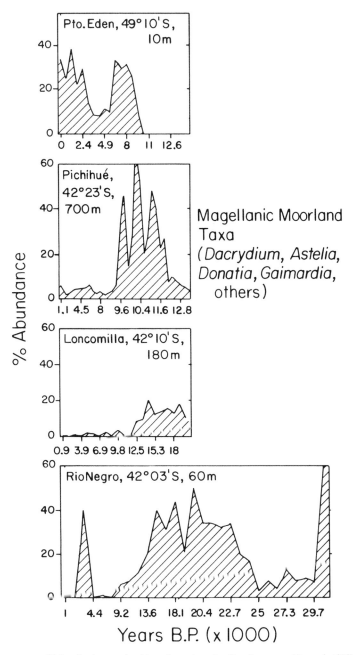

Fig. 4.   Importance of Magellanic moorland taxa in various fossil pollen assemblages in Chiloé Island (Rio Negro, Loncomilla, Pichihué) and in the Chilean Channels (Puerto Edén).

mediterranean as well as temperate regions in Chile during the full Glacial, forming a parkland. Subsequently, a rapid warming trend, associated with abundant rainfall favored the colonization of the north Patagonian forest at midlatitudes. The *Nothofagus* parkland extended southward, up to Puerto Edén at 50°S, at a time when sites located farther south were still dominated by Gramineae (Figs. 2 and 3). *Nothofagus* parkland colonized the southern tip of South America in the early Holocene, presumably in association with higher rainfall. In contrast, *Nothofagus obliqua* type (Fig. 2), which was present in central Chile (34°30'S) during the Pleistocene, expanded south along the Andes to midlatitudes (to 42°S) during the early Holocene, probably reflecting decreased precipitation during this period.

B. Climate during the Glacial–Holocene Transition

In this section, we present a possible scenario for the late glacial in southern South America. The timing of forest expansion in southern Chile during this period agrees with the timing of glacier regressions in the Lake District, as noted by a number of workers (Heusser, 1990a; Porter, 1981; Mercer, 1976, 1984; Clapperton, 1990). These authors proposed dates between 15,000 and 14,500 B.P. for the latest advances of glaciers in the area. A rapid and uninterrupted deglaciation followed in response to a warming trend that began about 13,000 B.P. This trend is supported by the record of SST inferred from a deep-sea core taken at 42°S,80°W in the subantarctic region (Salinger, 1981; Shackleton, 1978). Around 15,000 B.P., SSTs at that latitude were close to the glacial minimum, and by 10,000 B.P. they had increased to values near the Holocene maximum levels.

A similar trend is shown by two 65,000-year ice cores from Antarctica (Jouzel *et al.*, 1987, 1989). These cores show a coincident and pronounced increase in isotopic contents (^{18}O and deuterium) at 15,000 B.P., when temperatures were lowest. Maximum temperatures were reached around 9500 years ago. Salinger (1981) and Harrison *et al.* (1984) argued that changes in SST in the Southern Hemisphere preceded changes in the Northern Hemisphere by about 3000 years. The maximum Holocene warming in Chile occurred around 9500 B.P., about the same time as in New Guinea, Australia, and New Zealand. The warmest period in Europe and North America, in contrast, occurred only about 6000 years ago.

The different timing of deglaciation processes and the thermal maximum in the Northern and Southern Hemispheres can explain the increase in rainfall in Chile during the Glacial–Holocene transition. Because of the asynchronous ice melting in each hemisphere, the position of the thermal equator was displaced to the south during the late Glacial (see Climap, 1981), when temperatures were rising in the south but not in the north. As a consequence of this displacement, the South Pacific Anticyclone was greatly weakened.

This paleoscenario for the last deglaciation period is supported by geomorphic data and studies of paleosols in the semiarid region of Chile. Relict paleosols that originated during a more humid period have been described for the end of the late Glacial in the semiarid coast of Chile (27–33°S) (Veit, 1991). On the other hand, it has been documented that paleolakes in the Peruvian–Bolivian Altiplano covered an area about four times as large as their present extent from 12,500 to 11,000 B.P. (Hastenrath and Kutzbach, 1985). It was estimated that average annual rainfall must have been 30–50% higher than it is today. Increased summer rains, associated with a southward shift of the Atlantic Intertropical Convergence, may account for the wetter conditions in the Altiplano during the late Glacial (Kessler, 1985; Servant *et al.*, 1981; Servant and Villarroel, 1979).

IV. Summary

We review the palynologic data from 11 sites located along the west coast of southern South America (32–55°S) and propose a reconstruction of the Glacial and late Glacial paleoenvironments based on the dominant vegetation in each period. These scenarios are also supported by evidence from geological and paleobiological studies. The evidence for the full Glacial period indicates that wet, cooler conditions prevailed in south-central Chile and along the coast of central Chile. Extremely dry environments occurred in the inland northern Atacama Desert, extending to the high Andes. Several lines of evidence suggest that deglaciation rates were higher in the Southern Hemisphere than in the Northern Hemisphere and that late Glacial climates differed at different latitudes along the Pacific Coast of southern South America. Dry conditions in the late Glacial delayed the southward expansion of Magellanic moorland and forests. Forest expansion under wet and warmer conditions in the Lake District was rapid from refugia located on the coastal range. Aridity extended from the north to the coast of central Chile in the early Holocene.

Acknowledgment

This work has been supported by Fondecyt 860-88 and 91-0844. We are grateful to Dr. D. A. Axelrod for comments on the manuscript.

References

Ashworth, A., Markgraf, V., and Villagrán, C. (1991). Late Quaternary climatic history of the Chilean Channels based on fossil pollen and beetle analyses, with an analysis of the modern vegetation and pollen rain. *J. Quat. Sci.* **6**, 279–291.
Caviedes, C. (1972). Paleoclimatology of the Chilean littoral. *Iowa Geog. Bull.* **29**, 8–14.
Caviedes, C. (1990). Rainfall variation, snowline depression and vegetational shifts in Chile during the Pleistocene. *Climatic Change* **16**, 94–114.
Clapperton, Ch. M. (1990). Quaternary glaciations in the Southern Hemisphere: an overview. *Quat. Sci. Rev.* **9**, 299–304.
Climap Project Members. (1981). Seasonal reconstructions of the earth's surface at the last glacial maximum. Geological Society of America Map and Chart Series, MC-36, 1–18.
Fox, A. N., and Strecker, M. R. (1991). Pleistocene and modern snowlines in the central Andes (24–28°S). *Bamberger Geographische Schriften* **11**, 169–182.
Groot, J. J., and Groot, C. R. (1966). Pollen spectra from deep-sea sediments as indicators of climatic changes in southern South America. *Mar. Geol.* **4**, 467–524.
Harrison, S. P., Metcalfe, S. E., Street-Perrott, F. A., Pittock, A. B., Roberts, C. N., and Salinger, M. J.

(1984). A climatic model of the last glacial/interglacial transition based on paleotemperature and paleohydrological evidence. In "Late Cainozoic Paleoclimates of the Southern Hemisphere" (J. C. Vogel, ed.), pp. 21–34. A. A. Balkema, Rotterdam.

Hastenrath, S. L. (1971). On the Pleistocene snow-line depression in the arid region of the South American Andes. J. Glaciol. 10, 255–267.

Hastenrath, S. L., and Kutzbach, J. (1985). Late Pleistocene climate and water budget of the South American Altiplano. Quat. Res. 24, 249–256.

Heusser, C. J. (1966). Late-Pleistocene pollen diagrams from the Province of Llanquihue, Southern Chile. Am. Philos. Soc. Proc. 110, 269–305.

Heusser, C. J. (1974). Vegetation and climate of the southern Chilean Lake District during and since the last interglaciation. Quat. Res. 4, 290–315.

Heusser, C. J. (1981). Palynology of the last interglacial–glacial cycle in midlatitudes of southern Chile. Quat. Res. 16, 293–321.

Heusser, C. J. (1982). Palynology of cushion bogs of the Cordillera Pelada, Province of valdivia, Chile. Quat. Res. 17, 71–92.

Heusser, C. J. (1983). Quaternary pollen record from Laguna de Tagua Tagua, Chile. Science 219, 1429–1432.

Heusser, C. J. (1984a). Late Quaternary climates of Chile. In "Late Cainozoic Paleoclimates of the Southern Hemisphere" (J. C. Vogel, ed.), pp. 59–83. A. A. Balkema, Rotterdam.

Heusser, C. J. (1984b). Late Glacial–Holocene climate of the Lake District of Chile. Quat. Res. 22, 77–90.

Heusser, C. J. (1989a). Southern westerlies during the last glacial maximum. Quat. Res. 31, 423–425.

Heusser, C. J. (1989b). Late Quaternary vegetation and climate of the southern Tierra del Fuego. Quat. Res. 31, 396–406.

Heusser, C. J. (1989c). Climate and chronology of Antarctica and adjacent South America over the past 30,000 yr. Palaeogeogr. Palaeoclimatol. Palaeoecol. 76, 31–37.

Heusser, C. J. (1990a). Chilotan Piedmont Glacier in the southern Andes during the last glacial maximum. Rev. Geol. Chile 17, 3–18.

Heusser, C. J. (1990b). Ice age vegetation and climate of subtropical Chile. Palaeogeogr. Palaeoclimatol. Palaeoecol. 80, 107–127.

Heusser, C. J., and Rabassa, J. (1987). Cold climate episode of younger Dryas age in Tierra del Fuego. Nature 328, 609–611.

Johnston, I. M. (1929). "Papers on the Flora of Northern Chile". Gray Herbarium of Harvard University, Cambridge, Massachusetts.

Jouzel, J., Lorius, C., Merlivat, L., and Petit, J. R. (1987). Abrupt climatic changes: the Antarctic ice record during the Late Pleistocene. In "Abrupt Climatic Change" (W. H. Berger and L. D. Labeyrie, eds.), pp. 235–245. Reidel, Dordrecht.

Jouzel, J., Raisbeck, G., Benoist, J. P., Yiou, F., Lorius, C., Raynaud, D., Petit, J. R., Barkov, N. I., Korotkevitch, S., and Kotlyakov, V. M. (1989). A comparison of deep Antarctic ice cores and their implications for climate between 65,000 and 15,000 years ago. Quat. Res. 31, 135–150.

Kalin, M. T., Squeo, F. A., Armesto, J. J., and Villagrán, C. (1988). Effects of aridity on plant diversity in the northern Chilean Andes: results of a natural experiment. Ann. Missouri Bot. Gard. 75, 55–78.

Kessler, V. A. (1985). Zur Rekonstruction von spätglazialesn Klima und Wasserhaushalt auf dem peru-anisch- bolivianischen Altiplano. Z. Gletscherkd. Glazialgeol. 21, 107–114.

Markgraf, V. (1983). Late and postglacial vegetational and paleoclimatic changes in subantarctic, temperate and arid environments in Argentina. Palynology 7, 43–70.

Markgraf, V. (1989). Reply to C. J. Heusser's "Southern Westerlies during the Last Glacial Maximum." Quat. Res. 31, 426–432.

Martin, L., Absy, M. L., Fournier, M., Mourguiart, Ph., Sifedine, A., Turcq, B., and Ribeiro, V. (1992). Some climatic alterations recorded in South America during the last 7000 years may be expunded by long-term El Niño like conditions. Paleo ENSO Records, 187–192.

Mercer, J. H. (1976). Glacial history of southernmost South America. Quat. Res. 6, 125–166.

Mercer, J. H. (1984). Late Cainozoic Glacial variations in South America south of the Ecuador. In "Late

Cainozoic Paleoclimates of the Southern Hemisphere" (J. C. Vogel, ed.), pp. 45–58. A. A. Balkema, Rotterdam.

Mörner, N.-A. (1992). Present El Niño–ENSO events and past super-ENSO events effects of changes in the earth's rate of rotation. Paleo ENSO Records, 201–206.

Moreno, P., and Marshall, L. G. (1992). Mamiferos Pleistocenos del norte y centro de Chile en su contexto geográfico: una síntesis. Manuscrito.

Núñez, L., Varela, J., and Casamiquela, R. (1983). Ocupación Paleoindio en Quereo: Reconstrucción Multidisciplinaria en el territorio semiárido de Chile. Imprenta Universitaria, Universidad del Norte, Antofagasta, Chile.

Pittock, A. B. (1978). An overview. *In* "Climatic Change and Variability. A Southern Perspective" (A. B. Pittock, L. A. Frankes, D. Jenssen, J. A. Peterson, and J. W. Zillman, eds.), pp. 1–8. Cambridge University Press, Cambridge, U.K.

Porter, S. C. (1981). Pleistocene glaciation in the southern Lake District of Chile. *Quat. Res.* **16,** 263–292.

Rabassa, J., and Clapperton, Ch. M. (1990). Quaternary glaciations of the southern Andes. *Quaternary Sci. Rev.* **9,** 153–174.

Rundel, P. W., Dillon, M. O., Palma, B., Mooney, H. A., Gulmon, S. L., and Ehleringer, J. R. (1991). The phytogeography and ecology of the coastal Atacama and Peruvian deserts. *Aliso* **13,** 1–49.

Salinger, J. M. (1981). Paleoclimates north and south. *Nature* **291,** 106–107.

Servant, M., and Villarroel, R. (1979). Le Problème paléoclimatique des Andes boliviennes et de leurs piedmonts amazoniens au Quaternaire. *C. R. Acad. Sci. Paris* **288,** 665–668.

Servant, M., Fontes, J.-Ch., Argollo, J., and Saliège, J. F. (1981). Variations du régime et de la nature des précipitations au cours des 15 derniers millénaires dans les Andes de Bolivie. *C. R. Acad. Sci. Paris* **292,** 1209–1212.

Shackleton, N. S. (1978). Some results of the CLIMAP project. *In* "Climatic Change and Variability. A Southern Perspective" (A. B. Pittock, L. A. Frankes, D. Jenssen, J. A. Peterson, and J. W. Zillman, eds.), pp. 69–76. Cambridge University Press, Cambridge, U.K.

Thiede, J. (1979). Wind regimes over the late Quaternary southwest Pacific Ocean. *Geology* **7,** 259–262.

Veit, H. (1991). Jungquartäre Relief- und Bodenentwicklung in der Hochkordillere im Einzugsgebiet des Río Elqui (Nordchile, 30°S). *Bamberger Geographische Schriften* **11,** 81–97.

Villagrán, C. (1980). Vegetationsgeschichtliche und pflanzensoziologische Untersuchungen im Vicente Pérez Rosales National Park (Chile). *Diss. Bot.* **54,** 1–165.

Villagrán, C. (1985). Análisis palinológico de los cambios vegetacionales durante el Tardiglacial y Postglacial en Chiloé. *Rev. Chilena Historia Natural* **58,** 57–69.

Villagrán, C. (1988a). Late Quaternary vegetation of southern Isla Grande de Chiloé. *Quat. Res.* **29,** 294–306.

Villagrán, C. (1988b). Expansion of Magellanic moorland during the Late Pleistocene: palynological evidence from northern Isla de Chiloé, Chile. *Quat. Res.* **30,** 304–314.

Villagrán, C. (1990). Glacial climates and their effects on the history of the vegetation of Chile: a synthesis based on palynologycal evidence from Isla de Chiloé. *Rev. Paleobot. Palynol.* **65,** 17–24.

Villagrán, C. (1991a). Historia de los bosques lluviosos templados del sur de Chile durante el Tardiglacial y Postglacial. *Rev. Chilena Historia Natural* **63,** 447–460.

Villagrán, C. (1991b). Glacial, late Glacial and post-Glacial climate and vegetation of the Isla Grande de Chiloé, southern Chile (41–44°S). *Quaternary of South America and Antarctic Peninsula* (in press).

Villagrán, C. (1991c). Desarrollo de Tundras Magallánicas durante la transición glacial-postglacial en la Cordillera de la Costa de Chile, Chiloé. Evidencias de un evento equivalente al "Younger Dryas"? *Bamberger Geographische Schriften* **11,** 245–256.

Villagrán, C., and Varela, J. (1990). Palynological evidence for increased aridity on the central Chilean coast during the Holocene. *Quat. Res.* **34,** 198–207.

Villagrán, C., Kalin, M. T., and Marticorena, C. (1983). Efectos de la desertización en la distribución de la flora andina de Chile. *Rev. Chilena Historia Natural* **56,** 137–157.

Wells, N. (1986). "The Atmosphere and Ocean. A Physical Introduction." Taylor & Francis, Philadelphia.

C H A P T E R 16

Vegetation in Western North America, Past and Future

STERLING C. KEELEY HAROLD A. MOONEY

I. Introduction

The vegetation of western North America is remarkable in its diversity of species, growth forms, and community types. Along the west coast of North America, extending from Alaska to Baja California, plant communities occur representing, from north to south, tundra, tiaga, coastal coniferous forests, woodlands and chaparral, and finally desert scrub (Fig. 1). This vegetation gradient reflects the generalized continental trend of increasing temperature and decreasing precipitation from north to south. On a finer scale, particularly at the more southerly latitudes, there can be considerable variation in plant community types (Fig. 2) that is related to the imposition, by mountainous terrain, of localized climatic variability. The great mountain ranges of the West trend north to south, creating strong rain shadows on their easterly slopes as well as steep elevational thermal gradients. In addition to the complex precipitation pattern moving from west to east due to the mountain ranges, there is an increase in the yearly range of temperatures moving away from coastal regions. Superimposed on this complex climatic mosaic are unusual rock types such as serpentine, dolomite, limestone, and altered andesite that also exert a control over the distribution of plant life.

It thus is not surprising that there is such a diversity of species and communities in this region. For example, California alone has nearly 6000 species of flowering plants (Munz and Keck, 1959; Munz, 1968), of which nearly a third are endemic (Raven and Axelrod, 1978). A high number of these are threatened, endangered, or

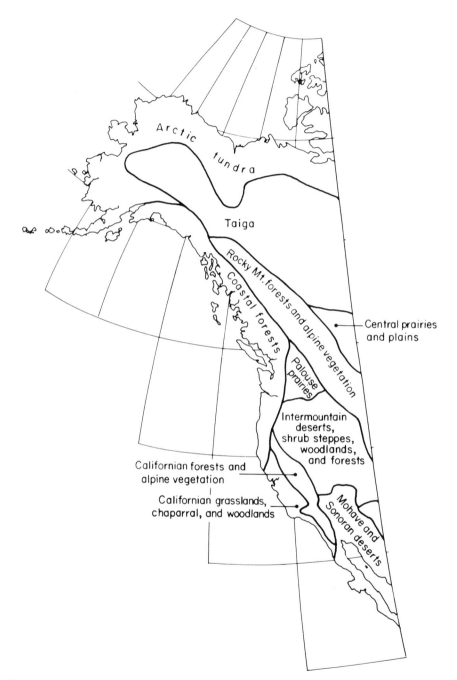

Fig. 1. Generalized vegetation distribution in the western United States reflecting the general climatic trend of decreasing precipitation and increasing temperature north to south and west to east (from Barbour and Billings, 1988).

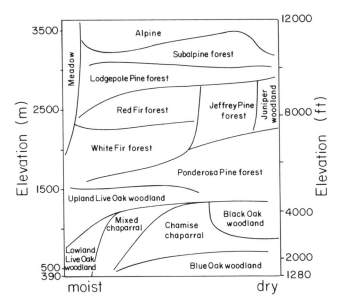

Fig. 2. Vegetation profile in the Sierra Nevada in Sequoia National Park. This profile exemplifies the complex vegetation patterning found throughout the western United States due to the sharp climatic gradients resulting from complex topography (from Vankat, 1982).

of limited distribution. In general terms this diversity is related to climatic variability driven at a fine scale by topographic and substrate variation.

We ask here what the future holds for the vegetation of the western United States. To do this we concentrate on the southwest and first look to clues from the past where there have been changes in climate analagous to those projected for the future.

II. Past Vegetation Change in Western North America

Because of the close correlation of vegetation and climate, records of past vegetation have been used to reconstruct past climates (e.g., Axelrod, 1967, 1988; Barnosky *et al.*, 1987; Brubaker, 1988; Betancourt *et al.*, 1990; M. Davis, 1986, 1989a; O. Davis, 1989; Thompson, 1988). Conversely, because of this climate–vegetation linkage, knowing how climate may change in the future makes it possible to make some predictions about potential vegetation change.

Data available from fossil pollen, macrofossils, and preserved plant parts in packrat middens provide records of vegetation, and hence climate, for a number of areas in western North America over the past 40,000 years and longer (Anderson, 1990; Axelrod, 1967, 1988; Betancourt *et al.*, 1990; Barnosky *et al.*, 1987; Brubaker, 1988; Thompson, 1988; Kutzbach, 1987; Kutzbach and Wright, 1986). These data indicate large climatic fluctuations, especially during glacial and interglacial periods. There is also evidence for substantial changes in temperature within the recent past, that is, within the past 10,000 years, the Holocene. This latter period is important in understanding vegetation response to climate because of the gradual warming that followed the Pleistocene glaciations and because of a period of increased temperatures, variously referred to as the Altithermal, Xerothermal, or Hypsothermal, that occurred for several millennia, between 9 and 3 ka, depending on the location (Anderson, 1990; Axelrod, 1988; Brubaker, 1988; M. Davis, 1989a; O. Davis, 1989). During this period temperatures were 2–4°C higher than today, in line with changes projected for the warming due to a doubling of CO_2 (Houghton *et al.*, 1990).

III. Vegetation Response to Past Climate Change

The paleobotanical record indicates several key features of species responses and therefore changes in vegetation in western North America during the past 20,000 years. Among these key points are (1) temperature changes are reflected in changes in species distribution, and even small temperature changes may have large effects; (2) species are independently distributed and may change range or distribution apart from community associates; (3) the rate at which species may move geographically or elevationally varies and is not necessarily the same from species to species; (4) community composition has not been static in the past and differs from modern plant community types even when common and dominant species are found in both past and present associations; (5) few species have become extinct; and (6) land use patterns can dramatically influence species distributions. In the following we illustrate a number of these points.

IV. Movement of Pinyon Pine in Relation to Climatic Change

Pinyon Pines (*Pinus,* subgenus *Haploxylon* section Cembroides) and pinyon–juniper associations have existed over large parts of western North America from before the Pleistocene (Axelrod, 1981b) and remain widely distributed throughout the region today (Betancourt *et al.*, 1990; Thompson, 1988; West, 1984) (Fig. 3). Because pinyons are found over such a large area and are well preserved in the fossil record, data on this species can serve to illustrate the kinds of geographic movements, change in elevation, community composition, and rate of change that characterize much of western vegetational change in the past 10,000 years and are presented here for that purpose. An abbreviated discussion of the historical movements of other vegetation types follows.

Pinus monophylla (single-needle pinyon) has been the predominant pinyon pine during the Holocene (Thompson, 1988). However, other pinyons, *P. edulis, P. cembroides,* and *P. juarizensis,* have also been important and remain significant at present. They have all changes in abundance since the Pleistocene (Thompson, 1988), but each species is important in plant communities in western North America (Table I).

The change in distribution and community composition typical of the past

Fig. 3.   Generalized distribution of pinyon pines in western North America (redrawn from Little, 1971).

10,000 years can be seen from fossils in packrat middens from the Sonoran and Mojave deserts and the Great Basin. In the Ajo mountains of the Sonoran desert of Arizona, for example, a pinyon–juniper oak woodland was present at 21–13 ka (Van Devender, 1990). In this fossil assemblage *Juniperus scopulorum* (Rocky Mountain juniper) occurred with Great Basin sage (*Artemisia tridentata*), *Atriplex*, and *Yucca brevifolia* (Joshua tree). A similar pinyon–juniper–oak community type that included *Y. brevifolia* and *Artemisia tridentata*-type sage and also included *Juniperus osteosperma* (Utah juniper) and *Quercus turbinella* was found at 12.7 and 12.5 ka in a limestone area of Arizona (Waterman Mountains) (Van Devender, 1990). By 11 ka (late Wisconsin) this community type dominated lower portions of the Arizona upland and intermixed with some type of woodland/chaparral (Van Devender, 1990). Between 11 and 10.3 ka single-needle pinyon had disappeared from the Sonoran desert; junipers persisted, however, until 9–8 ka (Van Devender and Spaulding, 1979; Thompson, 1988).

   In the Mojave pinyon–juniper woodlands were common from the Late Wis-

TABLE I.
Geographic Variation in the Composition of the Pinyon–Juniper Woodland

Colorado Plateau (eastern Utah, western Colorado, northern Arizona, northwestern New Mexico)
  Juniperus osteosperma
  Pinus edulis
  Juniperus monosperma
Great Basin (Nevada, western Utah, California east of the Sierra Nevada)
  Juniperus osteosperma
  Pinus monophylla
Mohave border (southern California)
  Pinus monophylla
  Juniperus californica
  Pinus juarezensis (P. *x quadrifolia*)
Pacific Northwest (eastern Oregon, southwestern Idaho, northeastern California)
  Juniperus occidentalis
Northern Rockies (Wyoming, northern Colorado Front Range, Montana, eastern Idaho)
  Juniperus osteosperma
  Juniperus scopulorum
Southern Rockies (southern Colorado, northern New Mexico)
  Juniperus monosperma
  Pinus edulis
Mogollon rim (central Arizona)
  Juniperus osteosperma
  Pinus monophylla
  Juniperus deppeana
  Cupressus arizonica
Sonoran Desert border (southern Arizona)
  Pinus cembroides var. *bicolor*
  Juniperus deppeana
  Pinus edulis
  Juniperus erythrocarpa
Edwards Plateau (central Texas)
  Juniperus pinchotii
  Juniperus ashei
  Pinus cembroides var. *remota*
Trans-Pecos-Chihuahuan Desert border (western Texas and southeastern New Mexico)
  Pinus cembroides var. *remota*
  Juniperus erythrocarpa
  Juniperus pinchotii

Source: West (1988).

consin (~ 15 ka) to 11 ka (Spaulding, 1990). The average upper elevational limit was approximately 1500 m, which is now the limit for many thermophilous species such as creosote bush (*Larrea divaricata*) of the southern deserts (Betancourt *et al.*, 1990). Between 13.2 and 11.7 ka pinyon–juniper woodland that included *Opuntia* and *Atriplex* in some areas replaced limber pine steppe to the north (Spaulding, 1990). Pinyon–juniper woodlands occurred at low elevations (700–850 m) in southern Nevada between 13.9 and 12.6 ka (Spaulding, 1990) and included oaks. Between 9.5

and 7.9 ka much of this woodland was lost throughout the Mojave (Spaulding, 1990) with the probable exception of high-elevation sites. *Juniperus osteosperma* ranged well into the Great Basin at 11 ka and continued to exist until 7.8 ka in the northern Mojave (Spaulding, 1990).

As pinyons vacated the desert lowlands after 11 ka, they extended north into the Great Basin and moved up in elevation in the southwestern mountains (Betancourt *et al.*, 1990). Single-needle pinyon did not arrive in the central Great Basin until 6.5 ka, but it was well established by 5 ka in the middle portion of its modern elevational range (Thompson, 1990). As pinyons moved into the Great Basin between 7 and 6 ka they formed communities with Pleistocene dominants such as limber pine (*Pinus flexilis*) and Rocky Mountain juniper. This was also the case in central Nevada from 6.5 to 6.1 ka (Thompson, 1990). At present, Utah juniper (*J. osteosperma*) grows with limber pine only as low as 2000 m in this area and single-needle pinyon is absent (Thompson, 1990). Records from 5.3 to 2.4 ka in central Nevada show a pinyon–juniper assemblage that included *J. scopulorum* (Rocky Mountain juniper) along with *P. monophylla* and *J. osteosperma* (Utah juniper). Rocky Mountain juniper no longer grows at this site, although the other two species are common. Data from 4 ka to present indicate little change in floristic composition of Great Basin woodlands except for the decline of Rocky Mountain juniper (Thompson, 1990).

Although pinyons left desert lowlands after 11 ka, the rate at which they achieved modern elevation ranges is unknown (few middens have been collected above 1500 m) (Betancourt *et al.*, 1990). In the Guadalupe Mountains of southern New Mexico pinyon (*P. edulis*) occurred at 2000 m by 10.9 ka, suggesting rapid colonization of the modern range. In the Great Basin pinyon pines have shifted at least 400 m in elevation over the past 10,000 years (Thompson, 1988). Data from the work of Cole (1985, 1986, 1990) in the Grand Canyon (Fig. 4) show the change in elevation of the pinyon–juniper woodland in this area and its relationship to other plant community types, including more common desert associations as well as alpine floras (Cole, 1985).

Pinyon pines have moved northward in what appears to be a relatively steady progression since the beginning of the Holocene (Fig. 5), although distribution is discontinuous because pinyons grow largely in rocky and upland areas. There is some evidence that *P. edulis* and *P. monophylla* are moving at present (Betancourt *et al.*, 1990), although it is not clear that the facilitating agent is changing temperature so much as effects due to fire and grazing. West (1984, 1988) surmised that half the area now occupied by pinyon–juniper woodland is of recent derivation. The northernmost site of *P. edulis* apparently became established about 400 years ago (West, 1984). Today its common associate, *J. osteosperma*, reaches northeast to the Big Horn Basin in Wyoming and could conceivably continue moving along the front range of the Rockies. West (1988) feels that past grazing has removed grass fuel that burned often enough to maintain a savannah-type community. Now, with loss of grasses from overgrazing, the establishment of nurse plants, the dominance of the soil by tree roots, allelopathic effects, and soil erosion, pinyons have increased their range both up- and downslope into grasslands and shrub steppes that are also being degraded. Fires, if

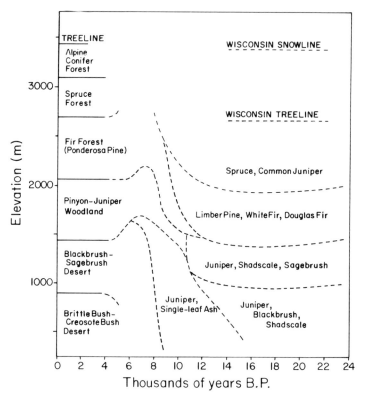

Fig. 4.   Zonation of vegetation in the Grand Canyon during the past 24,000 years (from Cole, 1985).

they should occur now, would result in destruction of large patches of pinyon, and reestablishment is uncertain. In areas of southern and Baja California R. Minnich (personal communication) has observed no seedling establishment in sparse pinyon stands on the edge of the Mojave 10 years after fire but saw much recruitment in a 55-year-old stand. In scattered areas of pinyon pine this may involve long-distance transport of seed and may be highly dependent on chance events.

Although movement appears to have been steadily northward during the warming of the Holocene in many areas of western North America, it also appears that the range of *P. monophylla* may have retracted in some areas. For example, the Mojave Desert vegetation may have extended much farther west during the Xerothermic (8–3 ka) than it does now (Axelrod, 1981b). Pinyons appear to have been distributed much more widely at that time at moderate elevations (<2000 m) (Axelrod, 1981b; O. Davis, 1989). Pinyon pine woodlands are confined to the eastern side of the Sierra Nevada in California today, but relict populations do exist on the west side of the Sierras near Yosemite and in unusual high-elevation locations near Carson Pass (Lake Tahoe area) and Bishop Creek somewhat more south (Axelrod, 1981). Axelrod (1981b) surmised that with earlier warmer temperatures populations were continuous

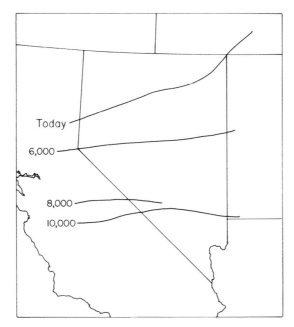

Fig. 5. The hypothesized northern limits of single-needle pinyon pine during the past 10,000 years. (Reprinted from Thompson, 1990, by permission of the University of Arizona Press, copyright 1990.)

from the lower southern Sierra Nevada, but these connections were cut with late Holocene cooling, leaving behind pockets of pinyon in protected sites.

Today pinyon pines grow on semiarid sites in the interior intermountain deserts and shrub steppes of the Great Basin, Mojave, and mountains of the Sonoran deserts (Fig. 3). *Pinus monophylla*, the most common species, is distributed in Nevada, Utah, and the middle to northern areas of Arizona and reaches to California, mostly on the eastern side of the Sierra Nevada extending south in Baja California. It overlaps slightly with the more easterly *P. edulis* but remains distinct (Mirov, 1967). The modern elevational range of *P. monophylla* is 1200–2300 m in California and 600–2900 m in the eastern end of its range (Thompson, 1988). Rainfall in much of the area is 25–50 cm, but the seasonality and availability vary greatly. As a consequence, the actual constituents of pinyon–juniper woodland vary greatly from area to area (Table I).

Growth and reproduction of *P. monophylla* (and its associates) are controlled chiefly by summer warmth, with a requirement for a 19.5–20°C mean July temperature (Axelrod, 1981). Axelrod (1981b) calculated that a rise in mean July temperature to 1–1.5°C would have been enough to enable pinyon and associated woodland species to extend northward in the Sierras during the Xerothermic. Today in alpine areas pinyon is confined to hot, dry, south-facing slopes with thin soils where maximum warmth is received during the growing season (Axelrod, 1981b). Relict populations such as those described for the high eastern Sierras can evidently persist in specialized microsites that are unlike surrounding areas. On the other hand, for Sonoran areas Thompson (1990) indicates that the elevational limits of pinyon pine, along with ponderosa and white pine, are correlated with limits of summer precipitation. *Pinus monophylla* and *Quercus gembelii* grow 300 m lower in mesic valleys there than elsewhere (Thompson, 1988).

The constitution of pinyon communities in the southwest and elsewhere has changed over time, geographically and elevationally, as have its associates. Figure 6 illustrates one chronological sequence from a packrat midden in the Whipple Mountains of California (Van Devender, 1990). The early prominence of pinyon pine, California juniper, and typical Mojave Desert species, *Yucca whipplei* and *Y. brevefolia* (Joshua tree) can be seen, along with an association of species that would have made a very different community type than is currently found. The sharp decline in pinyon and other Mojave species with increasing temperature and the present-day composition of the community indicate persistence of some species, but not within the same geographic area. Similarly, data from the work of Cole (1985) in the Grand

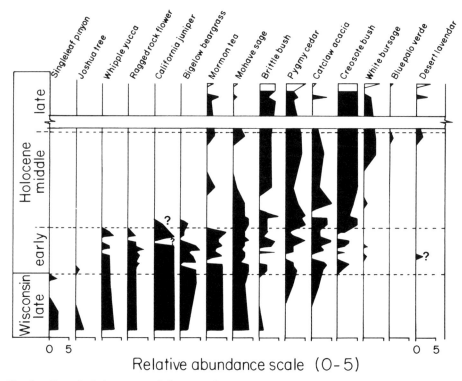

Fig. 6. Chronological sequence of plant macrofossils from packrat middens from the Whipple Mountains, California. (Reprinted from Van Devender, 1990, by permission of the University of Arizona Press, copyright 1990.)

Canyon (Fig. 4) show the kind of elevational variation that has occurred with increased warming. These changes in elevational range also have produced community associations unlike those found today. Despite the large changes in range, elevations, and sites, the pinyon pines and their common associates have persisted through long periods of time.

V. Other Community and Species Movements in the Past

As with pinyon pine, many other species in other parts of western North America show evidence of large-and small-scale movements during the past 10,000 years (see Anderson, 1990; Barnosky *et al.*, 1987; Betancourt *et al.*, 1990; Brubaker, 1988; Cole, 1985). The movement of species has also clearly been independent even if they remain within a recognizable community association. In addition, taxa found in these communities have persisted in spite of these movements and in spite of major vegetational changes (Table II).

One of the clearest examples of community dominants that have moved independently is provided from data on the modern dominants of the Mojave desert, *Ambrosia dumosa* (white bursage) and *Larrea divaricata* (creosote bush). These species did not move at the same rate, nor did they move together during the Holocene (Spaulding, 1990). *Ambrosia* migrated as far north as Death Valley, California, by 10 ka, from its earlier range in the southern Southwest, for example, (Van Devender, 1990), but *Larrea* did not appear until 8.2 ka. Similarly, in another northern site (Eureka View, California) *Ambrosia* was recorded at 6.8 ka whereas *Larrea* did not appear until 5.4 ka. Clearly, these two species did not share the same migrational history and therefore were not members of the same community types throughout the past. Data from a fossil site in the Whipple Mountains (Fig. 6) also show that these species were not found in association with the same species through time, in the same area. Today these species are also widespread in the deserts but, depending on topography and local variations in microclimate, may grow with each other in concert with typical Mojavean species such as *Y. brevifolia*, or with Great Basin species such as *A. tridentata*, or with low desert cactus species. For the most part, species appear to have migrated from areas that became unsuitable and established new habitats elsewhere in this region. Other species also shifted in geographic or elevational range but not necessarily at the same time or in the same way as the modern community dominants.

Species of coastal and inland California, Oregon, and Washington also show great persistence over time, changes in elevation and latitude, and varying community composition through time (Anderson, 1990; O. Davis, 1989a,b; Waring and Franklin, 1979; Barnosky *et al.*, 1987; Axelrod, 1967, 1981b, 1988).

Coastal California appears to have had a climate with fewer extremes and more mesic conditions than most areas since at least the end of the Pleistocene (Axelrod, 1981; Johnson, 1977). A cool mesic climate prevailed along the coast from about 14 ka, extending the range of *Pseudotsuga menzesii* and other mesic forest species 240–320 km south of their present range (Axelrod, 1981b). Closed cone pines existed

TABLE II.
Taxa Present in Various Regions during the Quaternary

Area	Present	Altithermal	Early Holocene
Northwest coastal	*Pseudotsuga*	*Pseudotsuga*	*Pseudotsuga*
	Tsuga	*Tsuga*	*Tsuga*
	Thuja plicata	*Thuja*	*Thuja*
	P. ponderosa	*Pinus*	*Pinus*
	P. contorta	*Lynothamnus*	*Lynothamnus*
	Lynothamnus	*Sequoia*	*Sequoia*
	Sequoia		
Southern California	*Artemisia*	*Artemisia*	Taxaceae
	Rosaceae	Rosaceae	Pinaceae
	Rhamnaceae	Rhamnaceae	Taxodiaceae
	Anacardiaceae	Anacar diaceae	*Quercus*
	Quercus	*Quercus*	
		Pinaceae	
Sonoran Desert	*Larrea*	*Larrea*	*P. monophylla*
	Ambrosia	*Ambrosia*	*Juniperus californica*
	Encelia	*Encelia*	*Yucca brevifolia*
	Succulents	Succulents	*Larrea*
	Riparian trees	Riparian trees	*Encelia*
Mojave Desert	*Larrea*	*Larrea*	*P. flexilis*
	Ambrosia	*Ambrosia*	*P. longaeva*
	Yucca brevifolia	*Y. brevifolia*	*P. longaeva*
	P. monophylla	*P. monophylla*	*Artemisia*
	J. osteosperma	*J. scopulorum*	*J. osteosperma*
	J. scopulorum	*J. osteosperma*	*Yucca brevifolia*
Great Basin	*Artemisia*	*Artemisia*	*Artemisia*
	Ephedra	*Ephedra*	*P. flexilis*
	P. monophylla	*P. longaeva*	*P. longaeva*
	J. osteosperma	*P. flexilis*	*Pseudotsuga*
		P. monophylla	*J. scopulorum*
		J. scopulorum	*J. osteosperma*
		J. osteosperma	

in a nearly continuous strip along coastal California at the end of the Wisconsin (11 ka) (Axelrod, 1967). Redwoods have persisted in some sites since well into the Miocene (Axelrod, 1967). With increased warming the range of Monterey and other closed-cone pines and Douglas fir moved northward. A 10-ka record of the San Bruno flora shows Douglas fir at a site 75 km south of its present distribution (Johnson, 1977; Axelrod, 1981b), with other records indicating this species at sites that now support drought deciduous coastal sage (Johnson, 1977). Along the coast moister and cooler conditions prevailed until 7 ka, as reflected by the fossil trees south of present limits (Johnson, 1977). The middle to late Holocene (8.5–4 ka) shows the existence of coastal sage and chaparral communities similar in composition to those found today (Baker, 1983). Data from the Santa Barbara channel (Baker, 1983) indicate near-modern vegetation by 7.8 ka, although records from 4 ka in southern California show redwoods in foothills with chaparral and foothill woodland associated higher up (Johnson, 1977), indicating at least some relict populations at that time. Axelrod (1981b) postulated that current vegetation in southern California is a mixture of xerically adapted species that are now in relict enclaves within more mesic sites (Fig. 7). These xerically adapted species spread with increasing dry hot climate during the Xerothermic (8–3 ka), but after 3 ka, when the climate returned to cooler conditions more like those of present coastal areas, they were restricted to isolated and disjunct localities within more mesic vegetation (Axelrod, 1981b).

Holocene vegetation records from inland areas of California are sparse (Baker, 1983). Huesser's 1978 study from the Santa Barbara area (which reflects nearby inland areas) indicates that plant associations similar to those now extant have existed since 12 ka. This correlates with work by Adam (1967), Casteel *et al.* (1977), and Casteel and Beaver (1978) for Clear Lake and interior California. Huesser (1978) reported that from 12 to 7.8 ka coniferous forest communities with Pinaceae–Taxodiaceae and Taxaceae species were important in upland areas. At lower elevations mixed conifer–oak forest prevailed from approximately 11 to 8.5 ka (early Holocene). From 7.8 to 5.7 ka lowland and cismontane communities were dominated by *Quercus* spp., along with members of the Rosaceae, Anacardiacea, Rhamnaceae,

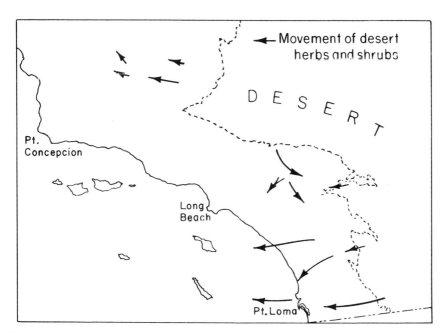

Fig. 7. Relictual distribution of species that presumably are a result of migration from more arid regions during the Xerothermic period (redrawn from Axelrod, 1967).

and Asteraceae including *Artemisia* (Adam, 1967; Baker, 1983; Huesser, 1978). Baker (1983) indicated that the progressive increase in chaparral and coastal sage pollen types indicated that *Artemisia* dominated coastal sage predominated in many areas in the middle Holocene and that chaparral became dominant in late Holocene. Vegetation approaching modern associations may have occurred around 7.8 ka (Baker, 1983). Pisias (1978) noted that there was a change in the vegetation of southern California at about 5.4 ka. Before this time pollen was characteristic of pine forest type, whereas after 5.4 ka the vegetation was more open and nonarboreal. From 5.6 to 4.9 ka there were maximum records of Asteraceae in Santa Barbara (Huesser, 1978). From 2.3 ka to the present chaparral and coastal sage formed the dominant vegetation types of the area (Huesser, 1978).

Anderson (1990) found marked changes in species abundance and forest structure with climate change during the past 10,000 years in the High Sierra of California. *Tsuga mertensiana* (mountain hemlock) and *Abies magnifica* (red fir) were more common at higher elevations at about 6 ka than they are at present and communities were more open and included montane chaparral species. O. Davis (1989) used pollen data to reconstruct Sierran vegetation on the west- and east-facing slopes during the Holocene. The data from 9 ka indicate a uniform distribution of community types, grassland, mixed montane with yellow pine forest, upper montane, and subalpine zone on both sides of the Sierras. This did not continue through 6 ka, however, when clearer differences in community distribution and type were found that are similar to those seen at present. Records for this area are poorer than for other regions but indicate that many of the species common today were also found during the Pleistocene and early Holocene. Modern community associations in this area, however, appear to be a phenomenon of the late Holocene (Anderson, 1990).

Vegetation of the Northwest has also varied over time and with proximity to the coast during the Holocene. Species such as Douglas fir (*Pseudotsuga menzesii*), western hemlock (*Tsuga heterophylla*), and spruce (*Picea sitchensis*), which are common components of the vegetation today, were also found in this region during the middle Holocene. *Sequoia sempervirens* was present in vegetation near the coast and somewhat inland during the Holocene, extending to near-modern southern ranges in Oregon and southern California. From 10 to 7 ka Douglas fir became a chief constituent of lowland areas with open forest savannahs (Barnosky *et al.*, 1987). During this same period prairies of Puget Sound expanded, suggesting a drier and sunnier climate than present (Baker, 1983). In north-central and northeastern Washington pollen cores indicate that an open tundra-like environment existed from 1.2 to 10 ka, with *Artemisia,* grasses, herbs, and haploxylon pines (Baker, 1983). These pollen assemblages have no modern analog (Baker, 1983). *Pinus ponderosa* and *P. contorta* appeared between 9 and 7 ka, when grasses and *Artemisia* moved farther north, displacing the border between grassland and forest northward (Baker, 1983). With subsequent cooling around 4 ka, pine species were replaced by *Abies lasiocarpus* in northeastern Washington and the treeline was lowered (Baker, 1983). Eastern Washington vegetation changed from spruce, pine, and alder to a temperate steppe of mixed pine and sagebrush similar to its modern appearance (Huesser, 1983).

Farther north in Alaska, as conditions warmed around 8 ka, a shrub tundra with birch–alder and willows formed (Barnosky *et al.*, 1987), with a spruce parkland between 9 and 7.5 ka, in a period warmer and drier than the present. Modern vegetation assemblages did not appear until recent times; however, as in most other cases already cited, they include species historically present in the region.

In summary, there have been substantial shifts in geographic and elevational distributions in response to climate change for pinyon pines, Douglas fir, or creosote with little evidence for extinction of plants in the western United States during the Quaternary (Thompson, 1988), a period of considerable temperature change. Anderson (1981) does report the extinction of one species, *Chrysothamnus pulchelloides,* during this period.

VI. Future Vegetation

We now turn our attention to making predictions of the nature of the vegetation of western North America sometime toward the end of the next century, when, given our current energy use patterns, our atmosphere will have a doubled concentration of CO_2 (Houghton *et al.*, 1990). Predictions based on global circulation models show the mean annual temperature of California increasing by 2 to 5°C (Fig. 8) (Smith and Tirpak, 1989). Ecosystem water balance will be greatly altered, particularly at higher elevations, where there will be less precipitation as snowfall and hence a reduced snowpack to feed spring and summer stream runoff. The predicted total amount of precipitation as well as its seasonal distribution varies by model. Some models predict an increase in summer precipitation that would have a profound effect on the present Californian summer drought-controlled ecosystems. These model predictions are based on averages of two 5×5 degree grid points that fall within California and thus are very crude. New approaches are being developed that will give greater local precision and will reflect the topographic diversity that so dominates the western North American local climate (Dickinson *et al.*, 1989). The fact remains, though, that these generalized predictions are of limited use for biologists in making derivative predictions about the impacts on particular plant distributions, as will be shown. The direction of the predictions in temperature change and moisture amount, however, enables us to make some statements about the general vegetation changes that can be expected. Also, there are climate predictions about rates of change with latitude that are useful for considering potential vegetation responses. A doubling of CO_2 will cause an uneven increase in temperature with latitude in the Northern Hemisphere (Manabe and Wetherald, 1980). Manabe and Wetherald (1980) predict about a 2° increase in mean temperature at low latitudes and as high as 8° in polar regions. The greater increase at high latitudes is due to a reduction in snow cover and an increase in poleward transport of latent heat. The net result would be a reduced latitudinal thermal gradient. They also predict a reduction in soil moisture at the midlatitudes between 37 and 47°. As noted by Trenberth in this volume (Chapter 3), the great impact on latitudinal temperatures of CO_2 doubling would not be matched in the Southern Hemisphere.

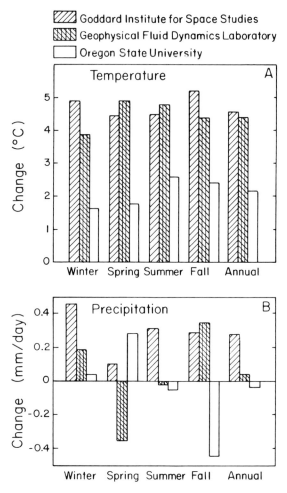

Fig. 8. Comparisons of predictions from different global circulation models for Californian temperatures and precipitation (from Smith and Tirpak, 1989).

VII. General Considerations for Making Predictions

Before discussing specific predictions of the changes that might be expected in the vegetation of the west coast of North America, we note some of the problems in making these sorts of predictions for anyplace.

A. Multiple and Changing Controllers

The controls on the distribution of any given organism are numerous and interactive. These include climate (temperature, moisture, and radiation—both means and extremes), soil type, and co-occurring biota such as pathogens, pollinators, seed vectors, competitors, and symbionts. All of these diverse agents can act to limit plant distribution at any given life stage, including germination, establishment, growth, and reproduction (Fig. 9). Furthermore, these limiting agents can be interactive. For example, factors limiting establishment one year may be quite different in another year. In a wet year conditions may be suitable for pathogenic fungi that destroy seeds, and in a dry year drought may have the same limiting effect on seedling establishment. Thus, the basic controls on establishment not only are diverse but also may differ in their effectiveness from year to year. Learning the controls on the distribution of any particular species in the presence of such complexity is one of our greatest challenges. However, it is not a new problem in plant geography. We really do not have complete information on the distributional biology of any noncrop species, nor do we have much prospect of getting it except for a few for which we can devote considerable resources. We do have fairly good information on the controlling climate, soil, and biotic agents for most crop species. However, for these the crucial establishment bottleneck is bypassed by cultural techniques.

B. Evolutionary and Dispersal Time versus Climate Change

All indications are that we will experience an unprecedented rate of climate change in the next century. Plants have, of course, always had to accommodate to changing climates in the past. They have done so by adapting in place and by migrating with the changing climate as noted earlier. The projected rate of change will, however, perhaps overwhelm the capacity of these mechanisms to adjust. M. Davis (1989b) has pointed out that the rate of climate change may exceed the ability of many plant species to migrate to keep up with their moving preferred climate.

It has been amply demonstrated that many plant species have evolved ecotypes with specific adaptations for the habitats in which they grow. One particularly

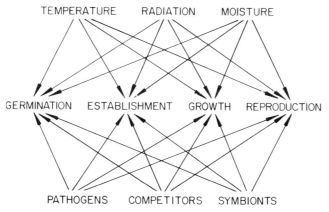

Fig. 9. Influence of climatic and biotic elements on life history events. This is a simplified illustration and does not show interactions of herbivores, for example.

prevalent example is the development of photoperiodic ecotypes along latitudinal gradients (McMillan, 1974; Mooney and Billings, 1961). Local races of a species are able to time their growth activities to match the most likely course of seasonal climatic change in any given locality. The development of these ecotypes has arisen over very long periods of time in the case of woody species. With a rapidly changing climate the fine tuning between local climate and plant behavior could well be disrupted.

Another disruption that will take place is between soil type and plant community type. The development of soil is interactive between climate and plant cover and takes place over thousands of years. There are examples of communities growing on soils that were formed under different climates and plant types. However, again it is a matter of the rate of change that is expected with the changing climate and the consequent rapid change in plant distribution that is anticipated. Plants will encounter soils quite different from those in which they evolved.

Thus, in assessing the impact of climate change on plant distribution the differing time scales of the changing climate, soil development, and the evolutionary response of herbs and trees must be considered (Fig. 10) in addition to differing dispersal rates of the organisms, as well as the differential buffering capacities of the resident organisms as discussed below. There is no question that the ecosystems we know today will be disassembled as the rate of climate change interacts with the differential dispersibility and adaptability of the constituent species.

C. Life Span and Climate Change

The west coast of North America not only has a great diversity of species and plant communities but also has a striking collection of plants that reach massive size and/or great age. The world's oldest living tree, the bristlecone pine, *Pinus longaeva*; the world's massive tree, *Sequoia sempervirens*; and one of the world's tallest trees, *Sequoiadendron giganteum,* all reside in California. The forests of the Pacific Northwest contain an abundance of dominant tree species that live for over a half-century and many that live over a millennium (Waring and Franklin, 1979).

It is obvious that the response to climate change will differ among species that live for millennia and those that live for only a single year. In the Sierran giant forests the successional cycle of *S. giganteum* operates on a different time scale than the "understory" mixed conifer species. In the White Mountains of California and Nevada the mature trees in the bristlecone pine forest have experienced over 4000 years of fluctuating climate. Successful reproduction is evidently episodic, occurring at long intervals (Billings and Thompson, 1957). Selection thus operates on an entirely different time scale for these trees than it does for co-occurring short-lived herbaceous species.

D. The New Flora

Beginning with the age of exploration, there has been a rapid change in the geographic distribution of the earth's biota. In particular, there has been massive transport of plant species from Europe and the Middle East to the New World. Many

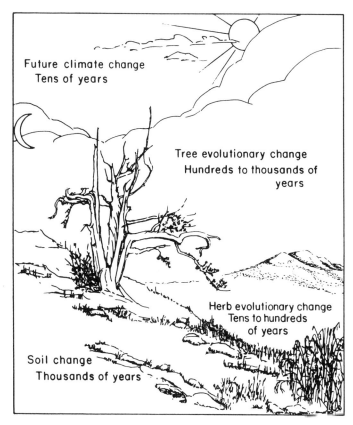

Future climate change
Tens of years

Tree evolutionary change
Hundreds to thousands of
years

Herb evolutionary change
Tens to hundreds
of years

Soil change
Thousands of years

Fig. 10. Differential rates of response of the physical and biotic components of an ecosystem to climate change. This simplified figure does not illustrate that different components of soil development have different time dimensions, nor does it capture the differential buffering capacities or dispersal potentials of the biotic parts of the system.

of these species are weedy types that have evolved in association with human-induced disturbance. *Bromus tectorum,* for example, was introduced into the Great Basin at the end of the 1800s and now occurs in nearly all western North American steppe vegetations (Mack, 1981; Thompson, 1988). In California, weedy Eurasian grasses dominate the interior valleys and have followed human disturbance into many areas. Approximately 25% of the California flora is now nonnative species (Mooney *et al.,* 1986). The biotic potential of these species and their interactions with native flora under present conditions are not fully known but could be significant. For example, the seeding of rye grass (*Lolium perenne*) in burns where native fire-annuals normally grow has been shown to retard significantly and prevent the growth of native species (Corbet and Green, 1965; Keeley *et al.,* 1981). Because of the shorter generation time and temperature and drought tolerance of many introduced species, these may be the players of the future.

E. The New Environments

Not only will there be new biotic players in the environments of the future but also the environments will be different, independent of the projected climate change. These environments will be enriched in CO_2 (Houghton *et al.*, 1990) as well as tropospheric ozone (Ashmore and Bell, 1991), both important gases metabolically. No doubt there will be accelerated landscape conversion due to increasing population pressures and increasing amounts of habitat disturbance due to human activities. Future scenarios, involving an altered climate, thus must consider not only the climate itself but also the new biotic players and the new environments.

VIII. On Making Predictions

What approaches do we have for predicting the responses of plants to new environments? Various techniques have been utilized. We assess a number of these in the following.

A. Climate–Vegetation Matching, Past and Present

One approach is to find an existing environment that matches the predicted new environment for a given locality. It is then assumed that the plants inhabiting the existing environment are those that will migrate to the new predicted environment. More specifically, for California it has been shown that the distribution of the major vegetation types correlates simply to mean annual temperatures and the mean annual range of temperatures (Fig. 11). It is thus easy to visualize how the predicted changes in mean annual temperature will affect community distribution, some to a greater extent than others. For example, utilizing these simple correlations of mean annual temperature with the distribution of alpine plant communities in the mountainous regions of southern California, it becomes obvious that some communities will disappear in the doubling CO_2 temperature scenarios, particularly because many of these communities exist on isolated mountaintops. As noted by Brown (Chapter 19, this volume), there will be no possibilities for these isolated populations to migrate northward to compensate for increasing temperatures.

Climate–vegetation matching has been utilized in an inverse manner by paleobotanists to determine past climates, as discussed earlier. Of course, from what the historical record shows us, as described earlier, these communities will not change just as beads on string but rather will assume some new compositional configurations because of the individualistic nature of the response of species to change. Then, regarding the details of response, it is quite clear that past relationships between climate and vegetation changes are not precise keys to the future for a number of reasons because of the new kinds of environments and the different species involved.

Holdrige (1947) devised a quantitative relationship between climate (temperature, precipitation, and evapotranspiration) and the world's vegetation types. This system has been utilized by Emmanuel *et al.* (1985) to predict changes in the world's vegetation distribution with the climate that would prevail under a doubled CO_2

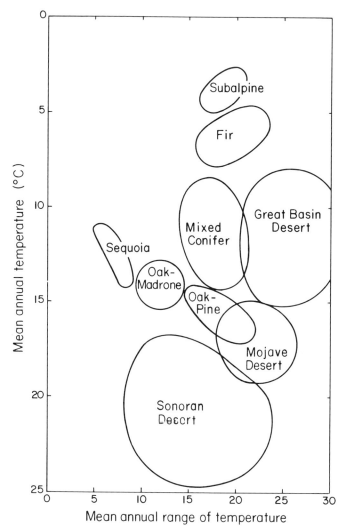

Fig. 11.   General relationship between the distribution of Californian plant communities and mean annual temperatures and range of temperatures (redrawn from Axelrod, 1979; Axelrod, 1981a).

atmosphere. More detailed climate, soil, and topographic features are being utilized in evolving global vegetation–environment models (Prentice *et al.*, 1992). These so-called static or equilibrium models are being supplanted by transient climate scenarios coupled with concomitant changing forest dynamics (Cohen and Pastor, 1991).

A comparable approach can be utilized to examine potential species distributions. The Nature Conservancy is examining all of the native plants of North America for their geographic range and elevational limits as well as their growth forms, reproductive potential, and dispersibility (Maddox and Morse, 1990). It is assumed that plants with a limited geographic range, and hence limited range of temperature and precipitation, will be most at risk from climate change.

B. Modeling Species Response to Climate Change

Future climates will be not only warmer but also enriched to a significant degree in CO_2 as well as a number of other atmospheric constituents that directly affect plant performance. This fact has led to attempts to make predictions of responses of the biota to global change at the species rather than the community level.

Knowledge of how the physiology of plants is affected by physical factors such as temperature, moisture, radiation, and CO_2 concentration has been consolidated into growth models for a number of crop species in particular. These growth models have been utilized to make predictions of the comparative response of species and functional groups to changing climates (Harwell and Hutchinson, 1985). Such approaches have the value of being able to incorporate factors that affect physiology other than those present in past climates, such as CO_2 enrichment or pollutant effects. We know that these factors differentially affect plant performance. These approaches can also be utilized in a more qualitative manner for species for which we have considerable cultural information, such as ponderosa pine (Fig. 12).

It is certainly worth the effort to develop comprehensive growth models for economically or socially important species to utilize in making future projections. There are difficulties, however. Most growth models incorporate only the physical environment–physiological relationships. Much of the detail that would be required to develop a full model incorporating the factors controlling growth and reproduction as shown in Fig. 9 is not presently available for any species. For crops, for which many of the controlling events are under cultural control, this detail is not necessary. But this detail is certainly required for plants in natural systems.

IX. What We Need to Do

We are obviously faced with an enormous challenge in making predictions of plant distribution and performance for the future even though we do have many clues from the past about the nature and direction of changes that may be anticipated. We should be able to increase our confidence limits considerably if we:

1. Improve our predictability of "forcing functions" controlling plant performance, including local climates and water balance, fire regimes, and pest cycles, and assess the nature of the new biotic players that will be present in future ecosystems (e.g., invading species). This is, of course, a large order.

2. Initiate experiments that will directly test plant performance in a natural ecosystem context under the new physical environmental conditions that we anticipate (e.g., enhanced CO_2 levels).

3. Acquire detailed life history–environmental relationships for economically or socially important organisms.

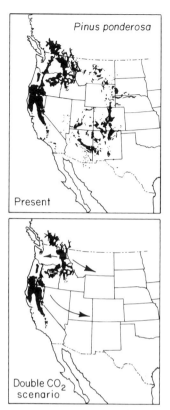

Fig. 12. Present and projected distribution of ponderosa pine under a doubled CO_2 scenario (redrawn from Little, 1971, and Leverenz and Lev, 1987). The arrows point to areas of reduced extent under doubled CO_2.

4. Develop new models that will link growth with competition as well as linking growth to dispersal and spatial community dynamics.

5. Develop generic models of the climatic response of functional groups that will aid in making generalized predictions of the kinds of community changes the global change will bring.

X. What We Can Say Now

Although our predictions cannot be precise at this time, we can still make a number of statements about the potential effects of global change on the vegetation of western edge of North America.

A. Moisture Regime

The predominating influence on the biota of the west coast is the summer drought, which exerts influence from the very southern part of California up through, to a lesser degree, Oregon and Washington. Nearly a third of the flora of California is composed of drought-evading annuals. The lack of summer precipitation is no doubt the reason why there are very few C_4 species in the coastal west. These plants are generally associated with warm temperatures, when water and hence growth is possible. The few C_4 grasses that occur in California are associated with vernal pools are possibly relics of a time when summer rainfall was prevalent in the region (J. Keeley, personal communication). Any climate change that would increase summer precipitation would favor C_4 species. However, countering this effect would be the potentially greater response of C_3 species over C_4 species to the fertilization effect of increasing CO_2.

The number of species found in California is positively correlated with the amount of rainfall and to a lesser degree local topographic diversity. Mean annual temperature is not a good predictor of species number as a whole, although it does explain the variance in the numbers of annual species that are not sensitive to total precipitation amount (Richerson and Lum, 1980). Thus, even at a very gross level of consideration, it can be seen how climate change would differentially affect the balance of growth forms found in California.

California is rich in plant endemics and relics. The greatest number of relics are found in the north coast and in certain desert regions (Stebbins and Major, 1965). These areas are respectively the wettest and driest in the state, yet they are still influenced to some degree by summer rainfall that characterized the region in the Miocene, some 25 million years ago (Axelrod, 1973). Thus, a change in precipitation seasonality alone could have a profound effect on vegetation patterning.

The Sierran snowpack will be altered by projected climate change with an earlier seasonal snowmelt (Smith and Tirpak, 1989). This, in turn, will influence plant distribution, because snowpack depth is strongly correlated with the distribution of high-elevation forest communities (Barbour *et al.*, 1991).

B. Fire Regime

Fire has played a major role in the evolution and distribution of the California flora. Climate is a principal determinant of fire frequency and severity. Swetnam and Betancourt (1990) have shown, for example, that the dry springs associated with the low phase of the Southern Oscillation are strongly correlated with extensive fires in the southwestern United States.

Fire return time varies latitudinally on the Pacific coast of North America. In Southern California, in the chaparral, fire frequency is on the order of 20 years (Keeley, 1982), whereas in the forests of central Oregon fire return intervals are somewhat over 100 years and in the central Cascades of Washington, over 400 years (Franklin *et al.*, 1992). With an increase in temperature and consequent increased evapotranspiration, coupled with increased human interaction with native vegetation, it is likely that fire frequency will increase along the latitudinal vegetation gradient. Franklin *et al.* note that this would result in a shift in dominance toward fire-sprouting hardwoods over nonsprouting conifers in the Pacific Northwest.

Not only fire frequency but also possibly fire intensity will increase. Westman and Malanson (1992) predict that under such a scenario the resprouting shrubs would be at a disadvantage due to death of root crowns.

C. Temperature

The implications of temperature change for vegetation change have already been noted. Barbour *et al.* (1991) stated that even "minor increases in temperature . . . will have a significant effect on vegetation zonation in the Sierra Nevada." Franklin *et al.* (1992) predict a total loss of the alpine vegetation in the Oregon Cascade Range and a significant decrease in the extent of subalpine communities; however, montane forests would increase in extent.

D. Oxidants

Atmospheric pollutants are affecting natural vegetation and ecosystem function to an ever greater extent in California. Forest damage first noted in the 1950s in the San Bernardino Mountains close to urban centers (Taylor, 1973) is now being observed in the comparatively remote forests of Sequoia National Park (Peterson *et al.*, 1989). Temperature increase will most likely increase ambient ozone levels (Malanson and Westman, 1991) with consequent detrimental effects on natural and agricultural systems.

E. Interactions

The many environmental changes predicted will have an adverse effect on individuals of the western vegetation that have so many species of unusual longevity (Waring and Franklin, 1979). These long-lived species will have reduced vigor and will become poor competitors, which will hasten their demise. Increase in pollutants and fire frequency will aid the process. Initially there will be a shift to species with short life spans. It will become a successional world. These will be prime conditions for an increase in the incursion of introduced weedy species, which, under present conditions, are increasing in extent because of accelerating land use modification by humans.

F. The "Other" Elements of Global Change

We have concentrated principally on the implications of climatic change for western vegetation. We would like to close with the thought that it is land use change that is the most immediate and larger threat to vegetation than the somewhat more distant projected climate change. Over 2 million persons are added to California every year with consequent high impact on California's natural resources. Even parks and reserves are being heavily affected by large numbers of visitors expecting modern conveniences. Many plant communities have already been drastically modified, including the grasslands, which have been totally altered by grazing activities, and the coastal marshes, which have been severely reduced in extent, as have the vast areas of riparian oak forest of the California Central Valley. The ecosystems of western

North America are destined to absorb yet greater impacts in the years ahead as the combined effects of human population growth and climate change really take hold.

XI. Summary

The vegetation and flora of the west coast of North America are extraordinarily rich. There are a great number of vegetation types with a high turnover from place to place, due in part to the diverse topography and the complex geological history of the region. Many of the vegetations are composed of species with vastly differing life spans. There are large numbers of endemic and endangered species, just as there is a large number of established alien species. There is little question that the predicted changes in climate, including fire regime, the composition in the air, and land use over the next century will alter the vegetation to a much greater extent than occurred during the climatic alterations of the past 10,000 years, even though these vegetation changes were rather impressive. As happened during the past, we can expect significant alterations in the changing distributions and dominance of species in an individualistic manner. The differential responses of varying life forms to change and the variety of those forms present ensure a new vegetative landscape in the years ahead, with losses of species, as well as of entire systems, and increased prominence of alien species.

References

Adam, D. P. (1967). Late-Pleistocene and recent palynology in the central Sierra Nevada, California. *In* "Quatrenary Paleoecology," INQUA Vol. 7 (E. J. Cushing and H. E. Wright, Jr., eds.), pp. 275–301. Yale Univ. Press, New Haven, Connecticut.

Anderson, L. C. (1981). A new species of fossil *Chrysothamnus (Asteraceae)* from New Mexico. *Great Basin Nat.* **40**, 351–352.

Anderson, R. S. (1990). Holocene forest development and paleoclimates within the central Sierra Nevada, California. *J. Ecol.* **78**, 470–489.

Ashmore, M. R., and Bell, J. N. B. (1991). The role of ozone in global change. *Ann. Bot.* **67**(Suppl. 1), 39–48.

Axelrod, D. I. (1967). Geologic history of the Californian insular flora. *In* "Proceedings of the Symposium on the Biology of the California Islands" (R. N. Philbrick, ed.), pp. 93–149. Santa Barbara Botanic Garden, Santa Barbara, California.

Axelrod, D. (1973). History of the mediterranean ecosystem in California. *In* "Mediterranean Type Ecosystems" (F. di Castri and H. A. Mooney, eds.), pp. 213–277. Springer-Verlag, New York.

Axelrod, D. I. (1979). Age and origin of Sonoran desert vegetation. *Occ. Pap. Calif. Acad. Sci.*, 1–74.

Axelrod, D. I. (1981a). Altitudes of tertiary forests estimated from paleotemperature. *In* "Geological and Ecological Studies of Qinghai–Xizang Plateau," pp. 131–137. Gordon & Breach Science Publishers, New York.

Axelrod, D. I. (1981b). Holocene climatic changes in relation to vegetation disjunction and speciation. *Am. Nat.* **117**, 847–870.

Axelrod, D. I. (1988). Paleoecology of a late Pleistocene Monterey pine at Laguna Niguel, southern California. *Bot. Gaz.* **149**, 458–464.

Baker, R. G. (1983). Holocene vegetational history of the western United States. *In* "Late-Quaternary Environments of the United States" (H. E. Wright, Jr., ed.), Vol. 2, pp. 109–127. Univ. of Minnesota Press, Minneapolis.

Barbour, M. G., and Billings, W. D. eds. (1988). "North American Terrestrial Vegetation." Cambridge Univ. Press, Cambridge, U.K.

Barbour, M. G., Berg, N. H., Kittel, T. G. F., and Kunz, M. E. (1991). Snowpack and the distribution of a major vegetation ecotone in the Sierra Nevada of California. *J. Biogeogr.* **18,** 141–149.

Barnosky, C. W., Anderson, P. M., and Bartlein, P. J. (1987). The northwestern U.S. during deglaciation; vegetational history and paleoclimatic implications. *In* "North America and Adjacent Oceans during the Last Deglaciation. The Geology of North America" (W. F. Ruddiman and H. E. Wright, Jr., eds.), Vol. K-3, pp. 289–321. Geological Society of America, Boulder, Colorado.

Betancourt, J. L., Van Devender, T. R., and Martin, P. S. (1990). "Packrat Middens, the Last 40,000 Years of Biotic Change." Univ. of Arizona Press, Tucson.

Billings, W. D., and Thompson, J. H. (1957). Composition of a stand of old bristlecone pines in the White Mountains of California. *Ecology* **38,** 158–160.

Brubaker, L. B. (1988). Vegetation history and anticipating future vegetation change. *In* "Ecosystem Management for Parks and Wilderness" (K. Agee and D. R. Johnson, eds.), pp. 41–61. Univ. of Washington Press, Seattle.

Casteel, R. W., and Beaver, C. K. (1978). Inferred Holocene temperature changes in the North Coast Range of California. *Northwest Sci.* **52,** 337–342.

Casteel, R. W. Adam, D. P., and Sims, J. D. (1977). Late-Pleistocene and Holocene remains of *Hysterocarpus traski* (tule perch) from Clear Lake, California, and inferred Holocene temperature fluctuations. *Quat. Res.* **7,** 133–143.

Cohen, Y., and Pastor, J. (1991). The responses of a forest model to serial correlations of global warming. *Ecology* **72,** 1161–1165.

Cole, K. L. (1985). Past rates of change, species richness, and a model of vegetational inertia in the Grand Canyon, Arizona. *Am. Nat.* **125,** 289–303.

Cole, K. L. (1986). In defense of inertia. *Am. Naturalist* **127,** 727–728.

Cole, K. L. (1990). Late Quarternary vegetation gradients through the Grand Canyon. *In* "Packrat Middens, the Last 40,000 Years of Biotic Change" (L. Betancourt, T. R. Van Devender, and P. S. Martin, eds.), pp. 240–258. Univ. of Arizona Press, Tucson.

Corbet, E. S., and Green, L. R. (1965). Emergency revegetation to rehabilitate burned watersheds in Southern California. Forest Service Research Paper PSW-22, Pacific Southwest Forest and Range Experiment Station, Berkeley, California.

Davis, M. A. (1986). Climatic instability, time lags, and community disequilibrium. *In* "Community Ecology" (J. Diamons and T. J. Case, eds.), pp. 269–284. Harper & Row, New York.

Davis, M. A. (1989a). Insights from paleoecology on global change. *Bull. Ecol. Soc. Am.* **70,** 222–228.

Davis, M. A. (1989b). Lags in vegetation response to greenhouse warming. *Climatic Change* **15,** 75–82.

Davis, O. (1989). Ancient analogs for greenhouse warming of central California. Appendix D: Forests. *In* "Potential Effect of Global Climate Change in the United States." Environmental Protection Agency Report 230-05-89-054, EPA, Washington, D.C.

Dickinson, R. E., Errico, R. M., Giorgi, F., and Bates, G. T. (1989). A regional climate model for the western United States. *Climatic Change* **15,** 383–422.

Emanuel, W. R., Shugart, H. H., and Stevenson, M. P. (1985). Climatic change and the broad-scale distribution of terrestrial ecosystem complexes. *Climatic Change* **7,** 29–43.

Franklin, J. F., *et al.* (1992). Effects of global climatic change on forests in northwestern North America. *In* "Global Warming and Biological Diversity" (R. L. Peters and T. E. Lovejoy, eds.), pp. 244–257. Yale Univ. Press, New Haven, Connecticut.

Harwell, M. A., and Hutchinson, T. C. (1985). "Environmental Consequences of Nuclear War." Wiley, Chichester.

Holdridge, L. R. (1947). Determination of world plant formations from simple climatic data. *Science* **105,** 367–368.

Houghton, J. T., Jenkins, G. J., and Ephraums, J. J. (eds.). (1990). "Climate Change: The IPCC Scientific Assessment." Cambridge Univ. Press, Cambridge, U.K.

Huesser, C. J. (1983). Vegetational history of the northwestern United States including Alaska. *In* "The

Late Pleistocene" (S. C. Porter, ed.), Vol. 1, "Late-Quaternary Environments of the United States," pp. 239–258. Univ. of Minnesota Press, Minneapolis.

Huesser, L. (1978). Pollen in Santa Barbara Basin, California: A 12,000 year record. *Geol. Soc. Am. Bull.* **89,** 673–678.

Johnson, D. L. (1977). The late Quaternary climate of coastal California: evidence for an ice age refugium. *Quat. Res.* **8,** 154–179.

Keeley, J. E. (1982). Distribution of lightning-and man-caused wildfires in California. Pacific Southwest Forest and Range Experiment Station General Technical Report PSW-58, pp. 431–437, U.S. Forest Service.

Keeley, S. C., Keeley, J. E., Hutchinson, S. M., and Johnson, A. W. (1981). Postfire succession of the herbaceous flora in southern California chaparral. *Ecology* **62,** 1608–1621.

Kutzbach, J. E. (1987). Model simulations of the climatic patterns during the deglaciation of North America. *In* "North American and Adjacent Oceans during the Last Deglaciation" (W. F. Ruddiman and H. E. Wirtght, Jr., eds.), Vol. K-3, pp. 425–446. Geological Society of America, Boulder, Colorado.

Kutzbach, J. E., and Wright, H. E. (1986). Simulation of the climate of 18,000 yr B.P.: results for North American/North Atlantic/European sector and comparison with the geologic record. *Quat. Sci. Rev.* **4,** 147–187.

Leverenz, J. W., and Lev, D. J. (1987). Effects of carbon dioxide-induced climate changes on the natural ranges of six major commercial tree species in the western United States. *In* "The Greenhouse Effect, Climate Change and U.S. Forests" (W. E. Shands and J. S. Hoffman, eds.), pp. 123–156. The Conservation Foundation, Washington, D.C.

Little, E. L., Jr. (1971). "Atlas of United States Trees. Conifers and Important Hardwoods." U.S. Government Printing Office, Washington, D.C.

Mack, R. N. (1981). Invasion of *Bromus tectorum* L. into western North America: an ecological chronicle. *Agro-Ecosystems* **7,** 145–165.

Maddox, D., and Morse, L. E. (1990). Plant conservation and global climate change. *Nature Conservancy Mag.* **40**(4), 24–25.

Malanson, G. P., and Westman, W. E. (1991). Modeling interactive effects of climate change, air pollution, and fire on a California shrubland. *Climatic Change* **18,** 363–376.

Manabe, S., and Wetherald, R. T. (1980). On the distribution of climate change resulting from an increase in CO_2 content in the atmosphere. *J. Atmos. Sci.* **37,** 99–118.

McMillan, C. (1974). Photoperiodic adaptation of *Xanthium strumarium* in Europe, Asia Minor, and northern Africa. *Can. J. Bot.* **52,** 1779–1791.

Mirov, N. T. (1967). "The Genus *Pinus*." Ronald Press, New York.

Mooney, H. A., and Billings, W. D. (1961). Comparative physiological ecology of arctic and alpine populations of *Oxyria digyna*. *Ecol. Monogn.* **31,** 1–29.

Mooney, H. A., Hamburg, S. P., and Drake, J. A. (1986). The invasions of plants and animals into California. *In* "Ecology of Biological Invasions of North America and Hawaii" (H. A. Mooney and J. A. Drake, eds.), pp. 250–272. Springer-Verlag, New York.

Munz, P. A. (1968). "Supplement to a California Flora." Univ. of California Press, Berkeley.

Munz, P. A., and Keck, D. (1959). "A California Flora." Univ. of California Press, Berkeley.

Peterson, D. L., Arbaugh, M. J., and Robinson, L. J. (1989). Ozone injury and growth trends of Ponderosa pine in the Sierra Nevada. *In* "Effects of Air Pollution on Western Forests" (R. K. Olson and A. S. Lejohn, eds.), pp. 293–307. Air and Waste Management Association, Pittsburgh, Pennsylvania.

Pisias, N. G. (1978). Paleoceanography of the Santa Barbara Basin during the last 8000 years. *Quat. Res.* **10,** 166–384.

Prentice, I. C., Cramer, W., Harrison, S. P., Leemans, R., Monserud, R. A., and Solomon, A. M. (1992). A global biome model based on plant physiology and dominance, soil properties and climate. *J. Biogeogr.* **19,** 118–134.

Raven, P. H., and Axelrod, D. I. (1978). Origin and relationships of the California flora. *Univ. Calif. Publ. Bot.* **72,** 1–134.

Richerson, P. J., and Lum, K.-L. (1980). Patterns of plant species diversity in California: relation to weather and topography. *Am. Nat.* **11,** 504–536.

Smith, J. B., and Tirpak, D., eds. (1989). "The Potential Effects of Global Climate Change on the United States." U.S. Environmental Protection Agency, Washington, D.C.

Spaulding, W. G. (1990). Vegetational and climatic development of the Mojave desert: the last glacial maximum to the present. *In* "Packrat Middens, the Last 40,000 Years of Biotic Change" (L. Betancourt, T. R. Van Devender, and P. S. Martin, eds.), pp. 166–199. Univ. of Arizona Press, Tucson.

Stebbins, G. L., and Major, J. (1965). Endemism and speciation in the California flora. *Ecol. Monogr.* **35,** 1–35.

Swetnam, T. W., and Betancourt, J. L. (1990). Fire–Southern Oscillation relations in the southwestern United States. *Science* **249,** 1017–1020.

Taylor, O. C., ed. (1973). Oxidant air pollutant effects on a western coniferous forest ecosystem. U.S. Forest Service Task B Report, Univ. of California.

Thompson, R. S. (1988). Western North America. Vegetation dynamics in the western United States: modes of response to climatic fluctuations. *In* "Vegetation History. Handbook of Vegetation Science" (B. Huntley and T. Webb, eds.), Vol. 7, pp. 415–458. Kluwer Academic Publishers, Boston.

Thompson, R. S. (1990). Late Quaternary vegetation and climate in the Great Basin. *In* "Packrat Middens, the Last 40,000 Years of Biotic Change" (J. L. Betancourt, T. R. Van Devender, and P. S. Martin, eds.), pp. 200–239. Univ. of Arizona Press, Tucson.

Van Devender, T. R. (1990). Late Quaternary vegetation and climate of the Sonoran desert, United States and Mexico. *In* "Packrat Middens, the Last 40,000 Years of Biotic Change" (J. L. Betancourt, T. R. Van Devender, and P. S. Martin, eds.), pp. 134–165. Univ. of Arizona Press, Tucson.

Van Devender, T. R., and Spaulding, W. G. (1979). Development of vegetation and climate in the southwestern United States. *Science* **204,** 701–710.

Vankat, J. L. (1982). A gradient perspective on the vegetation of Sequoia National Park, California. *Madroño* **29,** 200–214.

Waring, R. H., and Franklin, J. F. (1979). Evergreen coniferous forests of the Pacific Northwest. *Science* **204,** 1380–1386.

West, N. E. (1984). Successional patterns and productivity potentials of pinyon–juniper ecosystems. *In* "Developing Strategies for Range Management," pp. 1301–1332. Westview Press, Boulder, Colorado.

West, N. E. (1988). Intermountain deserts, shrub steppes, and woodlands. *In* "North American Terrestrial Vegetation" (M. G. Barbour and W. D. Billings, eds.), pp. 210–230. Cambridge Univ. Press, Cambridge, U.K.

Westman, W. E., and Malanson, G. P. (1992). Effects of climate change on mediterranean-type ecosystems in California and Baja California. *In* "Global Warming and Biological Diversity" (R. L. Peters and T. E. Lovejoy, eds.), pp. 258–276. Yale Univ. Press, New Haven, Connecticut.

CHAPTER 17

Global Change: Flora and Vegetation of Chile

MARY T. KALIN ARROYO   JUAN J. ARMESTO
FRANCISCO SQUEO   JULIO GUTIÉRREZ

I. Introduction

Chile is a narrow, mountainous, island-like territory, 736,532 km², spanning subtropical to subantarctic latitudes (17–55°S) on the western border of South America in the Southern Hemisphere. It is characterized by extraordinary climatic and vegetational diversity. Because of the interplay of topography and latitude, vegetation is highly variable. The Andes intercept westerlies in the south and monsoonal rain in the north. Climate is further affected by the drying effect of the cold-water Humboldt Current, and strong altitudinal effects. Vegetation types range from the sparse and episodically appearing plants of the Atacama Desert to some of the wettest temperate rain forests in the world, together with considerable extensions of deciduous forests. Despite these extremes, climate in Chile is generally strongly equable (in the sense of Sloan and Barron, 1990; Miller, 1976). This results from Chile's position on the South American landmass (inversely distributed latitudinally in relation to North America). There is a narrow land area from sea level to the main Andean divide, significant inland penetration of maritime airmasses occurs, and the Humboldt Current has a moderating effect on summer temperature.

Chile, like places at equivalent latitudes in western North America, will be affected progressively by the warming of the earth, attributed in large measure to the injection into its atmosphere of greenhouse gases (Berger and Labeyrie, 1987). What

239

are the future prospects for Chile and how well adapted are the Chilean flora and vegetation for abrupt climatic warming?

Because no biological research has specifically addressed global change in Chile, the principal objective of this Chapter is to set the stage for such an analysis. We will review (1) aspects of floristic composition and vegetation distribution, including the temperature and precipitation relationships of some vegetation types, and (2) responses of individual plant species. We will also point to some contrasts expected between the vegetation of Chile and that of western North America.

II. Chile and Global Change

Stouffer *et al.* (1989) predict that the Southern Hemisphere will warm more slowly than the Northern Hemisphere because of the higher ocean/land ratio. Moreover the latitudinal pattern of temperature increase will differ in the two hemispheres as a result of asymmetry of landmass distribution. Whereas in the Southern Hemisphere midlatitudes will warm more rapidly than high latitudes, the reverse trend is predicted for the Northern Hemisphere. These trends are illustrated in Table I, which provides

TABLE I.
Surface Temperature Increases for Localities in Chile Compared to 1958 Predicted from Zonally Averaged Values from Stouffer *et Al.* (1989)[a]

Latitude (°S)	Chilean locality	Year		
		1991	2030	2053
65[b]	Base Presidente	+0.4	+1.0	+1.5
	G. Gonzalez V.	(+1.5)	(+4.0)	(>+5.0)
55	Puerto Williams	+0.5	+1.2	+1.8
		(+1.4)	(+3.5)	(+4.9)
51	Puerto Natales	+0.6	+1.4	+2.1
		(+1.4)	(+3.4)	(>+4.0)
45	Coihaique	+0.8	+1.7	+2.4
		(+1.5)	(+3.2)	(>+4.0)
42	Castro (Chiloé)	+0.8	+1.9	+2.6
		(+1.5)	(+3.2)	(>+4.0)
36	Concepción	+1.0	+2.1	+2.9
		(+1.3)	(+3.1)	(>+4.0)
30	La Serena	+1.0	+2.3	>+3.0
		(+1.2)	(+2.9)	(+3.9)
23	Antofagasta	+1.0	+2.3	>+3.0
		(+1.1)	(+2.5)	(>+3.0)
18	Arica	+1.0	+2.3	>+3.0
		(+1.0)	(+2.3)	(>+3.0)

[a]Values in parentheses are for equivalent latitudes in the Northern Hemisphere. 2030 is predicted CO_2 doubling relative to 1958.
[b]On Chilean territory, West Antarctic Peninsula.

extrapolations from Stouffer *et al.* (1989) for some Chilean localities and their equivalent latitudes in the Northern Hemisphere. Of particular note, the temperature increase difference from the Northern Hemisphere might grow to as much as 3.5°C at latitude 65°S in 2053 (62 years from the time of this writing). What do these predictions mean for vegetation? Mean temperature at Constitución on the coast at 35°S in Chile is presently 13.9°C (di Castri and Hajek, 1976). Within around 62 years, according to present global circulation models (GCMs), this could be the temperature of Puerto Montt, 41°S. With direct migration this means that vegetation in central Chile would have to move about 6° of latitude south within 62 years! Northern rain forests would have to move about 3° of latitude south. Alpine vegetation would be forced to move 500 m upward in the same time!

A cautionary note is in order. The "predictions" given in Table I derive from a model in which only zonally averaged temperatures are provided. Chile's complex relief could strongly influence the numbers in Table I. Seasonal temperature distribution is not provided by this analysis. This Chapter hardly relies on these data, so they could have been omitted. They are included, however, to illustrate the *abruptness of global warming,* which to our minds should be the central point in any discussion of the effects of global change on organisms.

No precise information on precipitation patterns can be obtained at a meaningful scale from GCMs. However, some particularly long climatic records analyzed for Chile suggest that a trend toward decreased precipitation in southern to north-central Chile (on the western side of the Andes) is already occurring (see Aceituno *et al.,* Chapter 4, this volume). Analyses of meteorologic records for stations on the dry eastern side of the southern Andes in northern Patagonia by Weber (1951), however, failed to detect any trend toward decreased precipitation during the first half of the present century. The future precipitation evolution of Chile, and southern South America in general, thus is likely to be complex.

III. Flora and Vegetation

A. Flora

Floras are the result of a continuous assembly process dependent on migration and the local evolution of species. Because the individual taxa present in a given flora at any one time may have evolved under widely different evolutionary conditions, knowledge of the evolutionary development of a flora is an essential ingredient for assessing its capacity to respond to abrupt climatic change. Evolutionarily conservative taxa, reflecting their past histories of gradual change across diffuse climatic ecotones, are unlikely to be as well adapted to respond to rapid climatic warming as nonconservative species that have emerged rapidly under strong environmental pressures or across sharp climatic ecotones. Floras of mediterranean-type climates (e.g., Raven and Axelrod, 1978) usually contain a mixture of conservative and nonconservative taxa, reflected in a diverse life form mix, including many annuals. The floras of equable climates like that of Chile, however, also tend to show many ancient or relic evolutionary conservative groups (Axelrod *et al.*, 1992). The flora of Chile, as will be shown in the following, has certain imprints of an evolutionary history in an equable climate.

Within continental Chile's native flora of close to 4500 species (Marticorena, 1990) are many paleoendemic taxa. For example, restricted to Chile are the woody, monotypic Gomortegaceae, with a few populations in cool moist habitats in coastal forest in central Chile. The monotypic woody Aextoxicaceae of uncertain taxonomic affinity are mostly restricted to the Valdivian rain forest and coastal relic forests in the mediterranean-type climate zone. Similarly, the parasitic monogeneric Misodendraceae are found only on *Nothofagus* and only to a small extent in Argentina.

Sixty-five (Marticorena, 1990) or about 8% of the native vascular plant genera of continental Chile are almost certainly endemic, in contrast to the 3% endemic genera for the state of California (Raven and Axelrod, 1978). The Andes and the Atacama Desert have served to isolate many plant groups. Among the endemic genera, *Peumus, Desmaria, Notanthera, Gypothamnium, Pintoa, Lapageria, Lardizabala, Jubaea, Balsamocarpon, Metharme, Dinemandra, Lepidothamnus, Gomortega, Leontochir,* and *Pitavia,* concentrated in coastal forests and deserts, and mostly monotypic, are probably paleoendemics. Many more paleoendemic genera occur in the Chilean region defined to include a small extension of the southern rain forests across into Argentina (e.g., *Campsidium, Fitzroya, Pilgerodendron, Austrocedrus, Saxegothaea, Asteranthera, Philesia*). Such paleoendemics persist along with many neoendemic genera and species that have appeared as a result of severe geographic isolation. Chile is similar to subtropical islands like New Caledonia (14%; Thorne, 1989) with its many endemic genera.

Phytogeographic affinity is another important consideration in relation to global change. Darwinian selectionists tend to assume that all organisms are optimally adapted. However, according to a review by Travis (1989), there are few robust examples of direct optimizing selection in natural populations. Strong life history constraints could mean that optimal adaptation is rarely accomplished and adaptational lag is the norm. Adaptational lag is seen in flowering phenology (Kochmer and Handel, 1986), where flowering patterns fail to respond to pollinator pressures because of strong phylogenetic constraints. For example, plant species of the Asteraceae, regardless of geographic origin and the composition of the pollinating fauna, tend to flower toward the end of the flowering season. Genera of plants of austro–Antarctic affinity (e.g., *Azorella*) found today along the full length of the Andes might be poorly adapted for global warming. In contrast, taxa with a past history in warmer climates could show adaptational lag in the reverse direction, effectively being "preadapted" to global warming.

It is now becoming clear that the Chilean flora has very strong connections with the new World tropics. A striking number of vegetationally important and often dominant genera of south-central and central Chile can be found either disjunct to cooler tropical latitudes in eastern South America or to the northern Andes or fragmented around the cool highlands of Brazil and northern Argentina with a fewer outliers in Chile (M. T. K. Arroyo, unpublished data). Some genera fitting one of these patterns include *Araucaria, Azara, Blepharocalyx, Colliguaja, Crinodendron, Cryptocarya, Dasyphyllum, Deuterocohnia, Drimys, Eccremocarpus, Escallonia, Fuchsia* section *Quelusia, Greigia, Griselinia, Kageneckia, Lithrea, Myrceugenia, Myrcianthes, Podocarpus, Quillaja, Rhaphithamnus,* and *Quillaja*. These genera are further examples of evolutionarily conservative groups, and they emphasize again that the cool, moist equable conditions in central and southern Chile have provided a refugium for many genera that were probably distributed across southern South America in the late Tertiary (see Axelrod, 1979). Looking toward ancient connections across Antarctica, Chile also shares a number of woody rain forest genera (e.g., *Gevuina, Nothofagus, Eucryphia, Griselinia, Aristotelia, Pseudopanax*) and Magellanic moorland genera (e.g., *Astelia, Donatia, Gaimardia*) with Australia and/or New Zealand.

The many evolutionary conservative and relic woody groups and genera of the Chilean flora intermingle with neoendemic taxa, producing a rich and taxonomically diverse flora. The overriding effect of climatic equability, however, and a regionally cooler climate are clearly reflected in the overall life form spectrum (Table II). The entire vascular flora of continental Chile, which is strongly concentrated in arid areas and ones with a mediterranean-type climate, contains only 15.7% annuals (Arroyo *et al.*, 1990a). Central Chile, with a mediterranean-type climate, has proportionately more woody species (especially shrubs) and perennial herbs than California (Table

TABLE II.
Life Forms (%) in the Flora of Continental Chile, Central Chile (30–40°S), and California

	Continental Chile[a]	Central Chile[b]	California[c,d]
Native flora	$(n = 4408)^e$	$(n = 2760)$	$(n = 5046)$
Annuals[f]	15.7	15.7	28.6
Total flora	$(n = 4984)$	$(n = 3199)$	$(n = 5883)$
Annual herbs	19.8	20.9	32.6
Perennial herbs	61.0	59.5	52.0
Shrubs	17.0	16.8	12.0
Trees	2.2	2.8	3.4
All herbaceous	80.8	80.4	84.6
All woody	19.2	19.6	15.4

[a]Native annuals from Arroyo *et al.*, 1990a; the life form spectrum for the total flora of Chile was obtained from the Chilean Flora Data Base. The latter contains a checklist of all species from Chile together with data on life form.
[b]Data for native annuals and life form spectrum for total flora from 30 to 40°S from the Chilean Flora Data Base.
[c]Native flora; Raven and Axelrod (1978).
[d]Total flora: Richerson and Lum (1980).
[e]n = total number of species sampled. For California n = total species present.
[f]Other life forms not given for California. Differences for woody versus herbaceous are significant for Chile and central Chile versus California, respectively (G-tests).

II). Identical trends are seen in the northern Altiplano compared to the high cold deserts of North America (Arroyo *et al.*, 1988). The evidently slowly evolving Chilean flora could be strongly constrained in adjusting to global change. Nevertheless, the tropical roots of many of the woody groups, if some of these show adaptational lag, could counteract the negative aspects of a conservative evolutionary history.

B. Vegetation

How vegetation is assembled over the landscape is an important area of inquiry in relation to global change. Vegetation response will depend not only on the inherent migrational and adaptational capacities of species but also on how the dominant species are presently assembled and distributed over the vegetational landscape.

Some key features of vegetation in relation to global change are size of the dominant species' ranges, diversity, and the latitudinal and altitudinal differentiation of vegetation types. Range extent can provide an index of the scope of environmental conditions occupied by individual taxa and an indication of potential genetic variability. Vegetation with high alpha and beta diversity should have a greater chance of responding to global change because they have most "cards" to draw on at any one location ("lottery" effect). Latitudinal and altitudinal differentiation of vegetation would seem to be a decisive feature in any geographic region. With diffuse vegetation differentiation, there will always be outlying individuals of a given vegetation type in low frequencies in adjacent vegetation belts. Such outliers could render the large-scale migration required to keep up with rapid temperature and precipitation evolution largely superfluous (see Section IV). This kind of vegetation distribution would gain exaggerated importance in geographic areas like Chile, where human activities have already tended to create series of vegetation islands in the central and central-southern parts of the country (Fuentes *et al.*, 1990).

Most regional work on vegetation in Chile has been descriptive and phytosociological (e.g., Oberdorfer, 1960; Schmithüsen, 1956). Much of what follows is based on this slim base and our own observations and impressions.

1. Southern Chile. Southern Chile (south of 38°S), with high rainfall and cloudiness along the Pacific coast and a moderate to strong rain shadow on the eastern side of the Andes, supports a succession of closed-canopy, temperate-evergreen rain forests over 17° latitude, a belt of deciduous *Nothofagus*-dominated forest, treeless Patagonian steppe dominated by *Festuca* spp., low shrubs and hard cushion species

---

Fig. 1. Generalized scheme of major vegetation types of Chile based on Schmithüsen (1956), Pisano (1981), Donoso (1982), Veblen *et al.* (1983), Arroyo *et al.* (1988), and personal observations. 1, Absolute desert; 2, vegetated desert; 3, sclerophyllous forest and woodlands; 4a, deciduous *Nothofagus* forest; 4b, semideciduous forest with *N. obliqua*; 5, Valdivian rain forest; 6, north Patagonian rain forest; 7, deciduous Magellanic forest; 8, evergreen Magellanic forest; 9, Magellanic moorland; 10, high alpine. Hatched areas: permanent icefields. Isolated fog desert islands occurring along the coast north of 25°S and patches of deciduous *Nothofagus* forest and gymnosperm forest close to treeline in south-central Chile are not shown.

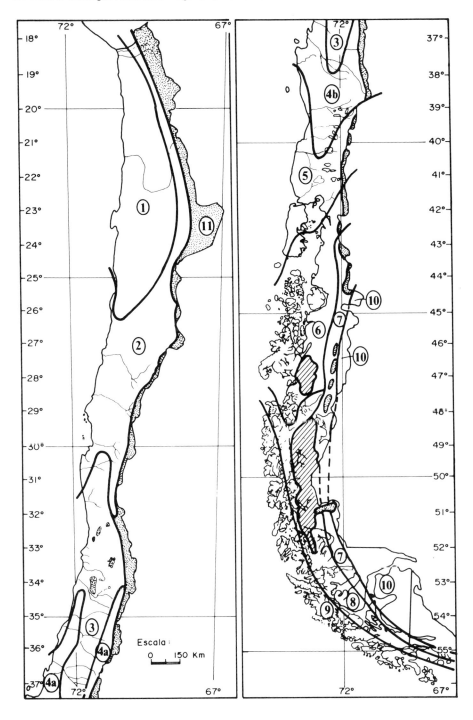

(Pisano, 1981; Arroyo *et al.*, 1989), and Magellanic moorland (Fig. 1). Relic stands of moorland can also occur on mountaintops of the Coast Range farther north (Villagrán, 1988). Fundamental differences from North American tundra are seen in a lack of permafrost due to the oceanic influence and a much smaller areal extent. Included in the moorland are hard cushion, sphagnoid, graminoid, and cyperoid types (Pisano, 1981; Moore, 19830. Gymnosperm dominance in the southern rain forests is restricted to discontinuous stands of *Fitzroya cupressoides* found above the Valdivian rain forest to the tree limit (700–1000 m) in the Andes and at higher elevations on the Coast Range from around 40 to 43°S (Rodríguez *et al.*, 1983).

From north to south, the evergreen forests show progressive physiognomic simplification with loss of vines and epiphytes, fewer strata, and floristic impoverishment (Veblen *et al.*, 1983). Alpha diversity is fairly high in the Valdivian rain forest (Table III). Most notably in the present context, the boundaries between these southern forest types are weak and their dominants can usually be found in adjacent types (Veblen *et al.*, 1983). Reflecting this, there is still much disagreement among authors as to where the vegetation ecotones should be drawn (see Schmithüsen, 1956; Pisano, 1981; Veblen *et al.*, 1983). Many tree species have large latitudinal ranges (Table IV).

TABLE III.

Alpha Diversity ($H' = -\Sigma\ p_i \ln p_i$) in Some Chilean Vegetation Types[a]

	H'	
Vegetation type	Mean	Range
Valdivian rain forest (40–42°S)—trees[b]	1.217	0.691–1.372
Deciduous *Nothofagus* forest (32–34°S)—all woody spp.[c]	1.631	0.543–2.079
Relic coastal rain forest (30–34°S)—all woody spp.[d]	1.612	0.888–2.473
Lowland sclerophyllous forest (33°S)—all woody spp.[e]	2.220	
Montane sclerophyllous woodland (33°S)—all woody spp.[f]	0.871	
Coastal matorral (32°S)—all woody spp.[g]	2.360	
Coastal desert (30°S)—all woody spp.[h]	2.250	
Desert scrub (18–28°S)—all spp.[i]	0.917	0.678–1.356
Alpine (18–28°S)—all spp.[j]	0.767	0.507–1.072

[a]p_i is relative abundance of the *i*th species. Abundance values are cover, unless otherwise indicated. Ranges indicate that measurements are available for more than one site in the vegetation type indicated.
[b]M. Riveros, unpublished, Puyehue; Armesto and Figueroa (1987), Talcán, Alao, Chaulinec, Isla Grande de Chiloé—abundance based on basal area.
[c]Casassa (1985), Termas de Chillán, Cerro Cantillana, Sierra Bella Vista, Quebrada del Roble, Cerro Campana.
[d]Pérez and Villagrán (1985), Fray Jorge, Talinay, Cerro Inés, Cerro Imán, Zapallar, Quebrada de Córdoba, Quebrada del Roble.
[e]Mooney *et al.* (1977), Fundo Santa Laura.
[f]Uslar (1982), Villa Paulina.
[g]Mooney *et al.* (1977), Papudo.
[h]Mooney *et al.* (1977), Potrerillos.
[i]Arroyo *et al.* (1988), Interior of Arica, Copiapó, Vallenar (2000–2500 m).
[j]Arroyo *et al.* (1988), Vallenar (4000–4500 m).

Reflecting the latter, a very high percentage of the tree species in southern Chile penetrate well into the mediterranean-type climate zone (Table IV).

With the fairly high diversity and physiognomic variability mentioned earlier, the temperate rain forests of Chile should have substantial flexibility for coping with global change. The Valdivian and north Patagonian rain forests includes a number of species whose ranges extend well to the south of the present limits of these forest types. In contrast, deciduous and evergreen Magellanic forests and Magellanic moorland do not and will be unable to rely on outlying individuals to the south, should climate be greatly affected. These last-mentioned vegetational belts, in fact, are likely to suffer considerable latitudinal contraction. For example, Magellanic moorland could be shunted off the edge of the South America continent in the next half-century.

The Chilean evergreen forests and moorland are characterized by low thermal oscillation and tend to occupy narrow ranges for mean annual temperature but have wide rainfall ranges (Fig. 2A). These vegetation types in Chile could be particularly sensitive to any increase in seasonality that might accompany global warming. In contrast, deciduous Magellanic forest and Patagonian steppe, adapted to broader thermal oscillation regimes, are distributed over a wider range for mean annual temperature and narrow precipitation ranges. These vegetation belts will perhaps be more sensitive to precipitation changes.

2. Central Chile. Central Chile (30–38°S) and inland areas of the Central Valley to around 40°S support a mediterranean-type climate with long dry summers and winter rainfall (di Castri and Hajek, 1976). The three major vegetation types—sclerophyllous forest and woodlands, semideciduous forest with *Nothofagus obliqua*, and strongly deciduous *Nothofagus* forest (Fig. 1)—intergrade into one another latitudinally (Villagrán and Armesto, 1980; Veblen *et al.*, 1983, Casassa, 1985; San Martin *et al.*, 1987). Large latitudinal ranges are again characteristic of the tree species, many of which penetrate well to the south of the mediterranean-type climate

TABLE IV.
Approximate Latitudinal Ranges of Tree and Treelets Occurring in Southern and Central Chile Deduced from Their Distributions Given in Rodríguez *et Al.* (1983) for the Geopolitical Regions of Chile[a]

	No. of species	Latitudinal range (°)			% Species in adjacent area[b]
		Mean	Min.	Max.	
Central Chile	87	8.4	0.8	21.8	59.8
Southern Chile	58	10.2	2.1	21.8	89.7

[a]The regions roughly follow the latitudinal gradient in Chile. Central Chile is considered as regions IV–IX. Southern Chile is considered as regions X–XII. Ranges expressed in degrees latitude were calculated from the midpoints of the northernmost and southernmost regions where a species is reported to occur.
[b]Adjacent area refers to southern Chile in the case of central Chile and vice versa in the case of southern Chile.

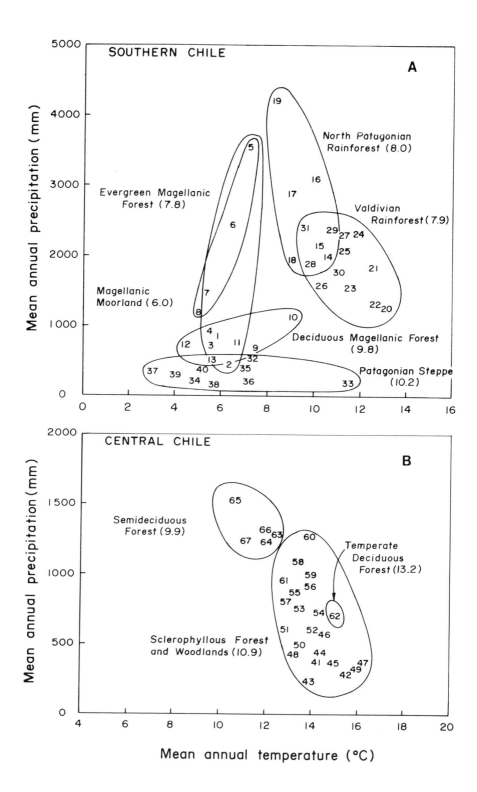

zone (Table IV). Somewhat sharper differentiation of vegetation belts seems to occur over elevation in comparison with the far southern rain forests. This difference could be related to the very gradual southward decrease in mean temperature in central Chile as a result of increasingly oceanic conditions that compensate for a southward decrease in temperature, which would be less felt inland in the Andes. Semideciduous forest, mostly found in the Central Valley, intergrades directly with Valdivian rain forest at higher elevations (Veblen *et al.*, 1983; Casassa, 1985) and with a progressive increase in *N. obliqua* or *N. glauca* becomes a strongly deciduous forest still containing many additional evergreen trees (*Azara* spp., *Maytenus boaria, Drimys winteri*). At its southern extreme semideciduous forest is interspersed with swamp forests characteristically containing *Tepualia stipularis* and many Valdivian rain forest elements (Ramírez *et al.*, 1983).

Sclerophyllous forest occurring up to the treeline at 31–32°S shows two major altitudinal forms, a lowland version dominated by *Lithrea caustica, Quillaja saponaria, Cryptocarya alba*, and *Peumus boldus* to 1600 m and a fairly discrete montane version with strong dominance of the small rosaceous tree *Kageneckia angustifolia* (Arroyo and Uslar, 1992) up to the treeline at 2000–2200 m. Lowland sclerophyllous vegetation is without doubt the most heterogeneous vegetation type in Chile. It is diverse for woody species (Table III), with some suggestion of higher diversity than in California. Diversity calculated for data supplied for a matched climatic site of chaparral vegetation in California by Mooney *et al.* (1977) is 1.47, compared to 2.22

Fig. 2. Mean temperature and precipitation for vegetation types (see Fig. 1) in southern (A) and central (B) Chile, excluding the alpine. Average temperature range (mean temperature of warmest minus mean temperature of coolest month) for localities within each vegetation type is given in parentheses. Climatic data from di Castri and Hayek (1975), di Castri and Hayek (1976), Pisano (1974, 1977, 1980), Zamora and Santana (1979), and Dollenz (1982). Only nonoverlapping localities are shown. Locality 5 was judged representative of two vegetation types.

(A) Evergreen Magellanic forest: 1, Río San Juan (53°S); 2, Puerto Navarino (54°S); 3, Isla Nueva (53°S); 4, Río Douglas (55°S); 5, Puerto Eden (49°S). Magellanic moorland: 5, Puerto Eden (49°S); 6, Evangelistas (52°S); 7, Orange Bay (55°S); 8, Diego Ramírez (56°S). Deciduous Magellanic forest: 9, Río Cisnes (44°S); 10, Coyhaique (45°S); 11, Río Paine (51°S); 12, Río Las Minas (53°S); 13, Puerto Williams (54°S). North Patagonian rain forest: 14, Quellón (43°S); 15, Futaleufú (43°S); 16, Melinka (43°S); 17, Puerto Aysen (45°S); 18, Cabo Raper (46°S); 19, San Pedro (47°S). Valdivian rain forest: 20, Lebu (37°S); 21, Contulmo (38°S); 22, Isla Mocha (38°S); 23, Puerto Dominguez (38°S); 24, Valdivia (39°S); 25, Punta Galera (40°S); 26, Frutillar (41°S); 27, Puerto Montt (41°S); 28, Maullín (41°S); 29, Punta Corona (41°S); 30, Pudeto (41°S); 31, Morro Lobos (42°S). Patagonian steppe: 32, Balmaceda (45°S); 33, Chile Chico (46°S); 34, Cerro Guido (50°S); 35, Cerro Castillo (51°S); 36, Punta Dungeness (52°S); 37, Laguna Blanca (52°S); 38, Oazy Habour (52°S); 39, Peckett Habour (52°S); 40, San Sebastian (53°S). (B) Sclerophyllous forests and woodlands: 41, Zapallar (32°S); 42, Jahuel (32°S); 43, Quintero (32°S); 44, Quillota (32°S); 45, Peñablanca (33°S); 46, Quilpué (33°S); 47, Colina (33°S); 48, San Antonio (33°S); 49, El Bosque (33°S); 50, Lo Espejo (33°S); 51, San José de Maipo (33°S); 52, Rengo (34°S); 53, San Fernando (34°S); 54, Curicó (34°S); 55, Molina (35°S); 56, Constitución (35°S); 57, Punta Carranza (35°S); 58, Panimávida (35°S); 59, Linares (35°S); 60, Los Angeles (37°S); 61, Angol (37°S). Temperate deciduous forest: 62, Cauquenes (35°S). Semideciduous forest: 63, Concepción (38°S); 64, Traiguén (38°S); 65, Cullinco (38°S); 66, Temuco (38°S); 67, Río Bueno (40°S).

for Santa Laura in Chile. Many locally distributed woody species requiring cooler and/or wetter conditions, succulents on xeric slopes, relic coastal stands of *Aextoxicon punctatum* rain forest (Peréz and Villagrán, 1985), and coastal swamp forest dominated by Myrtaceae that intergrade with each other contribute to high diversity. Such floristic mixing undoubtedly has been facilitated by central Chile's varied relief, very strong altitudinal gradients, contrasting microclimatic conditions on equatorial and polar-facing slopes (Armesto and Martinez, 1978), and the fact that mediterranean-type latitudes have always been subject to climatic oscillation because of strong dependence on the position of the polar front. Some vegetationally important species are distributed over 1000–1500 m (e.g., *Kageneckia oblonga, L. caustica, M. boaria, Q. saponaria*; Rodríguez *et al.*, 1983).

Despite the lower latitude, mean annual temperature range tends to be greater in central Chilean woody communities (Fig. 2B) than in the far south (Fig. 2A). This is a reflection of greater equability in southern Chile, producing a moderating effect on average temperature. Variation in precipitation for sclerophyllous vegetation is surprisingly high (Fig. 2B). It is important to appreciate that the temperature range for sclerophyllous vegetation is probably higher than indicated in Fig. 2B because there are few meteorological stations at the higher-elevation sites of occurrence.

3. Northern Chile. Inland northern Chile is under the influence of summer monsoonal rain to 25°S and the weak tail end of the frontal rainfall zone from 25 to 30°S. Coastal locations from 20 to 25°S receive low amounts of winter rainfall very sporadically. Intense aridity results from the combined action of the strong rain-shadow of the Andes to the east and the drying effect of the Humboldt Current to the west. The progressive southern weakening of monsoonal rainfall produces maximum aridity around 23–25°S, where the zonal vegetation is restricted to above 3000 m (Arroyo *et al.*, 1988). The intersecting east–west and north–south rainfall gradients produce a significant extension of absolute desert interrupted only on fog-bound summits along the coast (Johnston, 1929) and sporadic vegetation along some of the major water courses and edges of "salares."

Further reduction in rainfall in northern Chile could eliminate some of the fog islands close to the Pacific coast, where many endemic genera and local endemic species can be found (Johnston, 1929). The effect of increased aridity on the eastern margin of the desert can be predicted from detailed work in northern Chile, where it has been shown that a severe rainfall gradient from an average of 68 mm to 6 mm from 18 to 24°S, respectively, is associated with a reduction from 69 to 10 plant species from 2000 to 3000 m elevation (Arroyo *et al.*, 1988). At this same elevation on the eastern side of the Andes, where rainfall is higher today, there are 130 species of vascular plants. A reverse trend in monsoonal rainfall should alter the slope of the present species richness curve along the desert edge.

4. High Andean Chile. Alpine vegetation should be a sensitive indicator of global change, in that temperature plays a major role in determining vegetation distribution at high elevations. For the same reason, alpine vegetation is likely to be strongly affected by global warming. Effectively, alpine vegetation can be considered a thermometer of global change.

Chile shows perhaps the strongest development of alpine vegetation found at any temperate latitude. Alpine communities occur along the entire length of the Andes except in the very extreme south, where summits are low or still covered by ice fields (Arroyo et al., 1983) (Fig. 1). Mean annual temperature in the subalpine belts is around 9–10°C (Arroyo et al., 1990b). Treeline fails to exceed 2500 m and descends to sea level in the extreme south of the continent. There is no true treeline north of about 31°S. Generally the alpine vegetation in Chile is composed of two discrete phases: an arid, low-cover, zonal vegetation of low rounded shrubs, cushion plants, and perennial herbs developed in an altitudinal succession of three vegetation belts, interspersed with azonal high-elevation cushion bogs of continuous cover (Arroyo et al., 1981, 1983, 1988). Vegetation zonation is sharper than in the Chilean lowlands. Such alpine habitats are directly connected northward with tropical alpine vegetation types. The strong continuity of alpine habitats from subantarctic to tropical latitudes along the Andes is in contrast with western North America, where the mountain ranges tend to be more fragmented and insufficiently high at lower latitudes to provide continuous high-elevation habitats.

Such alpine habitats come under the influence of the monsoonal precipitation regime in northern Chile, where precipitation is received in the summer months as nonpersistent snow (Belmonte and Moscoso, 1985; Arroyo et al., 1988), and the polar frontal precipitation regime in central and southern Chile, where precipitation is received almost entirely in the winter as persistent snow (central Chile) or there is considerable precipitation during the summer months as well as snow during the winter (southern Chile) (Arroyo et al., 1988). The annual thermal oscillation is considerably lower in the northern subtropical Andes than in the central and southern Andes (Arroyo et al., 1990b). These precipitation and thermal patterns are at present reflected in highly contrasting phenological patterns in the two Andean climatic regions (Arroyo et al., 1990b). There is a bimodal flowering on either side of the middle to late summer rainy season in the north, versus flowering predominantly in the dry (central) or drier (southern Chile) summer months in the south. There are also latitudinal differences in life form composition (Arroyo et al., 1988).

Bog species in the Chilean Andes have broader latitudinal distributions than the zonal flora of the arid slopes (Arroyo et al., 1982b). High Andean bog species depend on accumulated ground water and richer soils (Arroyo and Squeo, 1990). Should ground water dry up or become less available, these plants locally would also be very sensitive to global change. Arroyo et al. (1988) demonstrated that reduced bog area as a result of strong aridity has been associated with a 60% decline in species richness from 18 to 24°S. This is a clear warning of the kind of future that can be expected for the large "bofedales" in the far north of Chile (an important resource for camelid grazing) if monsoonal rain were to be reduced in the near future. Unfortunately, too few meteorologic stations are available at high elevations to characterize the precipitation and temperature relationships of alpine vegetation.

Summarizing, each vegetation type in Chile is characterized by particular precipitation and temperature requirements. Some vegetation types occur over large precipitation ranges but small temperature ranges, whereas for others the situation is reversed. The degree of intergradation of the main vegetation types could turn out to be a key factor in the ability of any one vegetation type to cope with global change. Poorly differentiated vegetation types with outlying populations of dominant species in adjacent vegetation types should be better adapted to global warming than discrete vegetation types.

IV. Individualistic Responses

A. Positive *in Situ* Effects

Vegetation in Chile could show positive responses to global change. One possible response is CO_2 enrichment (Bazzaz, 1990; Mooney *et al.*, 1991). Enhancement of photosynthesis due to increased CO_2 is more likely in C_3 plants than C_4 plants, and greatest responses can be expected where other resources are not lacking (Bazzaz, 1990). In northern Chile *Portulaca philippii* is a C_4 species, while species of *Calandrinia,* another genus of Portulacaceae, are either C_3 or CAM (Arroyo *et al.*, 1990c). These species with different photosynthetic systems could be affected differently under global warming. Also, drought stress in plants grown at high CO_2 levels may be ameliorated because of more efficient water use. Direct effects of CO_2 enrichment should be looked for in vegetation types like the Valdivian rain forest, whereas an indirect response could be expected in arid communities. Northern outlying populations of *Austrocedrus chilensis* in the Andes and of *N. obliqua* in the Coast Range persist under conditions very similar to those seen for *Pinus aristata* in the Sierra Nevada, California, where increased CO_2 is thought to be responsible for larger growth rings (LaMarche *et al.*, 1984). However, the higher temperatures associated with increased atmospheric CO_2, if accompanied by reduced precipitation, could produce diminishing returns. Accurate knowledge of possible precipitation changes is vital to assess this possibility and its vegetational consequences in Chile.

In the Chilean Andes pollinators are highly sensitive to temperature and atmospheric conditions (Arroyo *et al.*, 1985). Likewise, flowering is strongly temperature dependent (Rozzi *et al.*, 1989). Increased pollination success could result from global warming in harsh, high-elevation communities such as those of the Chilean Patagonia. Here anthophilous insect visitation rates are about one-third of those in the central Chile Andes (Arroyo and Squeo, 1990). Optimal temperature for visitation in the Patagonian alpine communities, however, is higher than the most frequent daytime temperatures found in those communities (Fig. 3). Thus, with global warming, the Patagonian alpine could be more adequately serviced by pollinators. In the central Chilean Andes, to the contrary, visitation rates were found to decrease for abnormally high temperatures, presumably due to corporeal overheating (Arroyo *et al.*, 1985) (Fig. 3). Here upward altitudinal displacement of pollinators to cooler temperatures could be expected.

Significant changes in the dynamics of pollination can be expected in other

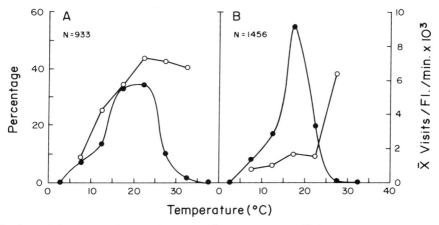

Fig. 3.   Relation between flower visitation rate (O) and temperature (●) in the subalpine belt in (A) central Chilean Andes (33°S) (after Arroyo *et al.*, 1985) and (B) Patagonian alpine (50°S) (F. Squeo and M. T. K. Arroyo, unpublished). N = number of 10-minute periods (during the daylight hours) over which flower visitation rates were quantified and temperature readings were taken. The curve for temperature gives the percentage of readings made during the 10-minute observation periods that correspond to a given temperature range. The ranges are 0–4°C, 5–9°C, etc.

Chilean communities as well. These include sclerophyllous forest, where there are striking microclimatic differences on north- and south-facing slopes (Armesto and Martínez, 1978; Del Pozo *et al.*, 1989), and Valdivian rain forest where reduced future cloudiness could increase pollination success. Such temperature effects on pollinators could eventually have noticeable effects on plant community composition. For example, bee-pollinated plants, presently in low frequency in the Patagonian alpine (Arroyo *et al.*, 1987), could increase in abundance in relation to dipteran- and lepidopteran-pollinated plants.

Many of the positive effects described could be offset by the negative effects of aridity on seed germination and seedling survival (see Section IVC) producing complex feedback. Moreover, changes in pollination rate must be counterbalanced against other likely effects of increased atmospheric CO_2 such as increased or decreased time to flowering (Bazzaz, 1990). These, together with the direct effects of temperature on phenology and flower longevity (Arroyo *et al.*, 1981) could offset positive effects of increased temperature on pollination rate by decoupling plants and their pollinators.

B. *In Situ* Adaptation

A plant species' ability to adapt *in situ* to abrupt climatic warming will ultimately depend on availability of genetic variability, generation time, breeding system, and recruitment possibilities. As recruitment is also relevant to migration, it will be considered at the end of Section IVC. In the following it is argued that the Chilean flora is generally not well adapted for rapid *in situ* adaptation.

Studies of genetic variation in Chilean plants are limited. Distributional patterns suggest a variety of situations. At one extreme, many forest dominants in Chile, as already mentioned, have very wide latitudinal and altitudinal ranges. Latitudinal clinal variation in seed size (Donoso, 1979) and in demographic strategy (Casassa, 1985) has been demonstrated in widely distributed *N. obliqua,* and clear-cut cases of altitudinal differentiation are known in other widely distributed species (e.g., *D. winteri, Embothrium coccineum, A. chilensis*). Whether such variation is genetic is largely unknown. On the other hand, some paleoendemic genera in the south-central coastal forests (e.g., *Pitavia, Gomortega*) and in the northern deserts (e.g., *Leontochir*) have very limited distributions and consequently might not have sufficient genetic variability to respond to rapid global warming. Several neoendemic annual genera (*Cyphocarphus, Microphyes, Araeoandra*) are also highly restricted in distribution. Of course, not all endemic genera in Chile are rare, the best example being *Peumus,* represented by a single widespread sclerophyllous forest species occurring over 10° of latitude in central Chile.

1. Longevity. Rapid population differentiation in many Chilean plant species will be hindered by longevity, which undoubtedly is a major constraint to rapid *in situ* adaptation in the Chilean flora. The life spans of many woody plants and high-elevation cushion plants are undoubtedly longer than 50 years. Extreme longevity characterizes some temperate rainfall trees species like *F. cupressoides.* In the latter species Bonisegna and Holmes (1985) reported an age of 1537 years for an individual in Argentina. A. Lara (personal communication) has evidence for much older trees of *F. cupressoides* in the southern Chilean forests. Such longevity implies that individuals will be forced to live through considerable climatic deterioration within a single generation! From time of establishment to first reproduction, significant changes in temperature and precipitation will already have occurred. Although phenotypic plasticity could save individuals over a lifetime, it is doubtful whether adaptive responses through selection at the progeny and later stages could occur quickly enough to keep up with the projected pace of global warming.

Of particular interest in this context are stump sprouters. Counterintuitively, we believe that long-lived stump sprouters might be poorly adapted for abrupt climatic warming in the long run. Any seeds sired or mothered by a very old, self-incompatible or dioecious stump sprouter (see Arroyo and Uslar, 1992) may have a genotype reflecting the temperature and precipitation conditions present when its mother or father was selected many years ago. Stump sprouting would seem to be viable long-term strategy only when climate changes at an intermediate pace, as would have been the case with development of the mediterranean-type climate as of the late Tertiary. In the next 50–100 years individuals of stump sprouting species may survive. The longer such individuals live, however, the less likely it is that their progeny will be able to adapt to the new temperature and precipitation environment. After a prolonged period of time, such species could undergo rapid local extinction. Assuming that global change is already under way, a simple way of testing this would be to compare the survival of progeny of artificially crossed seeds obtained between very old parents and between young parents under natural conditions. *Lithrea caustica,* a dominant, long-lived stump sprouter of the Chilean sclerophyllous forests and the very long-lived *F. cupressoides* in the southern rain forests of Chile would be ideal candidates for studies of this kind.

2. Breeding Systems. Many long-lived species in Chile will undoubtedly be additionally constrained for *in situ* adaptation because of their breeding systems. Self-incompatibility and dioecism are not conducive to rapid population differentiation (Hamrick *et al.,* 1979). These breeding systems are very common in long-lived woody species and in many perennial herbs and geophytes in the Patagonian alpine and montane sclerophyllous forest in Chile (Table IV). We now know that the Chilean montane sclerophyllous forest is about as rich in obligately outcrossed woody species as lowland tropical forest (Arroyo and Uslar, 1992). Ongoing work suggests that proportionately fewer obligately outbreeding species are found in the Valdivian rain forest than in drier montane sclerophyllous forest and Patagonian alpine vegetation considering given life forms (M. Riveros, unpublished). On this count, Valdivian rain forest might be in a better position to respond to global change than sclerophyllous vegetation and the high alpine.

The annuals of Chile should be best equipped to respond to global change by *in situ* adaptation given their short generation time and tendency for genetic self-compatibility favoring rapid population differentiation (Hamrick *et al.,* 1979). To date, all native Chilean annuals studied for breeding system have proved to be genetically self-compatible (Table V). Rapid chromosomal evolution and saltational speciation under extreme environmental conditions have been described in desert annuals (Lewis, 1966). This phenomenon could be expected in the Chilean coastal

TABLE V.
Breeding Systems in Two Vegetation Types in Chile

Vegetation type	% Dioecism in total flora	Hermaphrodites[a]	
		% SI	% SC
Montane sclerophyllous forest, 33°S[b] (n = 109)			
Woody species[c]	24.3	80.0	20.0
Perennial herbs	? 0	50.0	50.0
Annuals	0	0	100.0
Patagonian alpine, 50°S[d] (n = 311)			
Woody species	31.0	75.0	25.0
Perennial herbs	3.7	28.4	71.6
Annuals	0	0	100.0

[a]SI, genetically self-incompatible; SC, genetically self-compatible.
[b]Arroyo and Uslar (1990), 27 hermaphrodite species studied.
[c]Percent dioecism in trees considered alone is 57%. n = total number of species assessed for dioecism.
[d]Arroyo and Squeo (1990), 124 hermaphrodite species studied.

desert, where rainfall can be expected to deteriorate significantly in the coming years, leading to local increases in diversity. However, many of Chile's native annuals might be highly constrained in times of global warming. The destruction of Chile's woody cover over the past 300 years has fostered the introduction of close to 300 annual species of which no fewer than 80% are found in central Chile. Many of the introduced species (e.g., *Erodium circutarium, Medicago* spp., *Avena fatua, Filago gallica, Silene gallica*) already contribute heavily to the intertree cover of degraded sclerophyllous forest and have invaded coastal desert, where they show strong competitive advantage over the native annuals, corroborating the aggressive nature of weedy annuals from the mediterranean basin (Heywood, 1989). Vidiella and Armesto (1989) showed that introduced annuals in coastal desert are able to germinate with about 10 mm of artificially applied precipitation. The native annual species, in contrast, required over 40 mm of precipitation for germination. Further decreases in rainfall and increased temperature (causing higher evaporation levels), compounded by increased size of intertree gaps in sclerophyllous forest and gaps among shrubs in the desert, where there is already evidence of strong competition between some species (Gutiérrez *et al.*, 1987), could enhance the competitive edge of the introduced annuals, leading eventually to local extinction of native species.

C. Migration

The "simplest" option for any species to cope with rapid global warming, whether short- or long-lived, should be that of tracking its temperature and precipitation conditions through migration. The palynologic records show that migration has been a common response to past climatic change (e.g., Axelrod, 1978). There are numerous examples in the Chilean palynologic record of extensive north–south migrations (see Villagrán and Armesto, Chapter 15, this volume). Assuming the presence of the desired temperature and precipitation somewhere in the landscape, successful migration will depend on dispersal capacity, presence of pollinators in migration sites, and recruitment. For inherent migrational capacity in a rapidly changing temperature environment, both negative and positive features are seen in the Chilean vegetation.

1. Dispersal. Dispersal rate is likely to be the major obstacle for tracking temperature and precipitation in any vegetation. Animal dispersal is common and statistically overrepresented in the southern temperate rain forest of Chile in comparison with deciduous forests in eastern North America (Armesto *et al.*, 1987; Armesto and Rozzi, 1989) and in mediterranean-type climate communities in central Chile in relation to California (Hoffmann *et al.*, 1989). Animals can be expected to move seeds farther than wind in some cases and to direct seeds to safer sites than wind. However, many sclerophyllous plants have large seeds or fruits with poor mobility. The high alpine (Arroyo and Squeo, 1990) and desert floras in Chile have few animal-dispersed species, and in those studied in detail (*Chuquiraga oppositifolia*: Asteraceae and *Anarthrophyllum cumingii*: Papilionaceae) dispersal has proved to be very local (Rozzi, 1990).

Irrespective of good dispersal in many Chilean plants, the projected pace of global change makes it questionable that direct migrational displacement of vegetation belts is a viable way for Chilean vegetation to cope with global change. For a hypothetical central Chilean tree requiring, say, 5 years to reach reproductive maturity, about 12 dispersal advances could occur in 60 years. To cover the distances mentioned in Section II would entail hope of about 1/2 degree in latitude with successful recruitment each time! Although Chile has a number of migratory frugivorous birds (Armesto *et al.*, 1987), birds usually discharge seeds a few hours after ingestion (Ridley, 1930). *Intermediate-range dispersal,* as called for with global change, is more probable in epizoochorous species, where seeds may remain attached to a bird for extended periods. However, this kind of dispersal would tend to favor small-seeded herbaceous species and, in particular, many weedy species.

For alpine vegetation, without access to good dispersal mechanisms, direct upward displacement of vegetation belts by about 500 m seems highly dubious for the long-lived dominant cushion species like *Azorella* sp. and *Laretia acaulis* and shrubs like *C. oppositifolia, Nardophyllum lanatum,* and *A. cumingii.*

2. Pollination. The many long-lived obligately outbreeding species in the sclerophyllous forest, the Valdivian rain forest, and the high alpine are biotically pollinated (Uslar, 1982; M. Riveros, unpublished; Arroyo *et al.*, 1987). With global warming, their pollinators, whose generation times are short, could be expected to move faster than their nectar sources, leaving some plant species without pollination service. This kind of situation could make the northernmost temperate rain forests of Chile more vulnerable than their counterparts in western North America, where all gymnosperms and most angiosperm trees are wind pollinated. This constraint nevertheless will depend on the degree of pollinator specialization. In Chile generalist feeding (and nonspecialized pollination systems) are common in the Valdivian rain forest (M. Riveros, unpublished), sclerophyllous forest (Cody *et al.*, 1977), and high alpine (Arroyo *et al.*, 1982a). Specialist feeders and more specialized pollination systems tend to be more abundant in the desert (Cody *et al.*, 1977). Interrelationships between plants and animals determining functional pollination are thus likely to be most affected in the desert by global warming. Self-compatible annuals, usually not dependent on pollinators, should be least affected in all vegetation types.

3. Recruitment. Recruitment is a critical life history phase for both *in situ* adaptation and migration. In Chile, there are suggestions of wide differences in the abundance of recruitment between vegetation types. In sclerophyllous forest, some dominant woody species have large, short-lived propagules (e.g., *C. alba, P. boldus, Beilschmiedia* spp.) that require humid and shaded conditions for germination, today reduced because of anthropogenic influence (Fuentes *et al.*, 1990). However, seed banks depauperate because of degradation of seed sources may be a primary cause of lack of regeneration in some disturbed sites (H. Jimenez and J. J. Armesto, unpublished). Many self-incompatible species in sclerophyllous montane forests have very low fruit/flower ratios (Arroyo and Uslar, 1992). These need to be studied

as potential cases of inadequate pollination service caused by regional-scale anthropogenic effects. In the widespread weedy, highly self-incompatible *Acacia caven,*
which offers only pollen as a flower resource, Peralta *et al.* (1992) have proposed that
lack of enough nectar-producing plants in the same area to support consistent bee
populations could be the cause of extremely low fruit set. Recruitment thus seems to
be already critical in the sclerophyllous forest of Chile as a result of direct and indirect
intervention by humans. The present low seed resources are likely to be critical under
projected less favorable microclimatic conditions in the future.

Recruitment in midelevation southern rain forest, in contrast, is favored by open,
well-lit conditions (Armesto and Fuentes, 1988). However, such conditions are only
partially beneficial in low-elevation rain forests. Valdivian rain forest species also
show fairly high fruit/flower ratios, suggesting adequate pollination service (M.
Riveros, unpublished). Although no quantitative data have been collected, recruitment seems to be generally low in the high alpine. For the northern deserts, only
information on long-lived cacti are available (Gulmon *et al.*, 1979) and here seedling
mortality is extremely high. If climatic deterioration leads to ecologically significant
decreases in rainfall in Chile, recruitment is likely to be a second major constraint in
much of the Chilean flora for coping with global change.

V. Synthesis and Final Comments

Global change models suggest that Chile will experience a slower rate of warming
than the Northern Hemisphere. The accompanying rainfall pattern changes are unclear. When life history traits are looked at independently, it can be seen that there
are many strong constraints in the evolutionarily conservative flora of Chile for
responding to global change by *in situ* adaptation or by direct migration. Least well
equipped to respond to the projected changes are the coastal deserts, whose species
have quite high precipitation requirements for seed germination and where there are
many narrowly distributed endemics. Sclerophyllous forest species would also be
negatively affected because of strong emphasis on outbreeding and poor recruitment
possibilities, as would high alpine vegetation with poor dispersibility and much
obligate outcrossing.

Coastal desert and sclerophyllous forest native annuals would also face the threat
of being unable to compete with introduced annuals. The southern forest species
show fewer life history constraints.

This picture becomes modified when the distributional features of Chile's major
vegetation types are taken into account. That vegetation must move as a block to
respond to environmental change is a false notion deriving from limitations of the
palynologic method for detecting underrepresented species (see Birks, 1989). Outlying individuals probably have always played an important role in vegetation response to climate change except when totally devegetated surfaces undergo recol-
onization. Vegetation types, distributed in such a way as to contain many outlying
populations of the dominant species, could be considered as "resilient" to global
change. Considerable resilience to global change is visualized for the lowland woody

Chilean vegetation with poor latitudinal and altitudinal differentiation. Such resilience breaks down in the northern deserts, where many plant species have highly local ranges and exist in widely separated populations. Low resilience is also expected in areas of the high alpine where vegetation belts tend to show strong altitudinal differentiation. Paradoxically, the same factor that ultimately determines many of the strong life history constraints in the Chilean flora (an evolutionary history in an equable climate) is also responsible for the poor differentiation of vegetation types (Axelrod *et al.*, 1991) and hence its proposed resilience. The relative role of direct migration versus the *in situ* flourishing of outlying populations of dominant species is a critical issue for the vegetation responses to global change. We predict that outliers will play a more important role with rapid climatic change than they would with gradual climatic change.

Several features of Chile's topography could further offset the negative effects accruing from life history constraints. Steep mountainous topography as seen in Chile (about twice as steep in central Chile as in California) and strong slope aspect effects at some latitudes can produce local microclimate contrasts of a magnitude equivalent to many years of temperature increase under rapid climatic change. Such local microclimatic differences result in a rich array of local temperature refugia into which plants may migrate. Moreover, steep topography cuts the migration distance necessary to track a particular temperature condition. Plants with good mobility could use steep altitudinal gradients to great advantage. Species with sweepstakes dispersal mechanisms could be saved if their seed were widely enough distributed over the local microclimatic landscape.

If, moreover, in Chile's evolutionarily conservative flora with a high component of taxa with neotropical affinities there is hidden adaptational lag, potential positive *in situ* effects could accrue, further reducing the need for large-scale physical migration. The notion of adaptational lag, or the tendency for long-lived species in particular to be out of phase with their present environmental conditions, must be tested with solid ecophysiological data. A fascinating study, and one that would bring a new dimension to ecophysiology and pose novel theoretical questions, would be to compare genera of austral and neotropical origin in Chilean forests and in the Chilean Andes for tolerances in key ecophysiological traits.

Within Chile the temperate rain forests should be least affected by global change. This follows because greater climatic stability is projected for the higher latitudes where they occur, in such forests there is strong emphasis on animal dispersal, recruitment in general is good, and there is less emphasis on obligate outcrossing among the woody species than in other vegetation types. They could provide a useful control for assessing global change impacts within other regions of Chile and in western North America. The scientific value of studying temperate rain forests in Chile is thus outstanding. Conservation and protection should be encouraged with the development of a system of minimal-sized protected areas maintaining species over a large portion of their geographic range so that climatic change and species responses can be followed.

The dynamics of vegetation response to global change for temperate floras in equable versus continental climates is an important study area. Migrational distances required to track a given temperature condition might vary because of differences in the latitudinal and altitudinal thermal gradients. Vegetation persisting under a continental climate, where zonation is more marked, is likely to have less inherent resilience to global change than that growing in equable coastal regions and area of noncontinental climate. Interesting differences between regions like Chile and western North America are likely to have profound effects on the composition and life history properties of vegetation emerging under global change. Comparative analyses of vegetation structure, geographic ranges, and the temperature and precipitation tolerances of plant species in the two geographic areas are badly needed at this stage in order to assess comparative responses to global change.

Acknowledgments

Work supported largely by FONDECYT grant 88-1177 (M.T.K.A.) and FONDECYT 88-060 (J.J.A.). Latin American Plant Sciences grant 89-BINAC-4 (M.T.K.A.) and 90-Binac-4 (JJA) supported work in the Andes and coastal desert, respectively. Ongoing work on endemism in the Chilean flora is supported by the John D. and Catherine T. MacArthur Foundation and USAID-Biodiversity Support Program (M.T.K.A.). Thanks are extended to Peter H. Raven for providing literature and Cecilia Niemeyer for drawing the figures.

References

Armesto, J. J., and Figueroa, J. (1987). Stand structure and dynamics in the temperate rainforests of Chiloé archipelago, Chile. *J. Biogeogr.* **14,** 367–376.

Armesto, J. J., and Fuentes, E. R. (1988). Regeneration of main canopy and subcanopy trees in tree-fall gaps in a mid-elevation temperate rainforest in southern Chile. *Vegetatio* **74,** 151–159.

Armesto, J. J., and Martínez, J. A. (1978). Relations between vegetation structure and slope aspect in the mediterranean region of Chile. *J. Ecol.* **66,** 881–889.

Armesto, J. J., and Rozzi, R. (1989). Seed dispersal in the rainforest of Chiloé: evidence for the importance of biotic dispersal in a temperate forest. *J. Biogeogr.* **16,** 219–226.

Armesto, J. J., Rozzi, R., Miranda, P., and Sabag, C. (1987). Plant–frugivore interaction in South American temperate forests. *Rev. Chil. Hist. Nat.* **60,** 321–336.

Arroyo, M. T. K., and Squeo, F. A. (1990). Relationship between plant breeding systems and pollination. *In* "Biological Approaches and Evolutionary Trends in Plants" (S. Kawano, ed.), pp. 205–227. Academic Press, San Diego.

Arroyo, M. T. K., and Uslar, P. (1992). Breeding systems in a temperate mediterranean-type climate montane sclerophyllous forest in central Chile. *Bot. J. Linn. Soc.* **110.**

Arroyo, M. T. K., Armesto, J. J., and Villagrán, C. (1981). Plant phenological patterns in the high Cordillera de los Andes in central Chile. *J. Ecol.* **69,** 205–223.

Arroyo, M. T. K., Primack, R., and Primack, R. (1982a). Community studies in pollination ecology in the high temperate Andes of central Chile. I. Pollination mechanisms and altitudinal variation. *Am. J. Bot.* **69,** 82–97.

Arroyo, M. T. K., Villagrán, C., Marticorena, C., and Armesto, J. J. (1982b). Flora y relaciones biogeográficas en una transecta altitudinal en los Andes del norte de Chile (18°–19°S). *In* "El Ambiente Natural y las Poblaciones Humanas de los Andes del Norte Grande de Chile (Arica, Lat. 18°28'S)" (A. Veloso and E. Bustos, eds.), Vol. I, pp. 71–92. UNESCO, Muntevideo.

Arroyo, M. T. K., Armesto, J. J., and Primack, R. B. (1983). Tendencias altitudinales y latitudinales en mecanismos de polinización en los Andes templados de Sudámerica. *Rev. Chil. Hist. Nat.* **56,** 159–180.

Arroyo, M. T. K., Armesto, J. J., and Primack, R. B. (1985). Community studies in pollination ecology in the high temperate Andes of central Chile. II. Effect of temperature and visitation rates on pollination possibilities. *Plant Syst. Evol.* **49**, 187–203.

Arroyo, M. T. K., Squeo, F. A., and Lanfranco, D. (1987). Polinización biótica en los Andes de Chile: Avances hacía una síntesis. *In* "Ecologia de la Reproducción e Interacciones Planta/Animal" (E. Forero, F. Sarmiento, and C. La Rotta, eds.). *An. IV Congr. Latinoam. Bot.* **2**, 55–76.

Arroyo, M. T. K., Squeo, F. A., Armesto, J. J., and Villagrán, C. (1988). Effects of aridity on plant diversity in the northern Chilean Andes. *Ann. Missouri Bot. Gard.* **75**, 55–78.

Arroyo, M. T. K., Marticorena, C., Landero, A., Matthei, O., Miranda, P., and Squeo, F. A. (1989). Contribución to the high elevation flora of the Chilean Patagonian: a checklist of species on mountains on an east-west transect at latitude 50°S. *Gayana (Bot.)* **46**, 121–151.

Arroyo, M. T. K., Marticorena, C., and Muñoz, M. (1990a). A checklist of the native annual flora of continental Chile. *Gayana Bot.* **47**(3–4), 119–135.

Arroyo, M. T. K., Rozzi, R., Squeo, F. A., and Belmonte, E. (1990b). Pollination in tropical and temperate high elevation ecosystems: hypotheses and the Asteraceae as a test case. *In* "Mount Kenya Area: Differentiation and Dynamics of a Tropical Mountain Ecosystem" (M. Winiger, U. Wiesmann, and J. R. Rheker, eds.). Geograph. Bernensia, African Study Series, Vol. A8, pp. 21–31.

Arroyo, M. T. K., Ziegler, H., and Medina, E. (1990c). Distribution and delta 13 values of Portulacaceae species of the high Andes in northern Chile. *Bot. Acta Germany* **103**, 291–295.

Axelrod, D. (1978). History of the coniferous forests, California and Nevada. *Univ. Calif. Publ. Bot.* **70**, 1–62.

Axelrod, D. I. (1979). Desert vegetation, its age and origin. *In* "Arid Land Plant Resources" (J. R. Goodin and D. K. Northington, eds.), Int. Center for Arid and Semi-arid Land Studies, Texas Tech. Univ., Lubbock.

Axelrod, D. I., Arroyo, M. T. K., and Raven, P. H. (1991). Historical development of temperate vegetation in the Americas. *Rev. Chil. Hist. Nat.* **64**, 413–446.

Bazzaz, F. A. (1990). The response of natural ecosystems to the rising CO_2 level. *Annu. Rev. Ecol. Syst.* **21**, 167–196.

Belmonte, E., and Moscoso, D. (1985). Patrones fenológicos de 81 especies de precordillera y altiplano de la I Región, 18°–19°S, Chile. *Gema* **2**, 46–72.

Berger, W. H., and Labeyrie, L. D., eds. (1987). "Abrupt Climatic Change. Evidence and Implications." Reidel, Dordrecht.

Birks, H. J. B. (1989). Holocene isochrone maps and patterns of tree-spreading in the British Isles. *J. Biogeogr.* **16**, 503–540.

Boninsegna, J. A., and Holmes, R. L. (1985). *Fitzroya cupressoides* yields 1534-year long South American record. *Tree-Ring Bull.* **45**, 37–42.

Casassa, I. (1985). Estudio demográfico y florístico de los bosques de *Nothofagus obliqua* (Mirb.) Oerst. en Chile central. M.Sc. thesis, Univ. de Chile, Santiago.

Castri, F. di, and Hajek, E. R. (1975). "Bioclimatología de Chile." Dirección de Investigación, Univ. Católica de Chile.

Castri, F. di, and Hajek, E. R. (1976). "Bioclimatología de Chile." Editorial Univ. Católica Chile, Santiago.

Cody, M. L., Fuentes, E. R., Glanz, W., Hunt, K. H., and Moldenke, A. R. (1977). Convergent evolution in the consumer organisms of mediterranean Chile and California. *In* "Convergent Evolution in Chile and California: Mediterranean Climate Ecosystems" (H. Mooney, ed.), pp. 85–143. Dowden, Hutchinson & Ross, Stroudsburg, Pennsylvania.

Del Pozo, A. H., Fuentes, E., Hajek, E., and Molina, J. D. (1989). Microclima y machones de vegetation. *Rev. Chil. Hist. Nat.* **62**, 85–94.

Dollenz, O. (1982). Estudios fitosociológicos en las reservas forestales Alacalufes e Isla Riesco. *Anal. Inst. Patag. (Chile)* **13**, 145–152.

Donoso, C. (1979). Genecological differentiation in *Nothofagus obliqua* (Mirb.) Oerst. in Chile. *For. Ecol. Manage.* **2**, 53–66.

Donoso, C. (1982). Reseña ecológica de los bosques mediterráneos de Chile. *Bosque* **4**, 117–146.

Fuentes, E. R., Avilés, R., and Segura, A. (1990). The natural vegetation of a heavily man-transformed landscape: the savanna of central Chile. *Interciencia* **15**(5), 293–295.

Gulmon, S. L., Rundel, P. W., Ehleringer, J. R., and Mooney, H. A. (1979). Spatial relationships and competition in a Chilean desert cactus. *Oecologia* **44**, 40–43.

Gutiérrez, J. R., Aguilera, L. E., and Moreno, R. J. (1987). Intraspecific competition in *Atriplex. Rev. Chil. Hist. Nat.* **60**, 63–69.

Hamrick, J. M., Linhart, Y. B., and Mitton, J. B. (1979). Relationships between life-history characteristics and electrophoretically detectable genetic variation in plants. *Annu. Rev. Plant Syst.* **10**, 173–200.

Heywood, V. H. (1989). Patterns, extents and modes of invasion by terrestrial plants. *In* "Biological Invasions: A Global Perspective" (J. A. Drake, H. A. Mooney, F. di Castri, R. H. Groves, F. J. Kruger, M. Rejadnek, and M. Williamson, eds.), pp. 31–60. Wiley, New York.

Hoffmann, A. J., Teillier, S., and Fuentes, E. R. (1989). Fruit and seed characteristics of woody species in mediterranean-type regions of Chile and California. *Rev. Chil. Hist. Nat.* **62**, 43–60.

Johnston, I. M. (1929). Papers on the flora of northern Chile I. *Contrib. Gray Herbar. Harv. Univ.* **85**, 1–138.

Kochmer, J. P., and Handel, S. N. (1986). Constraints and competition in the evolution of flowering phenology. *Ecol. Monogr.* **56**, 303–325.

LaMarche, V. C., Jr., Graybill, D. A., Fritts, H. C., and Rose, M. R. (1984). Paleoclimatic inferences from long tree-ring records. *Science* **225**, 1019–1021.

Lewis, H. (1966). Speciation in flowering plants. *Science* **152**, 167–172.

Marticorena, C. (1990). Contribución a la estadística de la flora vascular de Chile. *Gayana (Bot.)* **47**(3–4), 85–113.

Miller, A. (1976). The climate of Chile. *In* "Climates of Central and South America" (W. Schwerdtfeger, ed.), pp. 113–145. Elsevier, Amsterdam.

Mooney, H. A., Kummerow, J., Johnson, A. W., Parson, D. J., Keeley, S., Hoffmann, A., Hays, R. I., Giliberto, J., and Chu, C. (1977). The producers—their resources and adaptive responses. *In* "Convergent Evolution in Chile and California: Mediterranean Climate Ecosystems" (H. Mooney, ed.), pp. 85–143. Dowden, Hutchinson & Ross, Stroudsburg, Pennsylvania.

Mooney, H. A., Drake, B. G., Luxmoore, R. J., Oechel, W., and Pitelka, W. C. (1991). Predicting ecosystem responses to elevated CO_2 levels. *Bioscience* **41**, 96–104.

Moore, D. M. (1983). "Flora de Tierra del Fuego." Anthony Nelson, England.

Oberdörfer, E. (1960). "Pflanzensoziologishe Studien in Chile—Ein Vergleich mit Europe." J. Cramer, Weinheim.

Peralta, I., Rodríguez, J., and Arroyo, M. T. K. (1992). Breeding system and aspects of pollination in *Acacia caven* (Mol.) Mol. (Leguminosae: Mimosoideae) in the mediterranean-type climate zone of central Chile. *Bot. Jahbr. Syst.* **114**(3), 297–314.

Pérez, C., and Villagrán, C. (1985). Distribución de abundancias de especies de boques relictos de la zona mediterránea de Chile. *Rev. Chil. Hist. Nat.* **58**, 157–170.

Pisano, E. (1977). Fitogeográfia de Fuego-Patagonia. I. Comunidades vegetales entre los latitudes 52° y 56°S. *Anal. Inst. Patag. (Chile)* **8**, 121–250.

Pisano, E. (1974). Estudio ecológico de la regional continental sur del área andino-patagónica. II. Contribución a la fitogeográfia de la zona del Parque Nacional "Torres del Paine." *Anal. Inst. Patag. (Chile)* **5**, 59–104.

Pisano, E. (1980). Distribución y características de la vegetación del archipiélago del Cabo de Hornos. *Anal. Inst. Patag. (Chile)* **11**, 191–224.

Pisano, E. (1981). Bosquejo fitogeográfico de Fuego-Patagonia. *Anal. Inst. Patag. (Chile)* **12**, 159–171.

Ramírez, C., Ferriere, F., and Figueroa, H. (1983). Bosques patanosos templados del sur de Chile. *Rev. Chil. Hist. Nat.* **56**, 11–26.

Raven, P. H., and Axelrod, D. I. (1978). Origin and relationships of the California flora. *Univ. Calif. Publ. Bot.* **72**, 1–134.

Richerson, P. J., and Lum, K. (1980). Patterns of plant species diversity in California: relation to weather and topography. *Am. Nat.* **116**, 504–536.

Ridley, H. N. (1930). "The Dispersal of Plants throughout the World." L. Reeve & Company, Ashford, Kent.

Rodríguez, R., Matthei, O., and Quezada, M. (1983). "Flora Arbórea de Chile." Editorial Univ. Concepción, Concepcion.

Rozzi, R. (1990). Períodos de floración y especies de polinizadores en poblaciones de *Anarthrophyllum cumingii* y *Chuguiraga oppositifolia* que crecen sobre laderas de exposición norte y sur. M.Sc. thesis, Univ. de Chile, Santiago.

Rozzi, R., Molina, J. D., and Miranda, P. (1989). Microclima y períodos de floración en laderas de exposición ecuatorial y polar de los Andes de Chile central. *Rev. Chil. Hist. Nat.* **62,** 75–84.

San Martin, J., Troncoso, A., and Ramírez, C. (1987). Fitosociología de los bosques de *Nothofagus antarctica* (Forst.) en la cordillera costera de Cauquenes (Chile). *Bosque (Chile)* **7,** 65–78.

Schmithüsen, J. (1956). Die raumliche Ordung der Chilenischen Vegetation. *Bonn. Geogr. Abh.* **17,** 1–86.

Sloan, L. C., and Barron, E. J. (1990). "Equable" climates during earth history? *Geology* **18,** 489–492.

Stouffer, R. J., Manabe, S., and Bryan, K. (1989). Interhemispheric asymmetry in climate response to a gradual increase in CO_2. *Nature* **342,** 660–662.

Thorne, R. F. (1989). New Caledonia. *In* "Floristic Inventory of Tropical Countries" (D. G. Campbell and H. D. Hammond, eds.), pp. 178–179. New York Botanical Garden, New York.

Travis, J. (1989). The role of optimizing selection in natural populations. *Annu. Rev. Ecol. Syst.* **20,** 279–296.

Uslar, P. (1982). Sistemas de reproducción de plantas: zona ecotonal entre la zona andina y el matorral esclerófilo de Chile central. Licenciatura thesis, Univ. de Chile, Santiago.

Veblen, T. T., Schlegel, F. M., and Oltremari, J. V. (1983). Temperate broad-leaved evergreen forests of South America. *In* "Ecosystems of the World" (J. D. Ovington, ed.), pp. 5–31. Elsevier, Amsterdam.

Vidiella, P. E., and Armesto, J. J. (1989). Emergence of ephemeral plant species from the north-central Chilean desert in response to experimental irrigation. *Rev. Chil. Hist. Nat.* **62,** 99–107.

Villagrán, C. (1988). Expansion of magellanic moorlands during the last glaciation: palynological evidence from northern Isla grande de Chiloé. *Quat. Res.* **30,** 304–314.

Villagrán, C., and Armesto, J. (1980). Relaciones florísticas entre las comunidades relictuales del Norte Chico y la zona central con el boque del sur de Chile. *Bol. Mus. Hist. Nat. Chile* **3,** 87–101.

Weber, T. F. (1951). Tendencia en las lluvias en la Argentina en lo que va del siglo. *Idia* **48,** 6–12.

Zamora, E., and Santana, A. (1979). Características climáticas de la costa occidental de la Patagonia entre las latitudes 46° 40'S y 56° 3'S. *Anal. Inst. Patag. (Chile)* **10,** 109–144.

C H A P T E R   18

North–South Comparisons: Vegetation

HAROLD A. MOONEY

The marked similarity of the vegetation of the coastal areas of Chile and the west coast of North America has long been noted. The array of ecosystems going from low to high latitudes of desert scrub, mediterranean scrub, to evergreen forests is exactly the same in general character. There are important differences, though, related presumably to dissimilarities in the climate equability and most certainly to differences in land use history. Within the mediterranean climatic region of Chile there appears to be less turnover of species with locality along either latitudinal or elevational gradients.

The central zone of Chile has had a long history of intensive land use, more so than the comparable climatic region in the Northern Hemisphere. This in part explains the dissimilar natures of the vegetations today. However, both the north and south have been invaded by large numbers of weedy species with origins in Eurasia, in fact, the very same ones. In a sense, then, the floras have converged in historical times at the specific level.

Both regions have a large number of endemics that will be threatened by human use of the landscapes, as well as by changing climate. Here, though, there will be considerable differences in the future. As noted in other chapters, the rate of climate change will be different between regions with the rate of change being faster in the north. Also, the rate of population change is greater in the north and consequently a much greater land use change is expected.

The mediterranean climate areas of both the Northern and Southern Hemispheres have high atmospheric oxidant loads. The effects of these toxicants are being seen increasingly in natural communities. Again, the vegetation of the north, with its greater population and industrialization, will be affected to a greater extent in the short run.

The convergent vegetations of the west coasts of the Northern and Southern Hemispheres have served as a great natural laboratory for the study of the evolutionary process. They will now serve as tests of the relative effects of the various components of global change on species distributions and interactions, since the rates of change of many of these components are expected to differ between continents.

Part VII: Animals

C H A P T E R 19

Assessing the Effects of Global Change on Animals in Western North America

JAMES H. BROWN

I. Introduction

In just the past 10,000 years, since the origin of agriculture, the human population has grown exponentially. This enormous increase, from a few million to over 5 billion, coupled with rapidly developing technology, is changing the globe. Much of the earth's surface has been converted to agriculture or transformed by pastoral, timber harvesting, water diversion, and mineral extraction activities. The waste products of human civilization are altering the water, air, and soil. The increased concentration of carbon dioxide caused by burning of fossil fuels threatens to change global climate.

All of these changes are having severe effects on the earth's biota. Many species have become extinct within the past few centuries, and many more are in imminent danger of disappearing. A few have benefitted from human-caused changes and dramatically increased their numbers and expanded their geographic ranges. There is good reason to believe that the full impact of existing conditions on the abundance, distribution, and diversity of organisms will not be realized for decades and even centuries. In the meantime, however, the pace of global change accelerates. It is critical to develop the ability to predict the impact of these changes on the biota.

The present chapter is a preliminary assessment of some effects of global change on the animals inhabiting western North America. The first part is very general. It

267

applies some basic principles of ecology and biogeography to the special features of the fauna and geography of western North America and uses this as a basis to infer effects of various kinds of environmental change. I deal primarily with the organisms and the region with which I am most familiar: the vertebrates inhabiting the temperate arid and montane zone between the Tropic of Cancer in the south and the mesic coniferous forests in the north, the Pacific Ocean in the west, and the Great Plains in the east. I avoid focusing exclusively on climate change. For one thing, the kinds and magnitude of predicted climate change are still controversial. In addition, other kinds of human-caused changes, such as water diversion and shifting patterns of human settlement and land use, have had major impacts on the animals of western North America in the past and are likely to continue to do so in the future.

The second part of the chapter considers the implications of a recent study by McDonald and myself (1992) which uses presently available data on the geographic distributions of one group of vertebrates (in this case montane mammals) to model the effects of one specific scenario of global change (in this case a 3°C climatic warming). This modeling exercise shows that it is possible to make relatively precise, quantitative predictions of biotic responses to abiotic change. The number of extinctions of populations predicted by the model is sobering.

II. The Organisms

Taken together, the vertebrates of the arid and mountainous habitats of western North America provide a good model system for assessing the impact of global and regional environmental change. There are several reasons for this:

1. The vertebrates include substantial taxonomic and biological diversity. The five major groups, fish, amphibians, reptiles, birds, and mammals, have unique characteristics that affect their capacities to respond to environmental change, and within each of these classes there are scores to hundreds of species, each of which has its unique set of biological attributes and ecological relationships.

2. The vertebrates occur in all of the major habitats, from the most arid deserts to the mesic mountaintops, including both terrestrial and freshwater environments.

3. The vertebrates exhibit enormous variation in their capacity to respond to environmental change, because of the diversity of their ecological requirements, life histories, and dispersal mechanisms.

4. The geographic distribution and ecology of the vertebrates of western North America have been relatively well studied. More is probably known about the factors affecting the abundance, distribution, and diversity of vertebrates than any group of organisms inhabiting this region.

5. Paleontologic and paleoecologic studies offer valuable insights into the effects of past environmental changes on the vertebrate fauna of western North America. Extinctions of populations and shifts in geographic distributions that accompanied the climatic cycles of the Pleistocene and the colonization of the region, first by aboriginal and then by European humans, are well documented. This historical

information provides some basis for predicting the effects of future environmental changes of comparable magnitude.

Rather than going into further detail here, I shall wait and use different kinds of vertebrates to illustrate the variety of responses that different elements of the fauna can be expected to exhibit to the same or different kinds of environmental change.

By focusing on vertebrates, the animals that are conspicuously ignored are the arthropods, especially the insects. Many of the points I will make with specific reference to vertebrates are probably also relevant to arthropods, because many of them exhibit similar patterns of abundance, distribution, and diversity and comparable (or even greater) variability in ecological requirements, life history traits, and dispersal capabilities. I will, however, mention one special feature of the life history of many insects that may make them particularly sensitive to certain kinds of environmental changes.

III. The Geographic Setting

Ultimately, the abundance, distribution, and diversity of organisms depend on the abiotic environment, especially on its spatial and temporal variation. The geology and climate of the earth's surface set the stage on which the play of life is acted out by the different kinds of organisms. Knowledge of the main features of this physical template at present and in the past is critical to any attempt to understand the composition of the present biota (e.g., MacArthur, 1972; Brown and Gibson, 1983), interpret the impact of past events, and predict the future response to human-caused environmental changes.

A little geography goes a long way in characterizing the predominant physical environmental features of western North America:

1. There is a latitudinal gradient in both the intensity and the annual variation of solar radiation and hence in both the average temperature and seasonality of the temperature regime. It gets colder and more seasonal as one goes north.

2. The land surface is dominated by multiple north–south ranges of mountains with intervening lowland valleys (Fig. 1).

3. The climate is dominated by westerly winds blowing inland after passing over the cold current that flows down the Pacific coast. These winds are initially cool, and they tend to be warmed when they pass over lowlands but cooled again each time they pass over mountains. Because air cools adiabatically as it rises and its capacity to hold water vapor varies directly with temperature, the lowlands are warm and arid while the mountaintops are cool and mesic. The wettest habitats typically occur on the western slopes of the mountains and the driest deserts occur in "rain shadows" in low valleys on the east sides of the mountains.

4. Because of the high topographic relief, much of the water that falls at higher elevations runs off and is collected in streams and rivers. These provide corridors of freshwater and adjacent mesic terrestrial habitats that run from the mountains through the arid lowlands to the ocean. An exception is the Great Basin, a vast area between

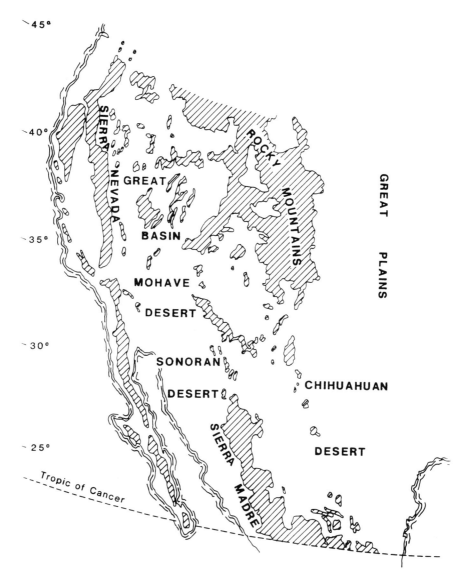

Fig. 1. Map of the part of western North America considered in this chapter, showing orientation of the major mountain ranges, lowland areas, and rivers.

the Sierra Nevada and Rocky Mountains, where there is no drainage to the sea; instead, the streams empty into shallow basins in the valleys, and the water evaporates, creating saline lakes or dry salt flats.

5. The climatic cycles of the Pleistocene—cool and wet during glacial periods and warm and dry during interglacials—profoundly changed the physical setting in the past (e.g., Wells, 1983; Van Devender *et al.*, 1987; Betancourt *et al.*, 1990). The present climatic regime has existed only for the past 10,000 years, since the beginning of the Holocene, and even during this period there has been considerable variation. Aquatic habitats were much more extensive in the past; the inland basins were filled with large freshwater lakes during the glacial periods, but evaporation during the interglacial periods caused them to become saline and often to dry up completely.

6. The influence of human activities in this region began with the colonization by aboriginal people about 20,000 years ago and increased with the increasing settlement of Europeans in the past few centuries. Many human impacts, such as desertification, deforestation, erosion, water diversion for irrigation, and soil salinization, were begun by Native Americans, but these have increased and new ones have been initiated with the increasing population and technology in the past two centuries.

The foregoing is an oversimplified description of the present and past physical setting of the biota of western North America. It captures only the major features of the variation and ignores many of the interesting details. The latter can be enormously important for certain organisms. A little feeling for this complexity can be given by a few examples:

1. In the southern part of the region, much of the climatic pattern is dictated by the seasonality of precipitation, which is concentrated in the winter in the west and in the summer in the east.

2. Two habitat types with distinctive biotas that have endemic species and genera of vertebrates are sand dunes and freshwater springs. These are distributed as widely dispersed islands, their locations dictated by specific features of geology and climate.

3. Although the mountain ranges nearly all run north–south (an exception is the Uinta Mountains), none extend continuously from the tropics to the boreal forest. In fact, many of the mountain ranges are effectively islands, with their cool, mesic, forested habitats at higher elevations isolated from other such habitats by the hot, arid desert and grassland habitats of the intervening valleys.

4. The responses of the vertebrate biota to Pleistocene and human-caused historical changes have often been complicated by specific local conditions that have caused the extinction of some populations and either inhibited or enhanced the dispersal of others.

Present and past relationships between the actors and the stage, the organisms and the physical setting, provide a basis for predicting responses to environmental changes in the future. I will develop a framework for making such predictions in two ways.

IV. General Responses to Future Environmental Change

First, I will make some very general points that should have wide applicability. Although these may seem obvious to many readers, they must be a starting point in any effort to apply basic ecological and biogeographic principles to predict the future consequences of human-caused environmental changes.

A. Because the region is predominantly arid, changes in the input and redistribution of water have major effects.

In talking about global environmental change there has been a tendency to equate global change with climate change and climate change with global warming caused by elevated concentrations of carbon dioxide and other greenhouse gasses. Although an increase in average environmental temperature would certainly have substantial impacts on the vertebrate animals of western North America, such a change is not likely to occur without concomitant change in the precipitation regime. Unfortunately, the state of the art in climate modeling is such that it is much more difficult to predict precipitation trends than temperature patterns.

Yet the amount and timing of precipitation are at present the most important factors affecting the biota of western North America. Changes in precipitation regime are likely to cause profound changes in the abundance, distribution, and diversity of not only vertebrates but all kinds of organisms. Indeed, many of the effects of altered precipitation on terrestrial vertebrates would be indirect, caused more by changes in vegetation than by effects of water availability on the animals themselves.

Both climatic changes that alter the input, runoff, and infiltration of precipitation and other anthropogenic changes in the redistribution of water (such as those caused by diversion of streams, damming of rivers, and depletion of aquifers by pumping) should also have major impacts. Effects on aquatic and amphibious animals will typically be direct, whereas those on other organisms will often be indirect, mediated primarily through changes in vegetation.

B. The elevational relief and the north–south orientation of mountains and valleys enable many species to shift their distributions rather than suffer extinction.

Many terrestrial vertebrates presently have geographic ranges that are restricted to certain elevational zones, climatic regimes, and vegetation types, but nevertheless span 5° or more of latitude. Because the high mountains offer steep gradients of climate and vegetation, animals can simply shift their elevational distribution to remain in a similar climatic regime and vegetation type in response to modest climate change. Because the major mountain ranges extend for hundreds and sometimes even thousands of kilometers in a north–south direction, animals can also find similar climatic and habitat conditions over a wide range of latitude. Southern, xerophytic species range far to the north in the bottoms of valleys, and boreal, mesophytic species range far to the south at high elevations in the mountains. Vertebrate species have shifted their latitudinal and elevational ranges dramatically in response to

Pleistocene climatic changes. They would almost certainly respond similarly to regional climatic warming and other kinds of environmental change (Fig. 2).

C. Species with small and/or fragmented geographic ranges are particularly susceptible to any kind of environmental change.

This point is the complement of the previous one. Not all species have wide ranges, and not all have ranges that follow the north–south orientation of the major mountains and valleys. Many vertebrates are narrowly endemic to local regions and to specific, highly restricted habitat types. These species already have narrow requirements, and they will be especially sensitive to any changes in environmental conditions. Compared to species that range widely up and down the valleys and mountains, these species will often be unable to remain in suitable environments by simply shifting their geographic ranges. Usually human-caused changes will be detrimental, threatening further reductions in suitable habitat, decreases in population size, and increased probability of local and total extinction (Fig. 2).

Not all environmental changes need be detrimental to narrowly distributed, insular species. The fishes and some of the amphibians that currently inhabit the isolated desert springs are relics of once much more widely distributed populations that occurred in the much more extensive aquatic habitats during the glacial periods of the Pleistocene. Many extinctions have occurred in the past century as humans have diverted the water from many of the remaining springs and streams. Additional extinctions can be expected if human water use increases or if the climate becomes drier. On the other hand, a climatic change that substantially increased precipitation and the availability of freshwater habitats could conceivably rescue many desert fishes and some amphibians from the brink of extinction.

Other examples of vertebrates that are endemic to particular kinds of insular habitats include the several species in the lizard genus *Uma,* which are restricted to small patches of aeolian sand dunes in the warm, southern deserts.

The distributions of some species have recently been reduced and fragmented by human activities. These will be particularly sensitive to any additional environmental change. For example, certain large mammals, such as bighorn sheep, pronghorn antelope, wolf, and grizzly bear, are currently restricted to small fragments of their formerly extensive geographic ranges by hunting, agriculture, urbanization, and other human activities that prevent the maintenance of populations and the dispersal of individuals. Any environmental changes that would cause extinction of these isolated populations would result in further drastic contractions in the geographic ranges of these species.

D. The impact of any kind of environmental change differentially affects different kinds of vertebrates depending on their individualistic ecological requirements.

By definition, each species is distinct. This distinctiveness is reflected in its unique requirements for abiotic and biotic conditions. Work on vertebrates (e.g., Graham, 1986; Brown and Kurzius, 1987) has confirmed the generality of Gleason's observation that each species is distributed across the landscape in a highly individ-

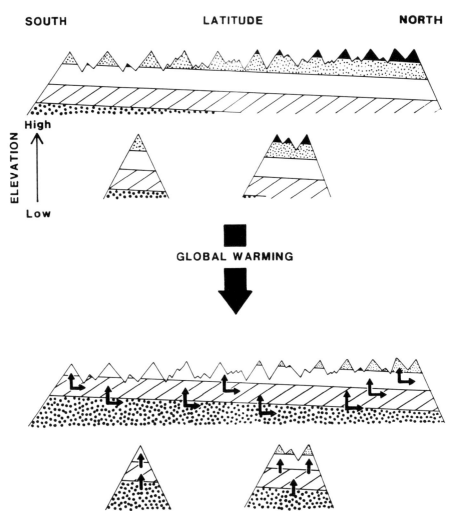

Fig. 2. Top: Diagram showing the present variation in elevational zonation of climate and vegetation with latitude in the Northern Hemisphere for a long continuous north–south mountain range and for two small, isolated ranges at different latitudes. Bottom: The northward and upward shifts in the zonation of climate and vegetation on the same mountains predicted to be caused by global warming. The small arrows show that species that occur along the extensive north–south cordilleras can respond similarly by dispersing northward and to higher elevations, but species endemic to the higher elevations on isolated mountains will be doomed to extinction unless they are capable of long-distance dispersal or rapid evolution.

ualistic fashion that reflects its ability to disperse into and/or maintain populations in sites that meet its specific requirements.

In many cases, this individualism makes it difficult to make sweeping gener-alizations about the effects of environmental change. Most human-caused environ-mental changes will inevitably have negative effects on the majority of species, because *any* change will tend to make it more difficult for any given species to meet its restrictive requirements. But some kinds of changes will also make the environ-ment more favorable for a minority of species, causing increases in their abundance and distribution (for some examples, see Brown and Gibson, 1983). Consequently, without much detailed information on ecological relationships, it will often be difficult to predict the direction let alone the magnitude and the specific effects of a particular anthropogenic change on any given species (but see the example of moun-taintop mammals later).

For example, human settlement of western North America has been concentrated along the permanent rivers. The general impact of concentrated human activities on the vertebrates of the aquatic and adjacent riparian forest habitats has been strongly negative; many native species have suffered great reductions in populations or been extirpated from large portions of their former ranges. However, a few species, such as the coyote, Canada goose, and house finch, have benefitted and increased their populations in the suburban and agricultural habitats.

Similarly, damming of rivers has contributed to the extinction or endangerment of many native fish species. On the other hand, the dams have created reservoir lakes. These represent new, human-created habitats that now support diverse communities of introduced, exotic fishes as well as thriving populations of some native vertebrates, such as waterfowl and beaver (for a general discussion of invasions by exotic vertebrates, see Brown, 1989).

E. Response to environmental change depends on life history characteristics.

Despite the inherent difficulty in predicting the response of individual species to any particular kind of environmental change, certain generalizations will hold. One is that organisms with certain kinds of life histories are more likely to be affected. Any feature of the life history that tends to cause reduced recruitment, slower population growth, and increased probability of local extinction will tend to increase sensitivity to most kinds of anthropogenic environmental change.

For example, long-lived species in which successful reproduction is episodic or juvenile survival is low tend to be very sensitive to small environmental perturbations that change the initially small recruitment rate to zero. An apparent example is the Colorado squawfish, *Ptychocheilus lucius,* a giant minnow endemic to the Colorado River and its major tributaries. Individuals live for many years, perhaps for several decades, and a few adults apparently still remain in the wild even though damming of the rivers, introduction of exotic species, and other human activities have appar-ently prevented any significant recent recruitment.

The apparent global decline of many amphibians (which also seems to be occurring in many regions of western North America) may reflect the complex life cycles of most amphibians rather than any specific environmental condition. The idea is that amphibians have to meet the specific requirements of both the aquatic larval and terrestrial adult stage of the life history. The reductions in population density and the extinction of local populations that appear to be occurring in many different species in many different habitats may indicate that human activities have altered some aquatic or terrestrial characteristic enough to affect negatively either the larval or the adult phase of the life cycle. Thus, in a sense, amphibians could be "miner's canaries," providing a clear warning signal of many different kinds of environmental degradation—changes which, if continued, would eventually affect many other kinds of organisms.

My colleague Fritz Taylor has pointed out that some arthropods may be particularly susceptible to global warming because their annual life histories are dependent on the relationship between temperature and photoperiod. Most temperate zone insect populations go through one (univoltine) or more (multivoltine) generations per year, and they have some life history stage capable of diapause during the winter when temperature conditions and food availability are unsuitable for being active. The critical cue for timing the life cycle, and especially for entering into and emerging from diapause, is photoperiod. Global warming threatens to uncouple the present relationship between temperature conditions suitable for activity and the photoperiod used to time diapause. Such changes must have occurred in the past, during the Pleistocene climatic fluctuations, but the rate of global temperature change was much less than that predicted to occur within the next century. Thus, although insects may have adapted genetically to the gradual climatic changes in the past, there may be insufficient time for at least some species to respond similarly to anthropogenic change.

F. Response to environmental change depends on dispersal capabilities.

Any kind of substantial, large-scale environmental change is likely to make some proportion of the present geographic range of a species uninhabitable, but it may also make other, presently uninhabited areas suitable for colonization. Thus, in the case of global warming, low-altitude and low-elevation sites may become too hot (either because of the direct effects of temperature or the indirect effects of temperature-mediated vegetation changes), but higher-latitude and upper-elevation sites that were previously too cold would presumably become suitable. The response of any species to availability of suitable conditions in new areas depends on its ability to colonize. The greater its capacity for movement, for temporary survival in intervening unfavorable environments, and for establishment of new populations, the more likely a species will be able to "track" an environmental change across the landscape.

Vertebrates exhibit an enormous spectrum of dispersal capabilities (Brown and Gibson, 1983) and thus can be expected to vary greatly in their response to large-scale environmental change. At one extreme are the birds, able to fly over large distances of unfavorable habitat and to breed and establish populations rapidly in new areas. As indicated by their frequency of occurrence on oceanic islands and their ability to occur in many habitat patches that are too small to support sustained populations, birds tend to be excellent colonists. At the other extreme are the fishes, most with no capability to move over land or to survive out of water for more than a few minutes, so that they require aquatic connections between habitats in order to disperse. The dispersal capabilities of mammals, reptiles, and amphibians are generally inter-mediate between those of birds and fishes.

The consequences of these differences in dispersal are dramatically different susceptibilities to most kinds of anthropogenic change. This is seen in the response of different kinds of vertebrates to recent environmental changes. Although several bird species have suffered severe contraction in geographic range during the past century, only one species, the California condor, has become extinct in western North America, and a number of native species (e.g., Canada goose, Anna's hummingbird, crow, mockingbird) have expanded their ranges to take advantage of favorable conditions created by human activities. In contrast, a large proportion of the native fish species of western North America have become extinct within the last century, many more are endangered, and nearly all have suffered substantial reductions in their geographic ranges (e.g., Moyle and Williams, 1990). Only a handful of native western North American fish species, most notably the rainbow trout, have succes-sfully colonized new areas, and their dispersal has almost invariably been aided by human transport.

G. Responses to past environmental changes offer valuable clues.

The historical changes in the physical setting had major effects on the biota. Species expanded and contracted their geographic ranges in response to the climatic cycles of the Pleistocene: boreal and mesic forms shifted southward and to lower elevations during glacial periods and then returned during interglacials; lowland and arid species exhibited the opposite pattern of range shifts. The colonization of western North America by humans, first by aboriginals about 20,000 years ago and then by European settlers in the past few centuries, had major effects. Initially these were primarily through biotic interactions (i.e., the role of humans as hunters), but impact of modern (and even aboriginal) humans has altered the abiotic environment through soil erosion, pollution, water diversion, and mineral extraction.

These historical changes in the environment can be viewed as biological ex-periments. The response of organisms to past perturbations can suggest how they will respond to anthropogenic changes in the future. For example, the Pleistocene–Holocene transition about 10,000 years ago was a time of global and regional climatic warming. Although the warming was much slower than that predicted to occur in the coming decades as a consequences of elevated concentrations of carbon dioxide and other greenhouse gases in the atmosphere, the magnitude of change is comparable. The paleontologic and paleoecologic record for western North America shows the magnitude of the resulting changes in vegetation and shifts in geographic ranges of vertebrates (e.g., Wells, 1983; Graham, 1986; Van Devender et al., 1987; Betancourt et al., 1990). Another major perturbation occurred with the arrival of aboriginal humans, which appears to have played some substantial role in the extinction of the vast majority of large mammals and a number of birds and other vertebrates (e.g.,

Martin and Klein, 1984). The most recent set of changes, caused by European settlement, are still in progress, but, as noted earlier, there have already been many extinctions in the fishes and dramatic reductions in the ranges of large mammals.

Probably the most useful insights to be obtained from the historical record concern the differential susceptibility of different groups of organisms to different kinds of environmental change and the mechanisms responsible for these differential responses. For example, Pleistocene conditions (especially the physical geography and vegetation of the last glacial period) have left an enormous imprint on the distributions of small terrestrial mammals, reptiles, amphibians, and fishes but have had much less effect on the ranges of large terrestrial mammals, bats, and birds. The reason is simple: the former groups have such limited dispersal capability that they have been unable to cross biogeographic barriers of unfavorable conditions to colonize new areas that became suitable when the climate and vegetation changed; on the other hand, birds, bats, and large mammals readily dispersed across the same barriers and are presently found in most areas that are suitable to maintain populations.

V. Predicting the Effect of Environmental Change: An Example

The individualistic environmental relationships of different species will often make it difficult to make simple predictions about the effects of any given environmental change on the vertebrate fauna as a whole or on any good-sized taxonomic or ecological group of species. But there is reason for guarded optimism—there are ways to circumvent some of the problems.

Consider the following problem. Global warming, if it occurs to the extent predicted by current climate models, should have a major impact on the vertebrates that are restricted to isolated mountaintops in western North America. As the temperature increases, the climatic and vegetational zones will shift to higher elevations. The species restricted to boreal conditions will see their habitat shrink, until eventually both the habitats and the species that depend on them are "pushed off the top of the mountain," so to speak. It is relatively easy to make this simple, qualitative prediction, but is there any way to predict more precisely which species are most likely to become extinct on which mountain ranges?

McDonald and I (McDonald and Brown, 1992) have made such predictions for the small mammals restricted to the upper-elevation forest, meadow, and streamside habitats on the isolated mountains of the Great Basin (Fig. 3). Each mountain range supports a number of boreal mammal species, isolated from other mountaintops by their inability to disperse across the sagebrush desert in the intervening valleys. These are survivors of much more widely distributed populations more than 10,000 years ago, during the Pleistocene, when the climate was cooler and wetter, and they and their habitats were broadly distributed across the Great Basin at much lower elevations (Brown, 1971; Grayson, 1987).

McDonald and I developed a simple, empirically based model. Following Mur-

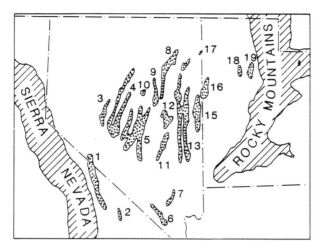

Fig. 3. Map of the Great Basin, showing the major isolated mountain ranges (identified by number in Table I) that rise from the desert valleys between the Sierra Nevada and Rocky Mountains. (Modified from Brown, 1971.)

phy and Weiss (1992; see also Peters, 1978), we assumed that the average temperature in the region will increase 3°C and that in response the lower border of woodland vegetation will be shifted upward 500 m from its present lower limit of approximately 2300 m. We used topographic maps to determine the present and expected future areas of boreal habitat on each mountain range. Since the occurrence of the small mammals on the mountain ranges is well documented, we used the present species–area relationship (Fig. 4) to estimate the area of boreal habitat required to support a given number of species and thus to predict the number of species that would be expected to become extinct on each mountain range when the habitat became reduced. Then, knowing the number of species predicted to become extinct, we were able to use the strong "nested subset" structure of the mountaintop mammal faunas (Table I; see also Patterson and Atmar, 1986; Patterson, 1987) to determine which particular species were most likely to be lost from each mountain range.

The results are sobering. The 19 isolated mountain ranges are predicted to lose between 9 and 62% of their present boreal small mammal species. Of the 14 species, 3 are predicted to become extinct on all the mountaintops where they presently occur, 2 are predicted to survive on all mountain ranges, and the remainder are expected to disappear from some mountain ranges but survive on others. These predictions should be taken with caution. They are only as accurate as the assumptions and logic of our model. Most important of these is that we assumed a particular scenario of climate change: 3°C increase in temperature and no significant change in the precipitation regime. However, if we wished to examine the effects of a different climatic regime, it would be a simple matter to make the appropriate changes in the model.

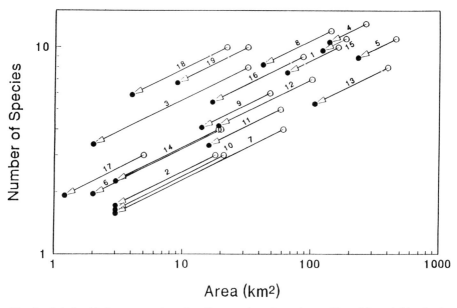

Fig. 4. Relationship between number of small mammal species and area of boreal forest habitat for the present distribution of small boreal mammals on Great Basin mountaintops (unshaded symbols) and after predicted extinctions of species owing to habitat reduction caused by climatic warming (shaded symbols). The arrows connect the two symbols for each mountain range (identified by number in Table I) and thus indicate the change in both the area of suitable habitat and the number of mammal species predicted to occur as a result of climatic change. (From McDonald and Brown, 1992.)

One other point of general interest emerges from this modeling exercise. We were able to use existing data on the distributions of the different mammal species among different mountain ranges to make some relatively precise predictions about which species were likely to become extinct of which mountaintops. It was not necessary to study the details of the population biology of each species in each locality. The predictions that we have made are probably at least as accurate as those that could be made from further field studies, unless an enormous amount of time and money were devoted to the project. This is not because the details of population biology, such as population sizes, habitat requirements, and genetic variation, do not affect the probability of extinction—they do. It is because the effects of these variables on extinction are reflected and integrated in the present distributions of species among mountaintops.

It should be noted that the extinctions predicted here are extinctions of local populations inhabiting individual isolated mountain ranges. All of these mammalian species also have populations inhabiting the Sierra Nevada to the west, the Rocky Mountains to the east, or both, and the species would be predicted to be able to survive the degree of global warming assumed here by shifting their ranges northward and to higher elevations on these much more extensive mountain ranges.

TABLE I.

Distribution of 14 Small Boreal Mammal Species among 19 Isolated Mountain Ranges in the Great Basin at Present (from Brown and Gibson, 1983) and after Predicted Extinctions due to Effects of Global Warming[a]

Species	Mountain Ranges																			Present number of ranges inhabited	Predicted number of ranges inhabited
	Toiyabe	Ruby	To-quima	White-Inyo	Snake	Oquirrh	Deep Creek	Schell	De-satoya	Stans-bury	White Pine	Spring	Grant	Dia-mond	Roberts Creek	Spruce	Sheep	Pana-mint	Pilot		
Eutamias umbrinus	X	X	X	X	X	X	X	X	X	X	X	X	X	X	X	X	X			17	17
Neotoma cinerea	X	X	X	X	X	X	X	X	X	X	X	X	X			X	X	X	X	17	17
Eutamias dorsalis	X		X	X	X	X	X	X	X	X	E	X	X	X		E	E	X	X	17	14
Spermophilus lateralis	X	X	X	X	X		X	X	E	X	X	X	E	E		E				14	10
Microtus longicaudus	X	X	X	X	X		X	X	E		X	E	E		X				E	13	10
Sylvilagus nuttalii	X	X	X	X	X	X	E	E	E	E		E						E		12	5
Marmota flaviventris	X	X	X	X	X	X	E	E	E	X	E									11	7
Sorex vagrans	X	X	X	X	X	X	E	E	E	X										10	7
Sorex palustris	X	X	X	X	E	X	E								E					8	5
Mustella erminea	X	E	E	X	E	E														6	1
Ochotona princeps	X	E	E	E										E						5	1
Zapus princeps	E	E				E					E				E					5	0
Spermophilus beldingi	E	E																		2	0
Lepus townsendii						E				E										2	0
Present number of species	13	12	11	11	10	10	9	8	8	8	7	6	5	4	4	4	3	3	3		
Predicted number of species	11	8	9	10	8	7	5	5	3	6	4	4	3	2	2	2	2	2	2		

[a]Species that were predicted to become extinct under the assumed scenario of climate and vegetation change are indicated by E, and species that were expected to persist are indicated by X.

However, our results have serious implications for the impact of global warming and certain other kinds of large-scale environmental change on the vertebrates and other organisms that are endemic to small, isolated habitats, including endangered species that are presently largely or entirely restricted to small biological reserves. Even relatively small changes in environmental conditions may be sufficient to make the present refuges of these species uninhabitable. Many species will be unable to disperse and colonize new habitats (even if they are available elsewhere) without human assistance. Thus, any kind of global, or even regional, environmental change threatens to undermine the effectiveness of present conservation strategies based on establishing a large number of relatively small reserves (see also Peters and Lovejoy, 1992).

Thus, one message of this study is pessimistic: many extinctions of isolated populations can be expected if projected levels of global warming occur. But another message is more optimistic: it may not be as daunting as it might seem at first to make simple models and use existing distributional data to assess the probable effect of certain kinds of anthropogenic environmental change on vertebrates and other organisms.

VI. Summary and Conclusions

To make a real effort to assess the possible or probable impacts of global environmental change on the animals, or even just on the vertebrates, of western North America would be a monumental task. The region is vast, the animals are diverse, and the possible effects of the various kinds of human-caused change are numerous.

In an effort to say something useful about this large, complex subject, I have tried to do two things. First, I have suggested some very general considerations that come from applying some fundamental ecological and biogeographic principles to the relationship between the physical geography of western North America and the biology of the vertebrate animals that occur there. Although these points may not be especially profound, they seem to provide a useful starting point for predicting the effects of different kinds of changes.

Then, to show how one might begin to go from the general to the specific and actually to predict the impact on a particular group of animals of a specific scenario of possible change, I have developed one concrete example. McDonald and I have developed a model to predict which species of small mammals are likely to go extinct on which Great Basin mountain ranges, assuming a particular degree of regional climatic warming.

Together, these two approaches suggest that animal ecologists and biogeographers have a great deal to contribute to understanding and mitigating the serious environmental changes caused by the growing human population. Just by applying data and concepts currently available, we can begin to develop a valuable, rigorous science of global ecology. Just as climatologists, oceanographers, and ecosystem ecologists are producing quantitative models to predict the physical patterns of global change, so should population and community ecologists and biogeographers begin to produce models to predict the biological consequences (see above and Pastor and Post, 1988; Peters and Lovejoy, in press). As in the case of the changes in the abiotic environment, however, the predicted impacts on the fauna are likely to be serious and difficult to avert.

Acknowledgments

This manuscript was prepared while I was on sabbatical leave at the CSIRO Centre for Arid Zone Research in Alice Springs, Australia. I thank G. Pickup, S. R. Morton, and the rest of the CAZR staff for their hospitality, F. Morton for the artwork, and A. Kodric-Brown and H. A. Mooney for helpful comments on the manuscript. One thing not available at CAZR was access to my references and to a well-stocked library; I apologize for the scant and rather biased citations. My sabbatical leave was supported by a John Simon Guggenheim Fellowship; my research on topics related to this chapter has been supported by the U.S. National Science Foundation, most recently by grants BSR-8718139 and BSR-8807792.

References

Betancourt, J. L., Van Devender, T. R., and Martin, P. S., eds. (1990). "Packrat Middens." Univ. of Arizona Press, Tucson.

Brown, J. H. (1971). Mammals on mountaintops: nonequilibrium insular biogeography. *Am. Nat.* **105,** 467–478.

Brown, J. H. (1989). Patterns, modes and extents of invasions by vertebrates. *In* "Biological Invasions: A Global Perspective" (J. A. Drake, H. A. Mooney, F. di Castri, R. H. Groves, F. J. Kruger, M. Remanek, and M. Williamsom, eds.), pp. 85–109. Wiley, Chichester, U.K.

Brown, J. H., and Gibson, A. C. (1983). "Biogeography." C. V. Mosby, St. Louis, Missouri.

Brown, J. H., and Kurzius, M. (1987). Composition of desert rodent faunas: combinations of coexisting species. *Ann. Zool. Fenn.* **24,** 227–237.

Grayson, D. K. (1987). The biogeographic history of small mammals in the Great Basin: observation on the last 20,000 years. *J. Mammal.* **68,** 359–375.

Graham, R. W. (1986). Response of mammalian communities to environmental changes during the Late Quaternary. "In "Community Ecology" (T. J. Case and J. Diamond, eds.), pp. 300–313. Harper & Row, New York.

MacArthur, R. H. (1972). "Geographical Ecology." Harper & Row, New York.

Martin, P. S., and Klein, R. G., eds. (1984). "Quaternary Extinctions." Univ. of Arizona Press, Tucson.

McDonald, K. A., and Brown, J. H. (1992). Using montane mammals to model extinctions due to global change. *Conserv. Biol.* **6,** 409–415.

Moyle, P. B., and Williams, J. E. (1990). Biodiversity loss in the temperate zone: decline of the native fishes of California. *Conserv. Biol.* **4,** 275–284.

Murphy, D. D., and Weiss, S. B. (1992). Effects of climate change on biological diversity in western North America: species losses and mechanisms. *In* "Global Warming and Biological Diversity" (R. L. Peters and T. E. Lovejoy, eds.), pp. 355–358. Yale Univ. Press, New Haven, Connecticut.

Pastor, J., and Post, W. M. (1988). Response of northern forests to CO$_2$-induced climate change. *Science* **234,** 55–58.

Patterson, B. D. (1987). The principle of nested subsets and its implications for biological conservation. *Conserv. Biol.* **1,** 323–335.

Patterson, B. D., and Atmar, W. (1986). Nested subsets and the structure of insular mammalian faunas and archipelagos. *Biol. J. Linn. Soc.* **28,** 65–82.

Peters, R. L. (1987). Effects of global warming on biological diversity: an overview. *In* "Preparing for Climate Change," pp. 169–184. Proceedings of the First North American Conference on Preparing for Climate Change: A Cooperative Approach. Washington, D.C.

Peters, R. L., and Lovejoy, T. E., eds. (1992). "Global Warming and Biological Diversity." Yale Univ. Press, New Haven, Connecticut.

Van Devender, T. R., Thompson, R. S., and Betancourt, J. L. (1987). Vegetation history of the deserts of southwestern North America: the nature and timing of the Late Wisconsin-Holocene transition. *In* "North America and Adjacent Oceans during the Last Glaciation" (W. F. Ruddiman and H. E. Wright, Jr., eds.), The Geology of North America, Vol. K-3, Geological Society of America, Boulder, Colorado.

Wells, P. V. (1983). Paleobiogeography of montane islands in the Great Basin since the last glaciopluvial. *Ecol. Monogr.* **53**, 341–382.

CHAPTER 20

Effect of Global Climatic Change on Terrestrial Mammals in Chile

LUIS C. CONTRERAS

I. Introduction

In the geological past, climate changes globally affecting the planet have been extensive. These changes have been caused by natural factors and have taken place over long periods of time at low rates. For instance, since the last glacial period global mean temperatures have varied <1°C in the past 7000–8000 years (see Ojima *et al.*, 1991). However, as a consequence of anthropogenic activities general models predict that global mean temperature may change from 2 to 5°C in the next 100 years. Global increase of temperature should bring changes in amounts and distribution of precipitation. These predictions have raised great concern about the effect that these global climatic changes may have on the ecosystems throughout the world.

Here I discuss the present relationship between the distribution of mammals and climate in Chile and the possible effects that a global climatic change may have on them.

It is well known that abiotic conditions are among the most important factors influencing the distribution and abundance of organisms, especially on large geographic scales. In the case of terrestrial environments, these factors are determined globally by climate, which at different temporal scales is not constant. Consequently, as a first approach, any consideration of the possible effect of global climatic change on terrestrial mammals should include correlations between climate and the distribution and abundances of mammals based on evidence from present and past conditions and an extrapolation of those relationships to future climatic scenarios. Second, this first approach is based on the idea that organisms in a community form unified assemblages, which we know does not occur in all cases. Furthermore, since species

285

differ in their biological characteristics, such as life history and dispersal capabilities, we should expect differential species responses to the same environmental change. Because of this, we should consider the response of single species to environmental changes. Third, we should take into account the fact that biotic factors may also affect the distribution of species.

II. Historical Setting of South American Mammals

The South American mammalian fauna has experienced dramatic changes through time because of its degree of connection to (or isolation from) other land masses, orogenic processes, or climatic changes. Briefly, after the late Cretaceous South America became more and more isolated until the Isthmus of Panama formed a land bridge connection to North America about 2 or 3 million years ago (MYA). The oldest mammalian taxa present in South America, at least since the late Cretaceous, are marsupials, xenarthrans, and notoungulates, with their different subgroups. Two new groups of mammals appeared during the Oligocene: platyrrhine monkeys and hystricognath (large) rodents, which probably came from Africa. At approximately 7 MYA faunistic interchange began between the Americas, leading to the great American biotic interchange about 2–3 MYA. Many "invaders" from North America entered South America. This group included mastodons, horses, tapirs, peccaries, camels, deer, shrews, rabbits, squirrels, cricetine rodents (small mice), canids, bears, raccoons, weasels, and cats. Differential extinction and diversification of these groups were common during all this time (Simpson, 1980; Reig, 1981; Pascual, 1984). At present there are 692 mammalian species in South America, most of which are rodents (41%) and bats (22%) (Reig, 1981; Honacki et al., 1982).

Until the Oligocene–Miocene, South America was relatively flat. However, during the Miocene and Pliocene an intense orogenic process gave rise to high and wide mountains in the western side of the continent. These conditions, together with oceanic currents, caused dramatic climatic changes, from relatively homogeneous during the Eocene to more heterogeneous up to the present (Simpson, 1983; Pascual, 1984).

III. Present Climatic Conditions in Chile

Presently, in northern Chile there is a strong decreasing east-to-west gradient of precipitation given by the rain shadow effect of the Andes mountains. In conjunction with the cold Humboldt Current and the Southern Pacific Anticyclone, this has resulted in the hot hyperarid Atacama Desert below 2500 m and the cold desert Altiplano above 3500 m. South of 30°S, an almost linear decrease of air temperature, an increase of precipitation, and a lowering of the snowline with increasing latitude are observed; the magnitude of the values varies according to distance from the coast, altitude, and topographic relief (di Castri y Hajek, 1976; Caviedes, 1990). In central and southern Chile the Andes intercept the prevailing humid southwestern winds coming from the Pacific Ocean, leading to a rain shadow effect with principal influences on the Argentinean side and minor effects on a few Chilean localities adjacent to and facing the Argentinean border. In central mediterranean Chile, climatic conditions are more variable within and between years than to the north or the south. Xeric conditions are ameliorated along the coast in the southern Atacama Desert by frequent clouds and fogs. Consequently, xeric conditions extend farther south into central Chile along the interior rather than along the coast. At high altitudes, along the Andes, precipitation decreases from the northern extreme to about 27°S, where precipitation starts increasing to the south (Arroyo *et al.*, 1988). Because of the steep slope of the Andes on the western side, there is a sharp contrast between low- and high-altitude climates, mainly given by the snowline.

IV. Present Distribution of Vegetation in Chile

Coincident with these climatic regimes, different vegetational formations exist in Chile. The northern xeromorphic vegetation zone corresponds to the barren Atacama Desert, transversed in some places by ravines coming from the Andes. The Andean vegetation extends southerly along the Andes to 36°S with an important maximal reduction of species and cover at about 27°S (Arroyo *et al.*, 1988). The xeromorphic vegetation from the north grades into the mesomorphic vegetation zone of the central Chile mediterranean area, which in turn grades into the hydromorphic vegetation zone south of 37°S. A Patagonian vegetation zone is found in some localities close to the Argentine border facing to the east, for example, Alto Bío Bío, Coihiaque, and the extreme southern end of Chile, with influences of the rain shadow effect (Schmithusen, 1956; Instituto Geográfico Militar, 1985; Troncoso, 1988; see also Arroyo *et al.*, Chapter 17, this volume).

V. Pattern of Mammalian Species Richness in Chile

At present, 95 species of terrestrial mammals are known in Chile. The most diverse orders are Rodentia (64%), Carnivora (14%), and Quiroptera (11%), which make up 89% of the Chilean terrestrial mammals. Marsupials, xenarthrans (edentates), and artiodactyls represent less than 6% of the total each (L. C. Contreras, in press). Species richness of Chilean nonvolant terrestrial mammals, excluding bats, are highest in the Altiplano and in areas with Patagonian vegetation in the south (21 to 27 species at different localities within the area) (Fig. 1) (L. C. Contreras, in press). The two areas with the lowest species richness are the Atacama Desert (≤9) and the temperate rain forest of southern Chile (12 to 14). The mediterranean, hygromorphic, and Andean life zones of central Chile show intermediate values of species richness (15–20).

VI. Mammalian Species Similarity among Areas in Chile

By cluster analysis of the mammalian fauna of 26 different areas covering all Chile, L. C. Contreras (in press) established six faunal groups that fit well to the vegetation

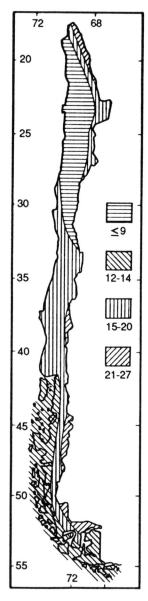

Fig. 1. Pattern of terrestrial nonvolant terrestrial mammal species richness in Chile. (From Contreras, in press.)

zones and climatic regimes mentioned above: the Altiplano of the northeastern extreme close to Peru and Bolivia (27 species and 8 endemics within this area), other northern areas including the Atacama desert and northern Andes (29 and 9), the Andes from central Chile (21 and 0), the mediterranean central Chile (23 and 3), the temperate rain forest (31 and 4), and southern localities with affinities to Patagonia (34 and 10) (for the same pattern in sigmodontine rodents see Marquet *et al.*, 1988).

VII. Possible Effects of Global Climatic Change on Mammal Associations in Chile

Assuming that mammalian species will respond quickly to a change in climatic conditions, the clear correspondence between the pattern of species richness and mammalian faunal similarities between areas with climate in Chile allows a first approximation to evaluating the impact of global climatic change on the life zones.

One of the most difficult aspects to do this is the reliability of predicting local effects of global climatic change in Chile (Fuenzalida *et al.*, 1989). There are several reasons for this. First, present models of global climatic change still have a considerable degree of uncertainty. Second, the global models operate at a large macroregional scale of hundreds of kilometers. The climatic regime of Chile is largely determined by the Andean Cordillera, which, as well as the country, has a smaller width than the resolution of global models. Thus, the effect of the Andes cannot be shown in the models.

Considering these problems, Fuenzalida *et al.* (1989) provided a tentative prediction of climatic change in Chile for the next 50 years based on the physical processes involved at the country level and compatibility with the global models. Regarding temperature, they predict a rise of 2 to 3°C in the north of the country to about 5°C at the southern end. The prediction of changes in precipitation is even more uncertain than that for temperatures; thus only qualitative estimates were provided. They postulate an increase and a southerly expansion of summer rains of the Altiplano of northern Chile, a reduction of precipitation in south-central Chile, and an increment of rains at the southern extreme of the country.

If we assume that the changes in climate may simply provide a geographic shift in the distribution of the present climatic regimes of Chile, and not the appearance of new climates, and consider the possible scenario indicated by Fuenzalida *et al.* (1989), we could expect the following changes:

1. If precipitation increases southward and perhaps westward throughout the Chilean Altiplano, we should expect a concomitant shift of the mammalian fauna in that area. Many species that presently have their southern and western limit of distribution in that area should move farther south and west; examples are *Chaetophractus nationi* (armadillo) and *Auliscomys boliviensis* (rodent). Other species, especially those presently restricted to the precordillera belt between 2500 and 4000 masl, may also be displaced to lower elevations; examples are *Octodontomys gliroides* (rodent), *Lama guanicoe* (camelid), and *Hippocamelus antisensis* (deer).

2. If an increment of precipitation in the Altiplano provides greater water availability in existing watershed areas, such as ravines and oases of the Atacama Desert, it is possible that some species currently inhabiting higher altitudes may reach those areas.

3. An increase in temperature and a decrease in precipitation in central Chile may result in a shift of xeric environments to the south. In turn, this change may cause a southern shift of the present latitudinal gradient of mammalian species richness between central Chile and the southern end of the Atacama Desert along the coast (Meserve and Glanz, 1978), as well as along the interior. Since these gradients are given by the drop-off of species going north, the shift to the south may lead to local extinction of species at their present northern range of distribution. This climatic change may also result in an increment of the area with present Patagonian influences close to the Argentinean border in central and southern Chile. If that is the case, we may expect an increment of mammalian species in those areas, since several species are currently found at localities near the frontier in the Argentinean Patagonia.

4. A further consequence of these climatic changes in central Chile may be a shift of the snowline toward higher elevations. In north-central Chile, this may have little effect other than an upward shift of the mediterranean and Andean mammalian fauna due to the higher elevation reached by the mountains in this portion of the Andean range (Caviedes, 1990). Contractions of the Andean environments may occur in southern Chile if the treeline shifts to higher elevations because the Andes exhibit lower altitude there. Consequently, some species currently restricted to Andean habitats, such as *Euneomys,* may suffer increased risk of local extinction (Pearson, 1987). However, if the snowline moves upward to a greater extent than the treeline, we may expect an increase of the area with Andean environments between these lines. A shift in the treeline and snowline to higher altitudes in central Chile may lead to contact between presently separated vegetational formations on both sides of the Andes. If this occurs, it may result in greater mammalian faunal similarity between Chile and Argentina in this area, as presently shown to occur farther south for cricetine rodents (Contreras, 1990).

VIII. Mammalian Species Differential Response to Changes in Climate

All of the possible changes that have been mentioned are based on rapid tracking of climate and vegetation changes by all the mammalian species associated with particular climatic regimes. Although shifts in the distribution of mammals associated with changes in climate have occurred recently in Chile, it is uncertain that they will reoccur because present global climatic change may be too rapid to be tracked closely by plants and animals. Perhaps some species may be able to track the changes more effectively than others. We know that in the recent past mammalian species have experienced differential success in geographic radiation and range extensions (Graham and Grimm, 1990). One species that has shifted its distribution in the recent past with changes in climate in Chile is the semisubterranean rodent *Aconaemys fuscus*. This species now lives in habitats characterized by bunch grass in Andean environments or low shrubby habitats with relatively high humidity at lower elevations. The northern limit of distribution of *A. fuscus* is 35°S along the Andes and 37°S along the Central Depression (Contreras *et al.*, 1987; Reise and Gallardo, 1989); however, 10,000 years ago, at the end of the last glacial period, it was found at least 1.5° farther north along the Andes and 3° farther north along the Central Depression. Other species that were found together with *Aconaemys* farther north are now extinct (Tamayo and Frasinetti, 1980; Simonetti and Saavedra, in press).

Global climatic change not only may bring a shift in the present distribution of the climatic regimes in Chile but also may produce new regimes. A most likely difference from present climatic regimes may be related not only to average values of temperature or precipitation but also to the variability of those factors. If this is the case, there would be greater reason to expect a differential species response to the changes. For instance, in places with high variability, species tolerating long periods of drought will prevail. This differential response may lead to new assemblages of species and consequently to new species-specific interactions, which in turn may affect species distributions and abundances. It should be noted that greater tolerance to drought depends not only on physiological adaptations, such as high renal capacity to concentrate urine, but also on ecological or behavioral adaptations. A good example is the subterranean rodent *Spalacopus cyanus*. This species has a very poor capacity to tolerate water deprivation (Cortés, 1985); however, *S. cyanus* is one of the species that is distributed from central Chile quite far north into the Atacama Desert along the coast, regardless of the low and highly irregular rainfall in this portion of its range (Contreras *et al.*, 1987). This seems possible because geophytes, which are the main food item of this rodent in the area (Contreras and Gutiérrez, 1991; Cox *et al.*, 1993; Contreras *et al.*, 1993), may capitalize on irregular pulses of precipitation and remain available to be eaten by *Spalacopus* for several years.

Because of the relative constancy of its food resources and its subterranean mode of life, *Spalacopus* demonstrates low population fluctuations despite the irregularity of rainfall. However, more generalized rodents in semiarid Chile may present strong population fluctuations leading to pest outbreaks in years with high rainfall (Fuentes and Campusano, 1985).

These two examples show how species with different modes of life and different life history characteristics may have differential responses to climatic changes. More generally, species with small, isolated populations, reduced population growth, and low dispersal capabilities will probably be most affected by climatic changes. It is difficult to indicate which species these may be because the present distribution of mammals in Chile is not sufficiently known and because not many species seem to have isolated populations. The main reason for the lack of isolation may be that, in general, the country as a whole is located on the west slope of a north–south oriented mountain range, with hills and areas with similar vegetation connected to each other. At present, the Atacama Desert is the only area where we know there are mammalian species with isolated populations; examples are *Cavia tschudii*, a 300-g guinea pig rodent, found close to the mouth of few ravines in the extreme north, and five other small mammal species found in the Pampa del Tamarugal (a system of dry salt lakes at 1000 m altitude). However, if climatic changes occur as indicated earlier, the Atacama Desert may receive larger amounts of water from precipitation or runoff, probably without detrimental effects on those species.

To better assess the possible effects of a global climatic change on the mammalian fauna in Chile, it is necessary to (1) predict local climatic changes more accurately, (2) understand the actual distribution of mammals and their correlation to climate better, and (3) improve our understanding of the autoecology of the species in order to evaluate their individual responses to these climatic changes.

Acknowledgments

I thank B. Lang and J. R. Gutiérrez for their comments on the manuscript and help with the English. This work has been funded by grants DIULS 120-2-35, FONDECYT 89/585, and 90/376. This is a contribution of the Arid Environment Study Program, Universidad de La Serena.

References

Arroyo, M. T. K., Squeo, F. A., Armesto, J. J., and Villagrán, C. (1988). Effects of aridity on plant diversity in the northern Chilean Andes: results of a natural experiment. *Ann. Missouri Bot. Gard.* **75,** 55–78.

Caviedes, C. N. (1990). Rainfall variation, snowline depression and vegetational shifts in Chile during the Pleistocene. *Climatic Change* **16,** 99–114.

Contreras, L. C. (1990). Cricetid species richness in the southern Andes: the effect of area. A critique of Caviedes and Iriarte (1989). *Rev. Chil. Hist. Nat.* **63,** 19–22.

Contreras, L. C. (In press). Biogeografía de mamíferos terrestres de Chile. *In* "Mamíferos de Chile" (A. Muñoz-Pedreros and J. Yañez Valenzuela, eds.).

Contreras, L. C., and Gutiérrez, J. R. (1991). Effects of the subterranean herviborous rodent *Spalacopus cyanus* on herbaceous vegetation in arid coastal Chile. *Oecologia* **87,** 106–109.

Contreras, L. C., Torres-Mura, J. C., and Yañez, J. L. (1987). Biogeography of octodontid rodents: an eco-evolutionary hypothesis. *Fieldiana Zool. N. S.* **39,** 401–412.

Contreras, L. C., Gutiérrez, J. R., Valverde, V., and Cox, G. W. (1993). Ecological relevance of subterranean rodents on arid coastal Chile. *Rev. Chil. Hist. Nat.* **66.**

Cortés, A. (1985). Adaptaciones fisiológicas y morfológicas de pequeños mamíferos de ambientes semiáridos. Tesis Magister, Facultad de Ciencias, Univ. de Chile.

Cox, G. W., Contreras, L. C., and Milewski, A. V. (1993). Role of fossorial mammals in community structure and energetics of mediterranean ecosystems. *In* "Ecology of Convergent Ecosystems: Mediterranean-Climate Ecosystems of Chile, California, and Australia" (M. T. K. Arroyo, P. H., Zadler, and M. D. Fox, eds.). Springer-Verlag, New York.

Di Castri, F., and Hajek, E. (1976). "Bioclimatología de Chile." Vicerrectoria Académica, Univ. Católica de Chile, Santiago.

Fuentes, E., and Campusano, C. (1985). Pest out-breaks and rainfall in the semi-arid region of Chile. *J. Arid Environ.* **8,** 67–72.

Fuenzalida, H., Bernal, P. A., Villagrán, C., Fuentes, E., Montecinos, V., Santibañez, F., Peña, H., Hajek, E., and Rutllant, J. (1989). "El Cambio Climático Global y sus Eventuales Efectos en Chile." Comisión Nacional de Investigación Científica y Tecnológica, Santiago.

Graham, R. W., and Grimm, E. C. (1990). Effects of global climatic change on the patterns of terrestrial communities. *Trends Ecol. Evol.* **5,** 289–292.

Honacki, J. H., Kinman, K. E., and Koeppl, J. W. (1982). "Mammal Species of the World: A Taxonomic and Geographic Reference." Allen Press, and Association of Systematics Collections, Lawrence, Kansas.

Instituto Geográfico Militar. (1985). "Atlas Geográfico de Chile para la Educación." Instituto Geográfico Militar, Santiago.

Marquet, P. A., Silva, S., and Contreras, L. C. (1989). Species richness of cricetine rodents on the Andean Pacific slope between northern Perú and south-central Chile. Pacific Science Association, VI International Congress, p. 41, Valparaiso, Chile.

Meserve, P. L., and Glanz, W. E. (1978). Geographycal ecology of small mammals in the northern Chilean zone. *J. Biogeogr.* **5,** 135–148.

Ojima, D. S., and Kittel, T. G. F. (1991). Critical issues for understanding global changes effects on terrestrial ecosystems. *Ecol. Appl.* **1,** 316–325.

Pascual, R. (1984). Late Tertiary mammals of the southern South America as indicators of climatic deterioration. *In* "Quaternary of South America and Antarctic Peninsula" (J. Rabassa, ed.), pp. 1–30. Balkema, Rotterdam.

Pearson, O. P. (1987). Mice and the post glacial history of the Traful Valley of Argentina. *J. Mammal.* **68,** 469–478.

Reig, O. A. (1981). Teoría del Origen y Desarrolo de la Fauna de Mammíferos de America del Sur. Monografie Naturae, Museo Municipal de Ciencias Naturales "Lorenzo Scaglia," Vol. 1, pp. 1–159.

Reise, D., and Gallardo, M. H. (1989). An extraordinary occurrence of the tunduco *Aconaemys fuscus* (Waterhouse, 1841) (Rodentia, Octodontidae) in the central Valley, Chillán, Chile. *Medio Ambiente* **10,** 67–69.

Schmithusen, J. (1956). Die Räumliche Ordnung der chilenische Vegetatio. *Bonn. Geogr. Abh.* **17,** 1–86.

Simonetti, J. A., and Saavedra, B. (In press). Reemplazando espacio por tiempo: Arqueofauna del estero el Manzano. *Anales Museo Hist. Nat. Valparaiso.*

Simpson, B. B. (1983). An historical phytogeography of the high Andean flora. *Rev. Chil. Hist. Nat.* **56,** 109–122.

Simpson, G. G. (1980). "Splendid Isolation." Yale Univ. Press, New Haven, Connecticut.

Tamayo, M., and Frassinetti, D. (1980). Catálogo de los mamíferos fósiles y vivientes de Chile. *Bol. Mus. Nac. Hist. Nat. Chile* **37,** 323–399.

Troncoso, A. (1988). Flora I y II. *In* "Enciclopedia Temática de Chile." Ercilla, Santiago.

C H A P T E R 21

North–South Comparisons: Animals

JAMES H. BROWN   LUIS C. CONTRERAS

Some similarities, but also pronounced differences, between North America and South America can be expected in the responses of animals to climatic and other kinds of anthropogenic environmental change.

It is not clear that the major environmental changes are likely to be similar between the two continents. Compared to North America, the different physical geography of South America and its distinctive temporal and spatial patterns of human settlement, land use, and industrial/technological development may cause substantial differences in the rates and magnitudes of many kinds of changes, including global warming.

As animal ecologists, we do not have the expertise to make predictions about changes in climate, land use, and vegetation. However, when the appropriate experts are willing to predict the extent of changes in abiotic conditions and vegetation on the two continents, it should be possible to assess the expected responses of the animals. Bearing this caveat in mind, it is possible to make some predictions based on characteristics of the physical settings and of the biotas of the two continents.

There are the important differences in the physical geography of North America and South America. South America narrows rapidly from the tropics to its tip at Tierra del Fuego, which, at just 55°S, is still in temperate latitudes. The Andes, the major mountain range, rises precipitously from the Pacific and runs predominantly north–south, leaving only a narrow strip of lowlands on its western (Chilean) side but much more extensive lowlands on the eastern (Argentine) side. By contrast, North America is moderately constricted at the tropics but widens rapidly in the temperate zone and extends far north to Arctic latitudes. There are multiple cordilleras running predominantly north–south; but compared to those in South America, the mountains are not so high, extent much farther inland, and are interrupted by extensive lowland areas.

The most conspicuous effect of these differences in physical setting on the vertebrate fauna is that the geographic ranges of most species in South America tend to consist of narrow strips running north–south along the Andes in the appropriate climatic and vegetational zone. This is particularly marked in Chile, but even in Argentina the contrast with the much more extensive (in both area and longitude) distributions of related taxa in North America is dramatic. Although South American vertebrates tend to be restricted to much smaller ranges, they have been able to respond to past environmental changes, including the climatic cycles of the Pleistocene, by shifting their elevational and latitudinal ranges (see Fig. 2 of Brown, Chapter 19, this volume).

There are at least three reasons why many South American species might be even more susceptible to environmental change—and especially to global warming—than their North American counterparts. First, the South American species already have smaller, narrower geographic ranges. Everything else being equal, this would suggest that they already tend to have smaller, less extensive populations and, hence, to be more susceptible to extinction.

Second, the fact that the ranges of South American species are restricted to narrow strips along the axis of the Andes makes them highly susceptible to fragmentation and subsequently increased probability of extinction. On the one hand, such fragmentation could occur because the pattern of climatic and vegetation change causes more reduction in some habitats than others (Contreras, Chapter 20, this volume). It could also occur because other kinds of anthropogenic change cause or contribute to the fragmentation. For example, the extensive human settlement and resulting conversion of natural habitats to urban, suburban, and agricultural ecosystems in the central depression of Chile are sufficient to cause a gap in the present ranges of several vertebrates or at least to restrict greatly population contact between the Andes and coastal ranges (e.g., medium-sized rodents like *Spalacopus cyanus* and *Octodon degus,* as well as felids and canids). This region of human-dominated landscapes is a potentially severe barrier to the dispersal of many animal species in response to climatic change.

Finally, although the vertebrates of temperate South America have survived repeated cycles of global cooling and rewarming during the Pleistocene, the rate and degree of global warming predicted to occur because of the greenhouse effect would cause an unprecedented situation. Any warming in the Southern Hemisphere will tend to push species southward and to higher elevations. But the ability of species to make these shifts is limited by the fact that toward the south the average elevation of the Andes decreases substantially and the continent narrows rapidly towards its tip at just 55°S. Thus, just as climatic warming would effectively push local populations of small mammals off the top of isolated mountains in North America (Brown, Chapter 19, this volume), so it would tend to push entire species off the South American continent.

We conclude that, given comparable kinds and magnitudes of environmental change, the effects on the vertebrate fauna might be even more severe in South

America than in North America. The diversity of most groups of vertebrates is already lower in temperate South America than in temperate western North America. For example, substantially more species (28) of nonvolant mammals occur on Brown's 20-ha experimental study site in the Chihuahuan Desert of southeastern Arizona (Brown and Nicoletto, 1991) than in all of central Chile (Contreras, Chapter 20, this volume)! Presumably, this much lower diversity in South America reflects higher extinction rates owing to the smaller land area and the impacts of past environmental changes, especially the climatic fluctuations of the Pleistocene. Now the South American fauna is faced with the threat of additional extinctions as a result of human-caused environmental change.

Reference

Brown, J. H., and Nicoletto, P. F. (1991). Spatial scaling of species composition: body masses of North American mammals. *Am. Nat.* **138,** 1478–1512.

Part VIII: Managed Systems and Human Impacts

C H A P T E R   22

Effects of Global Warming on Managed Coastal Ecosystems of Western North America

PAUL B. ALABACK   MICHAEL H. McCLELLAN

I. Introduction

An unprecedented integrated scientific effort is needed to learn the biological and economic implications of global climate change. The primary impetus for this work will be the need to develop a credible basis for adaptive strategies for managed ecosystems. Although it can be argued that uncertainties in general circulation models (GCMs) make immediate changes in natural resource policy premature, the potential magnitude of effects from climate change and the long-term and essentially irreversible nature of these changes dictate that scientific information on the mechanisms and technical information on adaptive strategies must be developed quickly (Tangley, 1988). As contrasted with natural ecosystems, more options may exist for adapting managed ecosystems to climate change. Adaptive strategies are urgently needed for managed ecosystems so that potentially catastrophic long-term consequences can be averted while reasonable options still exist.

The global warming problem, although theoretically formulated as early as 1827, has stimulated intense scientific interest only during the past few decades. Analyses of historical records and broad-scale ecological and economic models have received the most attention (e.g., Adams *et al.*, 1990; Boer *et al.*, 1990; Emmanuel *et al.*, 1985; Ritchie, 1986). Physiological experimentation and modeling have significantly increased the realism of global change models, especially for managed ecosystems

299

(Campbell, 1989; Eamus and Jarvis, 1989; Mooney *et al.*, 1991). Most efforts have focused on a coarse global scale or on one specific region or climatic zone. Few studies have attempted to make detailed comparisons along a climatic gradient or to compare trends in ecosystem response to analogous climatic gradients. An examination of the response of the climatic gradients in western North America and western South America to changes in climate could help develop both adaptive management strategies in these economically important regions and new scientific insights that could have broad-scale applications not only to the global warming problem but also to advancing our general understanding of how climate is related to landscape, ecosystem, and population level processes.

The climatic gradients along the Pacific coast are unique in providing two analogous gradients of similar scale, history, and geology while including nearly total contrasts in biota (Alaback, 1991; Mooney, 1977). The north–south orientations of these biomes and a long history of north–south migrations make these gradients a particularly tempting analog for climatic change. Study of these gradients should be useful in developing hypotheses about ecotonal dynamics and species thresholds and helping identify key climatic biogeographic or physiological variables that may govern biological responses to global climatic change. The differing responses of North America and South America also should provide insights about subtle distinctions in climate and human culture and their influence on landscape-level dynamics. The history of research and the strategic location of research institutions along the entirety of this gradient offer unique opportunities for collaboration, synthesis, and the integration of many scales and perspectives.

In this chapter we review the ecological literature on global climatic change with particular reference to managed systems along the climatic gradient of the west coast of North America. In addition, we synthesize information available from this diverse region to outline a range of probable consequences of climate change for this region, propose some adaptive strategies for managed ecosystems, and lay a foundation for and help stimulate future collaborative studies. We place heavy emphasis on northern forest ecosystems because of their presumed sensitivity to climate change and the relative difficulty of developing adaptive management practices to buffer them from climate change.

II. Present Conditions

A. The Climatic Gradient

Western North America from 30 to 60°N comprises diverse climatic conditions: from drought-ridden deserts to perpetually wet rain forests and tundra. For convenience of discussion we will divide the North American coastline into three principal sections: the north coast (50 to 60°N), where heavy precipitation occurs during most months; the Pacific Northwest (50 to 40°N), where summer droughts and fire are common; and the south coast (30 to 40°N), where extended droughts and fire can occur throughout the year. These regions include forest-dominated, mixed forest–

agriculture, and agriculture zones roughly corresponding to rain forest, seasonal forest, and mediterranean climatic zones (Alaback, 1991; Holdridge, 1947).

A key variable within this transect is precipitation, which steadily increases northward both in equability of distribution and in yearly accumulation (Fig. 1). Along this transect moisture changes from a scarce resource limiting plant productivity to an overabundant resource that is associated with declines in soil fertility (Stephenson, 1990; Ugolini and Mann, 1979). The general distribution of forest, grassland, and tundra vegetation types can be predicted along this transect simply by calculating site water balance (Stephenson, 1990). Along the south coast precipitation occurs primarily in the winter months except in the Sierra Nevada and Sierra Madre Occidental ranges and in the adjacent Sonoran Desert regions, where convective storms bring significant precipitation during summer.

Within the north coast the most ecologically important effects of precipitation are related more to its distribution than to its magnitude (Alaback, 1991). Along the north coast semipermanent low-pressure systems bring abundant precipitation throughout the year. Upwelling in the Pacific Ocean also increases moisture by producing fog during late spring and early summer. Fog drip interception is an important yet seldom quantified element of the hydrologic budget (Harr, 1982) and is also of key importance in defining the geographic limit of the most productive forest types, especially along the coastlines of northern California and the Pacific Northwest.

Throughout the region, but especially in the north, precipitation varies dramatically in relation to local topography and storm paths. Mountains, ice sheets, and river

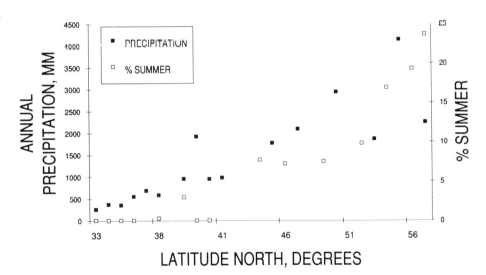

Fig. 1. Rainfall distribution and accumulation across the western coast of North America. Stations selected are within 10 km of the coast and without significant rain shadow effects. Locations are described in greater detail in Farr and Harris (1979).

corridors to the continent interior can also influence precipitation distribution. For example, within 5 km of Juneau, Alaska, annual precipitation can vary from 1300 to 2600 mm at sea level or 1300 to 4000 mm within a 1000-m elevational range.

An equally significant but more consistent gradient of temperature also characterizes this north–south transect (Fig. 2). This temperature gradient is closely related to forest yield and productivity and defines the northern limit of agricultural zones (Boer *et al.*, 1990; Farr and Harris, 1979). Because of persistent westerly winds and the warm Kuriosho ocean currents, the entire west coast has a distinctly maritime climate with moderate temperatures in both winter and summer. The moderating influence of the ocean is most constricted along the southern coast—often to within as little as 1 km—and is most pervasive along the northern coast, where moderated temperatures may extend 150 km inland. Fog often depresses spring and summer temperatures, particularly in the Pacific Northwest. Seasonal and diurnal variation is greatest in the north.

B. Natural Disturbance

The types of natural disturbance and their frequency, intensity, and scale also vary systematically along the Pacific coast. In the south, wildfire is of only local ecological significance because of discontinuous fuels and low frequencies of lightning-generating storms. Human settlement, however, has drastically elevated the economic and ecological importance of fires in many mediterranean and semidesert climatic zones. In the middle latitudes high storm frequencies result in highly frequent but mostly noncatastrophic fires in the Cascades and Sierras to moderately frequent but catastrophic fires in the Oregon and California Coast Range and the Washington Olympic Mountains (Hemstrom and Franklin, 1982; Morris, 1934). In

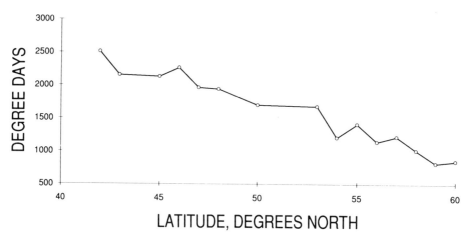

Fig. 2. Change in growing degree-days with latitude, western coast of North America. Stations selected are within 10 km of the coast and without significant rain shadow effects. Locations are described in greater detail in Farr and Harris (1979). (Adapted from Farr and Harris, 1979.)

general, fire in moist climates is less frequent but can be more catastrophic because of greater fuel accumulations (e.g., Davis *et al.*, 1980; Hemstrom and Franklin, 1982). Intensity of fire generally declines with latitude north of 40°N, except during extended droughts in forested regions, where heavy fuel accumulations lead to high fire temperatures and rapid fire spread. Fire ceases to be an ecologically important process when summer precipitation exceeds 10% of the annual accumulation and in coastal regions north of 50°N (Alaback, 1991). East of the coastal mountains low precipitation allows fire to remain an ecologically important process all the way north to the Arctic Circle.

In the Arctic and boreal regions fire is less intense because of cold soils and thick wet organic horizons, but it is extensive and moderately frequent. Fire maintains many ecosystem processes: fire accelerates permafrost thawing; increases decomposition rates; elevates soil temperatures and diversifies vegetation composition, productivity, and structure (Van Cleve *et al.*, 1986). Fire and insects are the most important disturbances that affect this climatic zone. In effect, fire maintains ecosystem productivity (Viereck *et al.*, 1983).

Wind is the most widespread and perhaps the most ecologically important form of disturbance in the northern portion of the west coast transect. Wind can cause catastrophic damage at infrequent intervals almost anywhere on the west coast, but windthrow is most common in forest zones where storms are most frequent: in the Pacific Northwest and the north coast. Noncatastrophic windthrow is frequent and pervasive throughout forested coastal zones but is ecologically most important in the north coast, where fire is rare. Windthrow can enhance ecosystem diversity by creating a wider variety of habitats and increasing light penetration through forest canopies, just as it does in the tropics. Soil disturbance accompanying windthrow reverses many effects of soil development and greatly increases soil heterogeneity (McClellan, 1991). Many major ecosystem types on the north coast can be distinguished by frequency and intensity of disturbance (Alaback, 1990). Wind damage is highly variable in both space and time but is most consistent in outer coastal regions (Harris, 1989; Ruth and Harris, 1979). In analogous climatic zones in Britain wind is considered the key factor in constraining commercial forest plantations (Cannell *et al.*, 1989).

Landslides and snow avalanches are also a major form of natural disturbance throughout the west coast. Road construction and logging greatly accelerate the rate of slope failures (Swanston and Swanson, 1976). The greatest risks generally develop where oversteepened slopes occur in climates in which high-intensity rainfall events are common (Swanston and Swanson, 1976). Loose, unstable substrates like expanding clays or volcanic ash also elevate the risk of slope failure. The California coast, northern California, the Oregon Coast Range, the Oregon Cascades, coastal British Columbia, and southeast Alaska are all considered highly susceptible to slope failures. Riparian ecosystems and fisheries resources are particularly susceptible to disturbance from slope failures.

Avalanches are also widespread throughout the northern coast and at high elevations in the Pacific Northwest and are an important form of natural disturbance.

The northern coast is among the world's most hazardous locales for avalanches because of the combination of heavy snow accumulation, mild temperatures (heavy, wet snow), and rapid warming in the spring. Frequency and destructiveness of avalanches are closely related to seasonal snowfall and temperature patterns, so they would probably be sensitive to even a subtle change in winter climate (Cooley, 1990). Also, human activities can drastically alter the frequency and destructiveness of avalanches. Forest clearing in run-out zones and increasing area in openings can magnify the destruction wrought by avalanches by increasing the distance affected by them, by promoting higher snow accumulations, and by intensifying the impact of rain-on-snow events in the spring (e.g., Simons, 1988).

C. Ecosystem Management and Climate

The climatic driving variables just discussed can be directly related to many functional ecosystem processes that affect resource management and other human activities. In the south coast and Pacific Northwest moisture is often limiting to plant growth (Emmingham and Waring, 1977; Zobel et al., 1976). East of the major cordillera moisture is limiting for most soil types north to the Arctic. Within the south coast and Pacific Northwest, leaf area and net primary production are directly predictable from available moisture data, except on extreme soil types (Gholz, 1982; Mooney et al., 1977). In areas with moderate to high rainfall nutrient availability can constrain net primary productivity (e.g., Axelsson, 1985; Binkley, 1983). From northern California (approximately 40°N) northward, frost also plays a key role in constraining net primary productivity and limiting species' ranges (Emmingham and Waring, 1977).

Where irrigation is available, optimal agricultural productivity occurs in the southernmost regions; otherwise it occurs near the coast in the south coast region. Optimal forest productivity occurs along the central coast (35 to 40°N) for redwood (*Sequoia sempervirens*), 45 to 48°N for Douglas fir (*Pseudotsuga menziesii*), Sitka spruce (*Picea sitchensis*), and western hemlock (*Tsuga heterophylla*) (Franklin and Waring, 1980). North of 48°N resource productivity is chiefly constrained by low temperatures, low nutrient availability, poor soil drainage, and possibly genetics (Farr and Harris, 1979; Ruth and Harris, 1979).

Along the northern coast, soil fertility and nutrient release decline with latitude (and precipitation) and form a major constraint on ecosystem productivity. Temperature, moisture, and plant chemistry are the key variables constraining decomposition and nutrient release (Alexander, 1988; Meentemeyer, 1978). Within the north coast, excess soil water and slow decomposition yield steadily increasing soil organic matter, thicker litter layers, more peatlands (muskegs), and less forest as latitude increases. Peatlands have been expanding since the Wisconsin glaciation as the climate has become warmer and wetter (Heusser, 1960; Ugolini and Mann, 1979). Local topography and glacial history also strongly influence the distribution and extent of peatlands, especially where impermeable layers develop in the soil or remain from glacial deposits. In the absence of disturbance many soils naturally develop impermeable layers that may ultimately lead to loss of site productivity and peatland formation (Ugolini and Mann, 1979). In the northernmost sites along the coast (>58°N) peatlands and alpine meadows or shrublands are the dominant vegetation type except on recently uplifted beach soils and on steep slopes (90% nonforest on the Chugach National Forest, 60% on the Tongass National Forest, and <10% on the Olympic National Forest). In contrast to comparable environments in northern Europe, few peatlands have been drained in the north coast region. The few experimental attempts to drain peatlands in southeast Alaska to increase land available for agriculture or forestry have met with only modest success. If the economics of peatland draining were to change, it could have significant consequences for regional hydrology and carbon emissions to the atmosphere.

Soils in the interior are similar to coastal soils in terms of increasing concentrations of carbon, thicker organic horizons, and increasing available moisture with increasing latitude. In contrast, however, nutrient cycling is strongly influenced by fire, which reduces litter accumulations and leads to increases in the rate of decomposition, nutrient release, and plant productivity (Van Cleve et al., 1986; Viereck et al., 1983). In addition, north of approximately 57°N permafrost can occur, constraining decomposition and nutrient release and leading to poor water drainage. Local topography, microclimate, and surficial geology determine the distribution of permafrost. These physical relationships are sufficiently understood that carbon and nitrogen cycling and nutrient availability of boreal forests can be precisely predicted from an ecosystem model driven by growing season climatic and soil litter layer parameters (Bonan, 1990).

Parallel to these changes in vegetation type and productivity, resource management strategies and human settlement follow a predictable pattern along this transect. Agriculture dominates the landscape along the southern coast. Agricultural land declines with latitude in the Pacific Northwest and north coast (Fig. 3), reaching its maximum northern extent at approximately 50 to 53°N in British Columbia (Dalichow, 1972). Forestry steadily increases in land area and economic importance from 40 to 50°N, after which it steadily declines in productivity and wood quality.

Product diversity declines rapidly northward, especially in the interior. Sawtimber, plywood, and specialty wood products are most common in the interior and the Pacific Northwest. Northward of approximately 50°N, pulpwood and chemical pulp are the dominant products. Northward and toward drier sites in the interior, fruits, vegetables, and grains such as corn and rice are replaced by wheat and rangeland.

Agriculture usually stops at 48–53°N except in unusual microsites associated with large river valleys and southern exposures (e.g., the Matanuska Valley in south-central Alaska). The northward extent of agriculture is generally predictable from accumulated growing degree-days. In Fennoscandia, for example, the northern limit of all arable crops is approximately 800 degree-days, with a 5°C threshold (Boer et al., 1990). Reliability of crop yields declines northward and toward the interior, where anomalous variation in growing season weather leads to crop failures from frost, flooding, or pests. Similar patterns occur in the southern extreme of the coastal transect, where droughts and high temperatures lead to high variation in crop yields from year to year. Air pollution, particularly ozone, is also an increasingly important

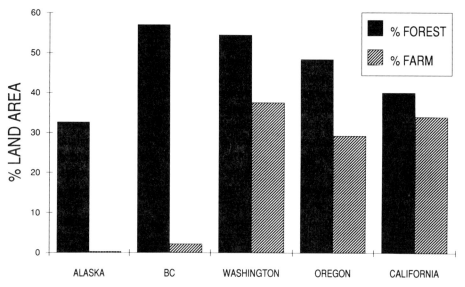

Fig. 3. Distribution of forest and agricultural land along the west coast of North America.

factor in constraining plant productivity, especially in the southern coast, and will be of even greater concern with climatic change.

D. Contrasts between Managed and Natural Ecosystems

Ecosystem management often strives to maximize the production or economic value of a few key resources, resulting in significant changes in ecosystem structure and function. Choosing the resources to manage for and the characteristics of those resources may be dictated as much by local economic conditions as by inherent ecological capabilities. Thus to examine the ecological implications of management requires consideration of both the social and economic context along with the physical and biological characteristics of the system.

Managed systems have several features that distinguish them from most natural ecosystems and that may have important implications for their response to climate change. Managed ecosystems have more frequent and more catastrophic disturbances than natural systems. Along the entirety of our transect, for example, forest trees tend to achieve much greater ages in pristine forests than in managed forest plantations. In addition, disturbance invoked for management (e.g., forest clearing, crop harvesting, chemical treatments, or weeding) generally results in less carryover from the previous mature ecosystem than would be expected following natural disturbance. For example, even the most intense and catastrophic wildfires, such as the Tillamook Burn in Oregon, left significant patches of unburned or lightly burned forest within the general burn area and many standing dead trees and logs that in turn strongly influenced the processes of seedling dispersal and establishment. In addition, natural disturbances tend to allow more carryover of below-ground plant propagules, associated microbes, mycorrhizae, and structure such as logs and snags than would be expected following management-invoked disturbance (Perry *et al.*, 1989).

These changes in disturbance regime have direct implications for ecosystem structure and function. More frequent disturbance favors the dispersal and establishment of early seral and weedy species. Fewer plant species dominate the upper canopy layer in managed ecosystems, and there tends to be a less equitable allocation of carbon between upper-canopy and understory vegetative strata. These structural changes generally imply a less diverse assemblage of animal species associated with these ecosystems; the species that do adapt are more subject to population outbreaks and crashes than those in corresponding natural ecosystems. Less carbon accumulates in managed forests because of more frequent disturbance or because of changes in ecosystem structure, unless forest plantations replace herbaceous or shrub-dominated ecosystems (Harmon *et al.*, 1990).

The landscape-scale structure of managed ecosystems differs from that of natural ecosystems. Patch sizes in natural landscapes tend to be larger and more diverse than in managed landscapes (Franklin, 1989; Franklin and Forman, 1990). Because of the heterogeneity of natural disturbance, landscape connectivity would be expected to be greater in natural landscapes. This implies more opportunities for the persistence of plant and animal populations following disturbance (Saunders *et al.*, 1991). Management is frequently highly selective in its effects on landscapes as well. Valley bottoms, stream corridors, lake shores, and well-drained, flat terrain all tend to be more intensively managed than hilly, mountainous, or high-elevation sites. This development pattern often constrains plant propagule and animal species dispersal across landscapes (Mader, 1990).

In the most economically valuable and most productive regions these contrasts between natural and managed landscapes are the most significant, particularly where population pressures are the greatest. Just as economic development has expanded over time, management intensity and ecological change have also increased. As the intensity, technology, and extensiveness of management have increased, there has been a growing concern about how maximization of productivity may be compromising long-term sustainability and resilience to natural disturbance (Battie, 1989; Perry *et al.*, 1989).

The effects of management on ecological processes can vary from barely detectable to overwhelmingly important, depending on the intensity of management. Forest management, for example, tends to be much less intensive and results in ecosystems that are much less distinct from their unmanaged counterparts than are most agricultural systems. At high latitudes in our coastal transect, forest management itself becomes less intensive, with less reliance on nursery-grown planting stock, limited use of herbicides, a small proportion of land area devoted to commercial activities, and less mechanization of management treatments.

Concerns over sustainability of management, soil fertility, maintenance of biodiversity, and related issues have gained increased recognition in recent decades. Development of management strategies for sustainability in light of increased costs and decreased yields over the short term has been difficult to implement on a wide-scale basis. The prospect of climate change provides a strong impetus for developing alternative management systems that emphasize resource sustainability.

III. Past Conditions

A. Presettlement

The pollen record is extensive and well known along the Pacific coast of North America, especially along the northern coast and interior (e.g., Brubaker, 1988; Heusser, 1960). Vegetation has been in a continual state of flux during the glacial and interglacial periods. As with other regions, these fluctuations demonstrate that plant species' distributions tend to vary in an individualistic fashion. Plant species respond differently to stress and vary in their speed of migration to suitable habitat areas. Life history strategies, reproductive mechanisms, and physiological tolerances vary between the dominant species and control migration rates. For example, many tree species in Alaska are physiologically adapted to grow much farther north or west than the limit of their present distribution (e.g., *Abies amabilis, A. lasiocarpa, Picea glauca*). In the Southern Hemisphere even more dramatic asymmetries between local populations and exotic species occur. Exotic tree species such as *Pinus contorta,* because of their adaptation to the current cool temperature regime, can extend timberline as much as 500 m or dramatically increase wood productivity over that of any native species (Hawkins and Sweet, 1989).

Many historical plant assemblages have no modern counterparts. For example, the present forest species mix in the Pacific Northwest has been in existence for no more than 8000 years. Western hemlock has been present in Prince William Sound for only 4000 years. Subtle changes in climate in southern areas correspond to dramatic changes in northern areas, including glacial surges or rapid recessions and rapid timberline movement. For example, during the Little Ice Age approximately 200 years ago Glacier Bay was filled with ice. At present ice remains only in the head of the bay, over 95 km from the mouth of the bay. Partly because of these massive fluctuations in glaciers and climate, species in the north tend to be habitat generalists and are the least likely to be uniquely adapted to current climatic conditions (Kellison and Weir, 1987).

B. Human Settlement

The deforestation and settlement of the west coast was one of the most rapid in history and has significantly contributed to atmospheric carbon (e.g., Harmon *et al.*, 1990; Harris, 1984). Forests in the Pacific Northwest have among the highest biomass accumulations of any forest type in the world, with typical values of 500–1000 Mg/ha and 2000–3000 Mg/ha on some exceptional sites (Franklin and Waring, 1980). Indigenous populations used fire extensively to clear land and improve habitat for wildlife. This created extensive grasslands in coastal valleys, on mountaintops, and throughout the interior (e.g., Johannessen, 1971; Sagan *et al.*, 1979). This extensive deforestation has also influenced local climatic conditions, generally increasing temperature extremes, aridity, and albedo (Hyams, 1952; Sagan *et al.*, 1979). When deforestation occurs near grassland or steppe ecotones, it may result in a long-term retreat of forests and soil degradation, even in the absence of subsequent human-caused disturbance (Hyams, 1952; Perry *et al.*, 1989; Veblen and Markgraf, 1988).

In the midlatitudes of the Pacific Northwest and northern California most of the remaining forests in valley bottoms and the most productive sites were cleared between the mid-1800s and the 1940s. The midelevation sites were mostly cleared during the economic expansion of the 1950s through the late 1960s (Sedell and Duval, 1985). Remaining high-elevation sites and the less productive or more remote sites have been logged from the 1980s through the present (Harris, 1984). In British Columbia and Alaska, logging of sea level sites began on an industrial scale in the 1950s and 1960s and peaked in the 1970s and 1980s (Sedell and Duval, 1985; M. Kirchhoff, Alaska Dept. of Fish and Game, personal communication). Logging harvest volume is expected to decline over the next four to five decades throughout the Pacific Northwest and north coast due to overharvesting and a resulting imbalance in forest plantation age structure (e.g., Beuter *et al.*, 1976).

Most large river systems of the Pacific Northwest and the south coast were diverted, channelized, cleared of debris, dammed, straightened, and otherwise modified since the turn of the century, which caused drastic changes in riparian ecosystems (Sedell and Duval, 1985). Since World War II many irrigation water systems and hydropower projects have been built on the largest river systems. Smaller-scale hydro projects continue to the present. In the Pacific Northwest hatcheries have largely replaced native salmon stocks since the 1970s. Careful regulation of water supplies to balance fisheries with electrical production has been attempted since the Bonneville Dam project in the 1950s. A drastic decline in fisheries has generally accompanied these effects on riparian systems.

Commercial forestry plantations were first established in the 1930s to 1960s. Problems with poor seed selection and poor regeneration techniques plagued early attempts at forest plantations. Since the 1970s, genetic improvement and refined nursery techniques have radically improved reforestation success. Intensive forestry—including thinning, fertilization, and pest control—has been implemented only in the past two decades and is practiced primarily on private industrial lands in the Pacific Northwest. Natural forest regeneration is relied upon rather than planting nursery stock for most lands within the north coast region. Little information is available on the genetic architecture of north coast forest species and how it is being influenced by forest practices. For example, since valley bottoms are cut first, most seed could come from subalpine or transitional higher-elevation forests, resulting in regeneration with less adapted stock than if seed were coming in primarily from adjacent low-elevation forests. Habitat fragmentation further complicates assessments of genetic- and population-level responses of crop, weed, and auxiliary species in managed landscapes.

C. Social Changes

Rapid settlement of the west coast paralleled social and economic changes in North America over the past four decades. The 1950s through the 1970s were the most rapid years of population growth and increase of energy use of any period in history. In our transect the most rapid population growth and radical transformation of landscapes occurred in the south coast, particularly on semiarid lands (Fig. 4). In 1980, 70% of the people on the west coast of North America lived in California, a post-1950 increase in population density of 224%, compared to 173% for Oregon and Washington. In the future, even greater growth is expected in the south coast (Fig. 5). The area with the greatest expected population expansion corresponds to the area with the greatest water supply problems and with the most severe air and water pollution and transportation problems, which would likely be exacerbated with climate change.

The past five decades have also included a dramatic change from a strictly agrarian society to an increasingly urbanized and industrialized population. Current trends suggest continued movement toward urbanization, high-technology industries, and information-based services that will encourage development and construction on the remaining flat lands and river valleys and further reduce the agricultural and forest land base throughout the region. The relative gap between land values in the south coast and the Pacific Northwest should increase pressure for rapid urbanization of valuable agricultural lands along the north–south trending valleys of Oregon, Washington, and British Columbia. These changes will put increasing pressure on land

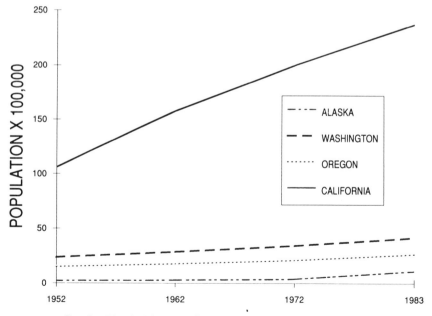

Fig. 4. Historical demographic patterns, west coast of North America.

Fig. 5. Population density and projected growth, west coast of North America and Chile.

owners to increase productivity and efficiency on a shrinking and less ecologically cohesive land base. Even in the absence of climate change, these managed landscapes will be increasingly stressed by human encroachment, competition for water resources, habitat fragmentation, air and water pollution, and pest and weed problems.

D. Relation of Managed Systems to Changes in Climate

The short period of human settlement in western North America prevents a detailed analysis of how changes in climate influenced these managed ecosystems. In similar climatic zones in northern Europe small variations in climate have had dramatic influences on agriculture and forestry. During the Middle Ages, for example, grains were grown at high latitudes in Scandinavia; conversely, crop failures and famine were widespread during the Little Ice Age. Interestingly, the period of maximum deforestation rates in Europe and North America corresponds to the onset of the Little Ice Age (Sagan *et al.*, 1979).

In the north coast, high year-to-year variation makes agriculture impractical in many areas that were formerly economically important. Drier periods in coastal Alaska also appear to have been associated with significant fire activity historically where fire is now rare. It is not known what proportion of these fires were human caused. Two centuries ago or more in the Pacific Northwest some tree species such as Douglas fir colonized coastal and high-elevation sites that are now dominated by cool-moist adapted species (Jan Henderson, USDA Forest Service, Olympic National Forest, personal communication). Natural regeneration following logging at midelevations in the Oregon Cascades can result in major shifts in dominant species. This suggests that the current climate differs from that of two to five centuries ago, when the dominant trees from the original stands were established. Significant fluctuations in economic activity and forest composition have occurred in response to global temperature fluctuations less than 1°C.

Since the late 1970s a 0.1°C per decade warming has been documented at long-term research sites in interior Alaska, providing a model of the effects of global warming on northern landscapes (Trenberth, 1990). This is similar to the predicted rate of warming in a global warming simulation [NASA Goddard Institute for Space Studies (GISS) GCM]. Agriculture is now possible on sites that were plagued with frequent crop failures until a few decades ago. Frequency of severe snow accumulation has also declined dramatically in the north coast, resulting in dramatic increases in large mammal populations that were formerly regulated by extreme winter conditions. Frequency of snow avalanches has also dramatically declined in coastal regions. Since the late 1800s severe winters have occurred at a frequency of about every 7–11 years in the north coast. The last major snowy winters in coastal Alaska occurred in the mid-1970s. Some authors suggest that this indicates global warming, but this warming in Alaska coincided with a strong cooling trend in the northern Pacific, while little change has occurred in other high-latitude regions. The recent warming in Alaska is most likely an anomalous shift in weather patterns that has been caused by trans-linkages with shifting tropical ocean circulation patterns (Trenberth, 1990).

IV. Future Conditions

A. Climate Models and Their Limitations

The GISS GCM predicts a 1 to 4°C warming within the next century and as much as a 10°C warming at high latitudes. As many other authors have noted, these predictions are difficult to use to study effects on ecosystems processes because of their imprecision and uncertainty, especially at regional scales. A key limitation of the models is their inability to predict variation in climate—the frequency of extreme weather events, including drought, flooding, and windstorms.

Precipitation and cloud cover can amplify or even reverse ecosystem effects of climate change. Few models are available that treat cloud cover realistically, and those that do generally predict smaller temperature increases. For this analysis we will assume the GISS model is a realistic approximation of future climate change, although it may tend to overestimate rate and magnitude of future temperature change.

With a doubling of CO_2, frequency of drought and moisture limitation of net primary productivity (NPP) would be expected to become more important in the southern and interior regions (Adams *et al.*, 1990; Boer *et al.*, 1990; Parry and Carter, 1989). At higher latitudes, especially in interior regions, temperature increases will be greater. Although not explicitly specified in the GISS model, these changes would imply an increase in susceptibility to fire and frequency of fire, because of more rapid drying of fuels and accumulation of fuel associated with forest decline or vegetation change. Increases in moisture and temperature may cause more intense convective activity, resulting in more lightning and thunderstorms. Decreased latitudinal temperature variation could lead to more extended hot dry periods and increased frequency of drought.

B. Ecosystem Functional Responses

1. Direct Effects of CO$_2$ Enrichment. Elevated CO$_2$ has been shown experimentally to result in higher rates of photosynthesis and lower respiration and water loss, leading to higher plant productivity in individual plants, but ecosystem-scale effects are still unknown (Mooney *et al.*, 1991). Because these effects are greater above 18°C, they would have the greatest influence in the southern, central, and interior regions of our coastal transect. CO$_2$ enrichment could lead to higher leaf area and to changes in plant chemistry (e.g., higher C/N ratio, more secondary compounds). Experimental evidence suggests that agricultural plants could become more palatable to insect pests, increasing control costs or decreasing yields (Campbell, 1989; Haynes, 1982). C$_3$ plants will benefit more than C$_4$ plants from CO$_2$ enrichment, thus potentially changing their competitive relationships (Patterson and Flint, 1980).

2. Temperature Effects. Other things being equal, increases in temperature should result in longer growing seasons, more efficient photosynthesis, greater crop yields, and a northward expansion of agriculture (Adams *et al.*, 1990; Boer *et al.*, 1990). In the southern and interior regions these yields will also be constrained by larger water losses or costs for increased irrigation. These costs may be compounded by the greater susceptibility of crop plants to insects and disease when afflicted by water stress. The beneficial effects of warmer temperatures will be greatest north of 48°N, where low temperatures now limit crop productivity. Warmer temperatures should increase rates of nutrient release, leading to increases in productivity, especially in the north where moisture is least limiting (e.g., Melillo, 1983; Pastor and Post, 1988).

Warmer temperatures will have some negative effects on plant growth and yield. Droughts and thermal stress will destroy agricultural crops or impede forest regeneration, especially in the south. Plant species that are near their southern limits of range or are near their limits of getting sufficient cooling to stimulate dormancy will be inefficient or even incapable of benefiting from longer growing seasons. Similarly, ovule sac budset may occur prematurely, resulting in flower, fruit, or seed crop failures. Dramatically warmer winters may increase winter desiccation not only in high-elevation areas but in some low-elevation sites as well.

3. Effects on Decomposition. Changing temperature and precipitation will influence decomposition rates directly by altering rates of chemical and physiological processes and indirectly by changing nutrient availability and litter chemistry. The response will be highly site specific, depending on the present soil temperature and moisture status and on the change, if any, in the local precipitation patterns. GCMs predict that some regions will become warmer and wetter, while others will become warmer and drier; the direction of change must be known to predict the response of a particular site or region.

Decomposition rates will increase in response to warming on most mesic sites, because the concurrent minor changes in the precipitation regime will not cause these systems to become either too wet or too dry to support decomposition. In contrast, slower decomposition may be expected on xeric sites east of the Cascade Mountains, where summer decomposition is drought limited and winter decomposition is limited by subfreezing temperatures. Winter warming will probably not compensate for the decomposition reduction caused by extended periods of summer drought likely to occur following a warming and drying trend. Decomposition at coastal and coastal montane sites continues through the mild winters of this region. Sites beyond the coastal fringe frequently experience summer drought sufficient to reduce decomposition in the upper litter layers and in small-diameter woody debris. Decomposition in this region will be enhanced if conditions become warmer and wetter, particularly if summer rainfall, fog, and cloud cover increase.

Coastal sites, particularly in British Columbia and southeast Alaska, contain large areas of poorly drained soils where decomposition is retarded by excess moisture. Some of these poorly drained sites would benefit from warmer and drier conditions, but it is more likely that climatic change will bring warmer and wetter conditions. Increased precipitation will cause flooding of some marginally drained sites, further reducing decomposition rates and possibly causing a shift from forest to peatland vegetation.

The uncertainty and low spatial resolution of the current generation of GCMs prevent reliable forecasting of the trajectory of change for specific sites, even with detailed knowledge of current site conditions. This presents a major challenge to resource managers attempting adaptive management schemes. The uncertainty is less where there is general agreement on the predicted precipitation response; for some areas, however, model predictions differ, particularly in zones separating regions of increasing precipitation from those of decreasing precipitation. Efficient adaptive management demands rainfall predictions with higher resolution and lower uncertainty.

Decomposition and nutrient availability interact in a positive feedback loop. Accelerated decomposition will generally enhance nutrient availability; if improved nutrition is reflected in lower carbon/nutrient ratios in plant litter, decomposition rates will increase more than might be expected from the changes in moisture and temperature alone. The converse is expected where decomposition is retarded by changing climatic conditions.

Decomposition may change dramatically where climate-induced changes in plant community composition engender changes in litter chemistry (Pastor and Post, 1988). For example, conifers with poorly decomposable litter might be replaced by hardwoods that produce more readily decomposed litter with a lower lignin/N ratio. Where precipitation is reduced, mesic species may be replaced by xeric species, which tend to have higher lignin/N ratios. This could greatly exacerbate the drought-induced reduction of decomposition and nutrient availability.

The projected doubling of atmospheric CO_2 further clouds the issue, potentially influencing rates of C fixation and water use efficiency. Increased C fixation in the absence of improved nutrient availability could yield plant tissues and litter with lower nutrient contents. For example, under enhanced CO_2, the C/N ratio of salt marsh plant tissues increased, but total N per unit area remained constant (Curtis *et al.*, 1989). With enhanced CO_2, increased water use efficiency may allow plant species to persist under increased drought stress, slowing the shift to more drought-adapted species with more decay-resistant litter.

C. Effects of Climate Change on Disturbance

Changes in the frequency and intensity of disturbance will likely be the earliest observed effect of climate change (Graham *et al.*, 1990; Overpeck *et al.*, 1990). In unmanaged systems, disturbance will hasten the rate of vegetation response to climatic change. This effect has been observed in both paleoecological studies and simulation studies. Frequent anthropogenic disturbances will mask the effects of climate-driven changes in disturbance frequency in many managed ecosystems but not in forests. Long-lived forest crops are more susceptible to damage by infrequent catastrophic events (wind, fire, pests, pathogens, frost) than are annual agricultural crops (Cannell *et al.*, 1989), so increased disturbance severity could reduce greatly the yield from managed forests.

Under the GISS global change scenario stronger thermal gradients across the coast would imply greater intensity and duration of windstorms. Also, increased precipitation could lead to more damaging and more frequent floods. These disturbances should lead to range extensions and increased abundance of weedy species and fire-adapted species and higher risks and reduced yields for forest and agricultural lands. Long-lived crops such as forest plantations will be particularly susceptible to these changes in disturbance frequency and intensity (Graham *et al.*, 1989; Peters, 1990). Emissions of CO_2 and loss of biomass will be most significant for wetland soils and forests, implying major changes in site productivity and land capability (Prentice and Fung, 1990).

Strong winds and thunderstorms are expected to be more frequent following global warming (Overpeck *et al.*, 1990). Wind is presently a major disturbing agent in forests of the northern coast, eastern North America, and Europe (Stephens, 1956; Harris, 1989). In the United Kingdom, for example, windthrow hazard limits the maximum tree height allowed in plantation forests and also limits the ability to thin young stands (Cannell *et al.*, 1989). Increased windthrow hazard will limit silvicultural options at most management intensities, because it affects stand improvement treatments, harvesting, salvage, and expected yield (i.e., return on investment).

Fire is not a major ecological force along the northern coast, but it could become so under warmer and drier conditions, causing significant changes in forest N cycles. Windthrow is currently the most common natural stand-replacing disturbance in this region. After a major windthrow event, it is unlikely that much N is lost through volatilization, leaching, or biomass removal, since most of the understory and advance regeneration remain intact and the wind-thrown trees remain close to their original position. In contrast, significant amounts of N can be lost to the atmosphere during biomass burning (Kuhlbusch *et al.*, 1991) or through leaching from devegetated sites. To maintain productivity in this new disturbance regime, any large N losses must be replaced through fertilization or natural N_2 fixation by locally adapted *Alnus, Lupinus,* or *Dryas* species.

Changes in disturbance will have significant effects on forest composition throughout the transect but especially in the midnorthern latitudes, where catastrophic disturbance is now infrequent (Overpeck *et al.*, 1990). In Britain, for example, almost 50% of the current commercial forest would suffer a decline in yield due to increased winds under a 2° warming scenario (Cannell *et al.*, 1989). In the Pacific Northwest and California changes in fire frequency or intensity would have a profound influence on forest structure and productivity, more so than direct physiological responses to altered climatic regimes (Dale and Franklin, 1989; Ritchie, 1986). Increases in disturbance in forests also would dramatically increase rates of change in forest composition, since the overall direction and structure of mature forests are often determined by climatic or stochastic events during stand establishment.

D. Species Geographic Shifts

Although it may seem logical to assume that vegetation zones would simply move northward as global warming progresses, it is unlikely that such a uniform response to global warming would result (e.g., Brubaker, 1988). Each species has its own genetic tolerance or level of adaptation to changing climate and ability to reproduce or disperse within a given time frame. In addition, heterogeneous soil conditions can drastically impede or increase the spatial complexity of vegetation responses to global warming. In northern Great Britain, for example, paleoecological evidence suggests that a time lag of up to 500–1500 years in forest response to global warming was at least in part due to edaphic restraints to vegetation colonization (Pennington, 1986). In Alaska a variety of recent postglacial environments also demonstrates the dramatic influence of local soil conditions on the rate of revegetation. Whereas till-derived soils may develop rapidly following glacial retreat, outwash gravel and other coarse alluvial deposits may take several centuries to develop fully.

Global warming should influence forest species distributions and agricultural crop mixes throughout the west coast transect. The most dramatic shifts will probably occur in the boreal zone, where many conifers will be replaced by early seral hardwood species and southern species will be able to colonize northern locales (Bonan *et al.*, 1990; Kullman, 1979; Pastor and Post, 1988). Many studies suggest that boreal forests will decline dramatically on a global scale, to be replaced by cool temperate forests on wet sites or by cool steppe on dry sites (e.g., Bonan *et al.*, 1990; Emmanual *et al.*, 1985; Kullman, 1979; Pastor and Post, 1988). In Alaska these new forests would be assumed to include mostly the hardwood species that now occur only as early seral fire species such as birch and aspen.

These changes in vegetation will be triggered primarily by changes in soil drainage, nutrient availability, lengthening of the growing season, and increases in disturbance. Effects and even direction of change will probably depend heavily on soil characteristics (Pastor and Post, 1988). For example, well-drained soils in boreal regions may become sufficiently dry to force vegetation type changes from forest to steppe, whereas permafrost areas will decline and may begin to resemble the well-drained sites prior to climate change (e.g., peatland to forest). The magnitude of these effects will vary depending on whether one assumes permafrost thaws directly in response to increased surface temperature or whether the increased insulative effect of permafrost drying arrests these changes on plant growth and yield (Bonan et al., 1990).

The spatial patterning of soil discontinuities could affect species dispersion and the overall direction of plant succession on individual sites. These edaphic constraints on native forests should pose significant restraints on agricultural expansion northward, where thin and poorly differentiated soils now occur, especially in the xeric climates of the interior.

Climate change will have the greatest effect on plants growing near the northern or southern limits of their distribution. Trees at the southern limit of their distribution may now receive winter chilling barely sufficient to ensure proper release from dormancy (budbreak) in the spring (Cannell et al., 1989; Murray et al., 1989; McCreary et al., 1990). Significantly warmer winters will fail to chill these trees adequately, thus delaying budbreak. This delay in growth may negate the benefits of the predicted longer growing season. Insufficient chilling may also increase plant susceptibility to late frosts, and delayed budbreak will reduce growth during the favorable moisture regime of early spring, potentially increasing the damage from summer droughts (McCreary et al., 1990). Trees that presently receive winter chilling in excess of their requirements should benefit from longer growing seasons, since they will be able to begin spring growth earlier (Murray et al., 1989).

Douglas fir, a tree of considerable ecological and economic value in the Pacific Northwest, experiences barely adequate winter chilling in southwest and coastal Oregon and may be susceptible to major growth disruption following warmer winters (McCreary et al., 1990). Few options are available to managers; preventive breeding will be difficult, because chilling requirements do not vary greatly between seed zones, and natural adaptation will be slow, because of Douglas fir's longevity and strong genetic control of budbreak (McCreary et al., 1990).

Ecotonal areas will most likely also show the most rapid changes in species composition (e.g., Bonan et al., 1990; Kullman, 1979; Pastor and Post, 1988). Because the whole north–south transect is marked by a continual loss of species northward (and few that occur only in the north), these changes should occur throughout our region. Species at their southern range limits will be stressed by moisture loss, periodic drought, fire, increased insect activity, increased disease infection and growth, altered competitive relationships, and less physiological fitness. Just as global warming historically produced savannahs or grasslands where continuous forest dominated before, in the future grasslands should expand westward and northward.

The Siskiyou Mountains on the border of California and Oregon are at the southern limit of range for many coastal tree and shrub species and are the sole locale for several rare species (Franklin and Dyrness, 1973; Whittaker, 1960). Many tree and shrub species are moisture limited, so they could easily exceed their physiological limits with a warmer, drier climate or with increased incidence or intensity of fire. Clearing of forested sites on southwest-facing slopes in this sensitive ecotonal area, for example, often results in nearly permanent conversion of forest to grassland (Perry *et al.*, 1989). The southern limits of wet coastal forest in northern California would be expected to be susceptible to retreat northward under the GISS global warming scenario.

Timberline ascent and associated glacial retreat may be rapid and could be the first effects of global warming along the northern coast (Kullman, 1979; Pelto and Miller, 1990). Although timberlines throughout the world are expected to rise with global warming—because they are controlled by available energy for growth, reproduction, and starch accumulation—the timberlines in the Pacific coast are expected to be particularly sensitive to global climate change. Coastal timberlines are mostly determined not by extreme cold temperatures but by heavy accumulations of snow that does not melt until late spring or early summer. During most of the winter these subalpine areas have temperatures close to 0°C. Even at 1°C warming could result in nearly total loss of the alpine zone on many mountain and island areas and a severe fragmentation of alpine zones on the mainland. Rapid glacial recession should contribute to rapid vegetation change in these northern sites. Currently, timberline varies from 2000 m above sea level (ASL) at the southern limit of the coastal rain forest zone to approximately 500–700 m ASL in southeastern Alaska to <300 m ASL in south-central Alaska.

The projected raising of timberline 300–500 m in Scandinavia over a similar latitudinal range would imply a significant forest expansion throughout the northern coast, greatly expanding opportunities for forest management. Avalanches are another important constraint on forest growth in coastal Alaska and British Columbia that could be considerably reduced with global warming.

E. Plant–Animal Relationships

One complex aspect of global climate change along this diverse north–south gradient will be the independent adaptation of different groups of plants and animals to climate and how this will affect their interrelationships. Tightly coupled plant–animal interactions would be expected to be sensitive to climate changes, more so than any individual species, because of between-species variations in the scope or speed of response. Insects and other small animals may respond more quickly to changes in climate than crop plants, resulting in increased plant and ecosystem stress and crop losses.

The direct effects of enhanced CO_2 on plants may result in altered chemistry that should in turn influence herbivores and diseases (Mooney *et al.*, 1991; Campbell, 1989; Haynes, 1982; Dennis and Shreeve, 1991). Elevated CO_2 tends to increase C/N ratios and decrease food quality for herbivores. Changes in plant secondary compounds further complicate the net effect of global change on plant chemistry. Foliage produced during warm, dry spring periods (more common with climatic warming), for example, has been shown to increase pest growth rates and survival over that of insects fed foliage grown during late cool spring periods (Campbell, 1989).

Currently approximately 33% of agricultural crops in the United States are lost due to insect, pathogen, nematode, and weed pests; thus any modest change in pest or weed relationships to crops would be of critical economic importance (Pimentel, 1977). In recent decades expenditures on pesticides, herbicides, and related measures to reduce losses have increased, while their efficiency and effectiveness have continued to decline. New pest control strategies that incorporate more detailed analyses of all possible control measures within the context of the ecological characteristics of both crop and pest species and the environment itself have the potential of being more effective than traditional strategies but require even greater expenditures on control. Crop losses are generally greater in warm climates, suggesting that global warming may exacerbate pest problems especially in the southern portions of our North American gradient (Haynes, 1982).

Weed species will likely expand their range and their effectiveness in competing with crop species following global warming. Where weed species are C_3 plants and crops are C_4 plants, CO_2 enrichment will differentially favor weed species (e.g., Patterson and Flint, 1980). Increases in aridity may favor weed species, especially at lower latitudes.

Insect populations are well known for their close ties to climatic conditions, especially when extreme conditions develop. Epidemic outbreaks of both forest and agricultural pest species are often associated with abnormally warm summers (for egg development) and mild winters (higher over-winter survival)—both predicted because of global warming along the Pacific coast. Much of the year-to-year or region-to-region variation in grasshopper density and crop damage, for example, can be modeled as a function of climate, in particular the heat–precipitation ratio (Gage and Mukerji, 1977). In northern Alaska, some early outbreaks of spruce budworm were reported following a decade of abnormally warm spring and winter seasons. During this anomalously warm period elevated rates of tree mortality from bark beetle attack also occurred. As is now understood, much of this problem was due to altered phenological patterns of the beetles and the trees: the insects were flying and attacking trees when air temperatures rose above freezing, but the soils were still frozen, reducing the trees' ability to repel attack.

Nematodes also would be expected to flourish after global warming because they, along with other soil organisms, tend to increase their population size and virulence in warmer and wetter climates. Precipitation increases generally enhance nematode survival and dispersion. More virulent southern species would be expected to colonize northern sites after climatic warming. By damaging root systems, nematodes often predispose host plants to infection by both insects and plant diseases (Powell, 1971).

Plant diseases and their virulence are also closely related to climatic conditions (Rotem, 1978). The spread of spores is favored by dry weather and stronger winds—

both of which may be more frequent in interior locations. Increased precipitation along the coast might increase infection by fungi, especially in agricultural populations. Plant diseases are particularly harmful when host plants are under severe stress (nutritional or lack of moisture), which would be expected to be more commonplace, especially in southern agricultural regions.

V. Adaptive Management

Considering the number and magnitude of probable injurious effects of global climate change on managed ecosystems, abatement strategies or adaptive strategies clearly need to be developed, even before we fully understand or can precisely predict future climate change. Development of serious alternatives will require substantial time and scientific effort that, if not initiated promptly, may lead to a severe constraint on decision making and policy options in the future, assuming a dramatic change in climate does occur. In this section we present a few basic principles that could be applied to developing these strategies and offer some specific examples of how these might be employed.

A. Land Management Philosophy

Resource management usually has simplified ecosystems and restructured them to optimize the production of a specific resource under a narrowly defined set of environmental conditions (e.g., Perry and Maghembe, 1989). Prevailing forest management practices have been largely driven by the criteria of maximum wood fiber production per unit time and maximum yield at any point in time, with little consideration of sustainability or ecological integrity (Clary, 1986; Franklin, 1989; O'Toole, 1988; Perry and Maghembe, 1989). There is growing recognition of the value of forests for noncommodity uses such as maintenance of biological diversity, climate control, buffering nutrient export from agricultural systems, maintenance of water and air quality, recreation, aesthetics, soil erosion control, pest (or predator) control, education, and water management. However, the inability of society to assign economic values to such uses has allowed the sacrifice of some of these functions without full consideration of their societal benefits (Clary, 1986; Krutilla and Fisher, 1975).

Genetic breeding programs and agricultural and forestry practices have been developed to optimize productivity under the existing climatic and soil regime, presumably at the expense of adaptability to widely varying climatic regimes and ecosystem diversity. Natural ecosystems, in contrast, have evolved or at least can persist through widely varying climatic regimes and under a wide range of plant–animal interactions. The fitness of these species may be more a measure of their persistence through environmental extremes than of their adaptation to any given condition.

If new management systems or philosophies are to be developed to address global climate change, then an intermediate path of ecosystem management is needed that incorporates the essential characteristics of natural ecosystems while maintaining as much as possible the high productivity of currently managed sites. To implement this altered land management philosophy will require developing decision criteria that better incorporate the ideas of ecological sustainability, resilience, climate change, and their interactions (Battie, 1989; Ehrlich, 1989). In forestry, perhaps the greatest opportunity exists to develop new philosophies for land management that will not only improve forest resilience to climate change but also deal more effectively with other critical ecological issues (e.g., biodiversity, long-term site productivity, integrated pest management, and noncommodity resource management).

Mature forests are generally more resilient to climate change than newly regenerating forests. Once established, the mature canopy of a dense forest creates its own microclimate and can to a certain degree buffer climatic change, in contrast to open environments. The most environmentally sensitive stage in the development of a forest is generally the stand establishment phase, when new seedlings are established by planting nursery stock or by natural seeding, sprouting, or release of residual plants from the original mature forest (Connell and Slatyer, 1977; Oliver, 1981). For many plant species the niche can be defined more easily in terms of reproductive strategies than by the environmental or competitive interactions of the mature plant (Grubb, 1977). Thus, even a small change in climate could result in dramatically shifting species composition or changes to the trajectory of plant community development over long periods of time through its effects on plant reproductive biology (e.g., Egler, 1954; Lertzman, 1989). To manage forests for climate change there should be a distinct advantage to maximizing the proportion of mature forest.

To mitigate climate change, the mature stage in forest development needs to be prolonged by changing management emphasis toward high-value wood products that require longer rotations or by de-emphasizing clear felling and other traditional even-aged management practices. Better maintenance of "ecological legacies" or structural and biological components from the previous generation of forest can play a critical role in maintaining ecosystem resiliency and diversity (Harmon et al., 1986; Perry et al., 1989). Although absolute fiber yield may be reduced by altered management practices, ecological, long-term productivity, social, and other benefits from increased persistence of mature forest may significantly improve the overall value of these forests to society. Development of small-patch or uneven-age management techniques could make the development of more complex multispecies management more practical. Increasing the genetic diversity of ecosystems by managing for multiple dominant species and breeding for increased heterozygosity also should have benefits for ecosystem resilience and buffering against climate change (Kellison and Weir, 1987). The maintenance of more intact mature forest should serve as a major benefit for plant and animal migration or dispersal by providing suitable habitats or at least migration corridors.

B. "New Forestry"

A fundamental tenet of "new forestry" or "new perspectives in forestry" or "ecosystem management" is that land management decisions need to be based on much broader criteria than have been used in the past (Franklin, 1989). Besides specific changes in forest cultural practices, this requires changes in how decisions are made in both a landscape and temporal context. A major debate emerging in forest landscape management, for example, is the issue of dispersion of cutting units.

In recent decades the U.S. Forest Service has had a national policy of dispersing cutting units throughout the landscape to minimize erosion and harm to wildlife and fisheries. However, this cutting pattern has caused rapid fragmentation of old-growth or mature forest patches, resulting in rapid loss of bird and mammal species (Franklin and Forman, 1987; Harris, 1984). Proponents of new forestry are now proposing that it would be less ecologically destructive if cutting units were aggregated together, assuming less destructive cutting practices can be developed and that *a priori* attention is paid to maintenance of migration corridors and preservation of natural habitat patches within the matrix of a managed landscape.

In agricultural landscapes, many management objectives and principles of new forestry or new resource economics should apply as well. Resilience and sustainability of agricultural systems could be enhanced with greater attention to maintenance of landscape connectivity and biological function. A wider range of species or varieties, combined with a cohesively functioning landscape, could reduce some of the pest and weed problems expected after climate change. Breeding strategies geared more toward heterozygosity, persistence, and resilience in the face of more unpredictable climatic regimes could decrease crop losses, especially during climatic extremes (Tangley, 1988).

Many commonly employed land use planning tools seem unable to address the key issues raised by climate change. In particular, the use of natural plant assemblages to define ecological characteristics of sites assumes a static environment. A more dynamic model of ecological capability is needed (Kimmins, 1990). Yield models also make many assumptions about a constant climate. More realistic models of yield in relation to climate change, and of ecosystem change in response to management, could greatly ease decision making for land use planners.

VI. Summary

Managed ecosystems reflect the economic and social condition of the people who manage them more than they reflect absolute limits of environmental conditions. Social, technological, and economic factors have largely dictated a rapid progression from low-intensity to high-intensity land management during recent decades. Although concern about sustainability and the long-term consequences of these new more intensive cultivation strategies has grown over time, the prospect of climate change gives a new sense of immediacy and relevance to the development of alternative management models.

As compared with natural systems, considerable latitude exists in developing adaptive strategies for agronomic systems over a wide range of global warming scenarios. Northward expansion of agriculture and an increase in yield are likely at high latitudes. Along the southern coast, availability and cost of water will determine the severity of global warming consequences. Changes in the relationships of pests and pathogens to their hosts will increase the uncertainty of predictions of ecosystem response to climate change.

Although intensive breeding programs may be able to help develop appropriate stock for short-lived agricultural species, forest species will be much slower to adapt. We would expect ecotonal areas, especially at the southern forest and mediterranean boundaries, to be highly sensitive to climatic changes and show a response before many other ecosystem types. Species or ecosystems sensitive to climate change need to be identified to monitor effects of climate change.

Development of a diverse array of management strategies, especially on a land-scape scale, will be required to minimize negative effects of climate change. Whereas current trends in ecosystem management emphasize maximizing resource prod-uctivity under a narrowly defined set of environmental conditions, management of ecosystems for climate change may require a fundamental shift toward maximizing resilience and stability over a wider range of environmental conditions. In particular, management strategies that emphasize landscape diversity, long-term sustainability of productivity, and landscape connectivity need to be explored. New forestry and similar initiatives in forestry and agricultural economics need to be more fully evaluated so that they can be incorporated into a cohesive strategy for adapting to climate change. Management of both crop and noncrop species and their inter-relationships need to be better emphasized in management planning.

References

Adams, R. M., Rosenzweig, C., Peart, R. M., Ritchie, J. T., McCarl, B. A., Glyer, J. D., Curry, R. B., Jones, J. W., Boote, K. J., and Allen, L. H., Jr. (1990). Global climate change and US agriculture. *Nature* **345,** 219–224.

Alaback, P. B. (1990). Dynamics of old-growth temperate rainforests in southeast Alaska. *In* "Proceedings of the 2nd Glacier Bay Science Symposium, September 19–22, 1988, Gustavus, Alaska" (A. M. Milner and J. D. Wood, eds.), pp. 150–153. National Park Service, Alaska Region, Anchorage, Alaska.

Alaback, P. B. (1991). Comparative ecology of temperate rainforests of the Americas along analogous climatic gradients. *Rev. Chil. Hist. Nat.* **64,** 399–412.

Alexander, E. B. (1988). Rates of soil formation: implications for soil-loss tolerance. *Soil Sci.* **145,** 37–45.

Axelsson, B. (1985). Increasing forest productivity and value by manipulating nutrient availability. *In* "Forest Potentials: Productivity and Value" (R. Ballard, P. Farnum, G. A. Ritchie, and J. K. Winjum, eds.), pp. 5–37. Weyerhaeuser Science Symposium No. 4. Weyerhaeuser Company, Tacoma, Wash-ington.

Battie, S. (1989). Sustainable development: challenges to the profession of agricultural economics. *Am. J. Agric. Econ.* **71,** 1083–1101.

Beuter, J. H., Johnson, K. N., and Scheurman, H. L. (1976). Timber for Oregon's tomorrow, an analysis of reasonably possible occurrences, Res. Bull. 19, Oregon State Univ., Forestry Research Laboratory, Corvallis.

Binkley, D. (1983). Ecosystem production in Douglas-fir plantations: interaction of red alder and site fertility. *For. Ecol. Manage.* **5,** 215–227.

Boer, M. M., Koster, E. A., and Lundberg, H. (1990). Greenhouse impact in Fennoscandia—preliminary findings of a European workshop on the effects of climatic change. *Ambio* **19,** 2–10.

Bonan, G. B. (1990). Carbon and nitrogen cycling in North American boreal forests. II. Biogeographic patterns. *Can. J. For. Res.* **20,** 1077–1088.

Bonan, G. B., Shugart, H. H., and Urban, D. L. (1990). The sensitivity of some high-latitude boreal forests to climatic parameters. *Climatic Change* **16,** 9–29.

Brubaker, L. B. (1988). Vegetation history and anticipating future vegetation change. *In* "Ecosystem Management for Parks and Wilderness" (J. K. Agee and D. R. Johnson, eds.), pp. 41–61. Univ. of Washington Press, Seattle.

Campbell, I. M. (1989). Does climate affect host-plant quality? Annual variation in the quality of balsam fir as food for spruce budworm. *Oecologia* **81,** 341–344.

Cannell, M. G. R., Grace, J., and Booth, A. (1989). Possible impacts of climatic warming on trees and forests in the United Kingdom: a review. *Forestry* **62,** 337–364.

Clary, D. A. (1986). "Timber and the Forest Service." Univ. of Kansas Press, Lawrence.

Connell, J. H., and Slatyer, R. O. (1977). Mechanisms of succession in natural communities and their role in community stability and organization. *Am. Nat.* **111,** 1119–1144.

Cooley, K. R. (1990). Effects of CO_2-induced climatic changes on snowpack and streamflow. *Hydrol. Sci. J.* **35,** 511–522.

Curtis, P. S., Drake, B. G., and Whigham, D. F. (1989). Nitrogen and carbon dynamics in C_3 and C_4 estuarine marsh plants grown under elevated CO_2 *in situ. Oecologia* **78,** 297–301.

Dale, V. H., and Franklin, J. F. (1989). Potential effects of climate change on stand development in the Pacific Northwest. *Can. J. For. Res.* **19,** 1581–1590.

Dalichow, F. (1972). "Agricultural Geography of British Columbia." Versatile Publications, Vancouver.

Davis, K. M., Clayton, B. D., and Fischer, W. C. (1980). Fire ecology of Lolo National Forest habitat types. Gen. Tech. Rep. INT-79, USDA Forest Service, Missoula, Montana.

Dennis, R. L. H., and Shreeve, T. G. (1991). Climatic change and the British butterfly fauna: opportunities and constraints. *Biol. Conserv.* **55,** 1–16.

Eamus, D., and Jarvis, P. G. (1989). The direct effects of increase in the global atmospheric CO_2 concentration on natural and commercial temperate trees and forests. *Adv. Ecol. Res.* **19,** 1–55.

Egler, F. E. (1954). Vegetation science concepts. 1. Initial floristic composition—a factor in old field vegetation development. *Vegetatio* **4,** 412–417.

Ehrlich, P. R. (1989). The limits to substitution: meta-resource depletion and a new economic–ecological paradigm. *Ecol. Econ.* **1,** 9–16.

Emanuel, W. R., Shugart, H. H., and Stevenson, M. P. (1985). Climatic change and the broad-scale distribution of terrestrial ecosystem complexes. *Climatic Change* **7,** 29–43.

Emmingham, W. H., and Waring, R. H. (1977). An index of photosynthesis for comparing forest sites in western Oregon. *Can. J. For. Res.* **7,** 165–174.

Farr, W. A., and Harris, A. S. (1979). Site index of Sitka spruce along the Pacific coast related to latitude and temperatures. *For. Sci.* **25,** 145–153.

Franklin, J. F. (1989). Toward a new forestry. *Am. For.* **95,** 37–44.

Franklin, J. F., and Dyrness, C. T. (1973). Natural vegetation of Oregon and Washington, General Technical Report PNW-8. USDA Forest Service, Pacific Northwest Forest and Range Experiment Station, Portland, Oregon.

Franklin, J. F., and Forman, R. T. T. (1987). Creating landscape patterns by forest cutting: ecological consequences and principles. *Landscape Ecol.* **1,** 5–18.

Franklin, J. F., and Waring, R. H. (1980). Distinctive features of the northwestern coniferous forest: development, structure, and function. *In* "Forests: Fresh Perspectives from Ecosystem Analysis" (R. H. Waring, ed.), Proceedings of the 40th Annual Biology Colloquium, pp. 59–86. Oregon State Univ. Press, Corvallis.

Gage, S. H., and Mukerji, M. K. (1977). A perspective of grasshopper population distribution in Saskatchewan and interrelationship with weather. *Environ. Entomol.* **6,** 469–479.

Gholz, H. L. (1982). Environmental limits on aboveground net primary production, leaf area, and biomass in vegetation zones of the Pacific Northwest. *Ecology* **63,** 469–481.

Graham, R. L., Turner, M. G., and Dale, V. H. (1990). How increasing CO_2 and climate change affect forests. *Bioscience* **40,** 575–587.

Grubb, P. J. (1977). The maintenance of species richness in plant communities: the importance of the regeneration niche. *Biol. Rev.* **52,** 107–145.

Harmon, M. E., Franklin, J. F., Swanson, F. J., Sollins, P., Gregory, S. V., Lattin, J. D., Anderson, N. H., Cline, S. P., Aumen, N. G., Sedell, J. R., Lienkaemper, G. W., Cromack, K., Jr., and Cummins, K. W. (1986). Ecology of coarse woody debris in temperate ecosystems. *Adv. Ecol. Res.* **15,** 133–302.

Harmon, M. E., Ferrell, W. K., and Franklin, J. F. (1990). Effects on carbon storage of conversion of old-growth forests to young forests. *Science* **247,** 699–702.

Harr, R. D. (1982). Fog drip in the Bull Run municipal watershed, Oregon. *Water Resour. Bull.* **18,** 785–789.

Harris, A. S. (1989). Wind in the forests of southeast Alaska and guides for reducing damage, Gen. Tech. Rep. PNW-GTR-244. USDA Forest Service, Pacific Northwest Research Station, Portland, Oregon.

Harris, L. D. (1984). "The Fragmented Forest." Univ. of Chicago Press, Chicago.

Hawkins, B. J., and Sweet, G. B. (1989). Evolutionary interpretation of a high temperature growth response in five New Zealand forest tree species. *N.Z. J. Bot.* **27,** 101–107.

Haynes, D. L. (1982). Effects of climate change on agricultural plant pests. *In* "Environmental and Societal Consequences of a Possible CO_2-Induced Climate Change," DOE/EV/10019-10, Vol. II, Part 10, pp. 1–35. U.S. Department of Energy, Washington, D.C.

Hemstrom, M. A., and Franklin, J. F. (1982). Fire and other disturbances of the forests in Mount Rainier National Park. *Quat. Res.* **18,** 32–51.

Heusser, C. J. (1960). "Late-Pleistocene Environments of North Pacific North America." Special Publication No. 35, American Geographical Society, Washington, D.C.

Holdridge, L. R. (1947). Determination of world plant formations from simple climatic data. *Science* **105,** 367–368.

Hyams, E. (1952). "Soil and Civilization." Thames and Hudson, London. [Reprinted 1976, Harper Colophon, New York.]

Johannessen, C. L. (1971). The vegetation of the Willamette Valley. *Ann. Assoc. Am. Geogr.* **61,** 286–302.

Kellison, R. C., and Weir, R. J. (1987). Breeding strategies in forest tree populations to buffer against elevated atmospheric carbon dioxide levels. *In* "The Greenhouse Effect, Climate Change, and U.S. Forests" (W. E. Shands and J. S. Hoffman, eds.), pp. 285–293. The Conservation Foundation, Washington, D.C.

Kimmins, J. P. (1990). Modelling the sustainability of forest production and yield for a changing and uncertain future. *For. Chron.* **66,** 271–280.

Krutilla, J. V., and Fisher, A. C. (1975). "The Economics of Natural Environments: Studies in the Valuation of Commodity and Amenity Resources." Johns Hopkins Univ. Press, Baltimore, Maryland.

Kuhlbusch, T. A., Lobert, J. M., Crutzen, P. J., and Warneck, P. (1991). Molecular nitrogen emissions from denitrification during biomass burning. *Science* **351,** 135–137.

Kullman, L. (1979). Change and stability in the altitude of the birch tree limit in the southern Swedish Scandes, 1915–1975. *Acta Phytogeogr. Suec.* **65,** 1–121.

Lertzman, K. (1989). Gap-phase community dynamics in a sub-alpine old growth forest. Ph.D. dissertation, Univ. of British Columbia, Vancouver.

Mader, H. (1990). Wildlife in cultivated landscapes: introduction. *Biol. Conserv.* **54,** 167–173.

McClellan, M. H. (1991). Soil carbon and nutrient dynamics of windthrow chronosequences in spruce-hemlock forests of southeast Alaska. Ph.D. thesis, Oregon State Univ., Corvallis.

McCreary, D. D., Lavender, D. P., and Hermann, R. K. (1990). Predicted global warming and Douglas-fir chilling requirements. *Ann. Sci. For.* **47,** 325–330.

Meentemeyer, V. (1978). Macroclimate and lignin control of litter decomposition rates. *Ecology* **59,** 465–472.

Melillo, J. M. (1983). Will increases in atmospheric CO_2 concentrations affect decay processes? *In* "Annual Report of the Ecosystems Center," pp. 10–11. Woods Hole Marine Biology Laboratory, Woods Hole, Massachusetts.

Mooney, H. A. (ed.) (1977). "Convergent Evolution in Chile and California. Mediterranean Climate Ecosystems." US/IBP Synthesis Series 5. Dowden, Hutchinson & Ross, Stroudsburg, Pennsylvania.

Mooney, H.A., Drake, B. G., Luxmoore, R. J., Oechel, W. C., and Pitelka, L. F. (1991). Predicting ecosystem responses to elevated CO_2. *Bioscience* **41,** 96–104.

Morris, W. G. (1934). Forest fires in western Oregon and western Washington. *Oregon Hist. Q.* **35,** 313–339.

Murray, M. B., Cannell, M. G. R., and Smith, R. I. (1989). Date of budburst of fifteen tree species in Britain following climatic warming. *J. Appl. Ecol.* **26,** 693–700.

Oliver, C. D. (1981). Forest development in North America following major disturbances. *For. Ecol. Manage.* **3,** 153–168.

O'Toole, R. (1988). "Reforming the Forest Service." Island Press, Washington, D.C.

Overpeck, J. T., Rind, D., and Goldberg, R. (1990). Climate-induced changes in forest disturbance and vegetation. *Nature* **343,** 51–53.

Parry, M. L., and Carter, T. R. (1989). An assessment of the effects of climatic change on agriculture. *Climatic Change* **15,** 95–116.

Pastor, J., and Post, W. M. (1988). Response of northern forests to CO_2-induced climate change. *Nature* **334,** 55–58.

Patterson, D. T., and Flint, E. P. (1980). Potential effects of global atmospheric CO_2 enrichment on the growth and competitiveness of C_3 and C_4 weed crop plants. *Weed Sci.* **28,** 71–75.

Pelto, M. S., and Miller, M. M. (1990). Mass balance of the Taku Glacier, Alaska 1946–1986. *Northwest Sci.* **64,** 121–130.

Pennington, W. (1986). Lags in adjustment of vegetation to climate caused by the pace of soil development: evidence from Britain. *Vegetatio* **67,** 105–118.

Perry, D. A., and Maghembe, J. (1989). Ecosystem concepts and current trends in forest management: time for reappraisal. *For. Ecol. Manage.* **26,** 123–140.

Perry, D. A., Amaranthus, M. P., Borchers, J. G., Borchers, S. L., and Brainerd, R. E. (1989). Bootstrapping in ecosystems. *Bioscience* **39,** 230–237.

Peters, R. L. (1990). Effects of global warming on forests. *For. Ecol. Manage.* **35,** 13–33.

Pimentel, D. (1977). Ecological basis of insect pest, pathogen and weed problems. *In* "The Origin of Pest, Parasite, Disease and Weed Problems" (J. M. Chenet and G. R. Sagar, eds.), pp. 3–31. Blackwell Scientific Publications, Oxford.

Powell, N. T. (1971). Interaction of plant parasitic nematodes with other disease causing agents. *In* "Plant Parasitic Nematodes" (B. M. Zackerman, W. F. Mai, and R. A. Rhodes, eds.), Vol. II, pp. 119–136. Academic Press, New York.

Prentice, K. C., and Fung, I. Y. (1990). The sensitivity of terrestrial carbon storage to climate change. *Nature* **346,** 48–51.

Ritchie, J. C. (1986). Climate change and vegetation response. *Vegetatio* **67,** 65–74.

Rotem, J. (1978). Climatic and weather influence on epidemics. *In* "Plant Disease" (J. G. Horsfall and E. B. Cowling, eds.), Vol. II, pp. 317–337. Academic Press, New York.

Ruth, R. H., and Harris, A. S. (1979). Management of western hemlock-Sitka spruce forests for timber production, General Technical Report PNW-88. USDA Forest Service, Pacific Northwest Forest and Range Experiment Station, Portland, Oregon.

Sagan, C., Toon, O. B., and Pollack, J. B. (1979). Anthropogenic albedo changes and the earth's climate. *Science* **206,** 1363–1368.

Saunders, D. A., Hobbs, R. J., and Margules, C. R. (1991). Biological consequences of ecosystem fragmentation: a review. *Conserv. Biol.* **5,** 18–32.

Sedell, J. R., and Duval, W. S. (1985). Water transportation and storage of logs. Influence of forest and rangeland management on anadromous fish habitat in western North America, No. 5. Gen. Tech. Rep. PNW-186, USDA Forest Service, Pacific Northwest Experiment Station, Portland, Oregon.

Simons, P. (1988). Après ski le déluge. *New Sci.* **117**(1595), 49–52.

Stephens, E. P. (1956). The uprooting of trees: a forest process. *Soil Sci. Soc. Am. Proc.* **20,** 113–116.

Stephenson, N. L. (1990). Climatic control of vegetation distribution: the role of the water balance. *Am. Nat.* **135,** 649–670.

Swanston, D. N., and Swanson, F. J. (1976). Timber harvesting, mass erosion, and steepland forest geomorphology in the Pacific Northwest. *In* "Geomorphology and Engineering" (D. R. Coates, ed.), pp. 199–221. Dowden, Hutchinson & Ross, Stroudsburg, Pennsylvania.

Tangley, L. (1988). Preparing for climate change. *Bioscience* **38,** 14–18.

Trenberth, K. E. (1990). Recent observed interdecadal climate changes in the northern hemisphere. *Bull. Am. Meteorol. Soc.* **71,** 988–993.

Ugolini, F. C., and Mann, D. H. (1979). Biopedological origin of peatlands in south east Alaska. *Nature* **281,** 366–368.

Van Cleve, K., Chapin, F. S., III, Flanagan, P. W., Viereck, L. A., and Dyrness, C. T., eds. (1986). "Forest Ecosystems in the Alaskan Taiga." Springer-Verlag, New York.

Veblen, T. T., and Markgraf, V. (1988). Steppe expansion in Patagonia? *Quat. Res.* **30,** 331–338.

Viereck, L. A., Dyrness, C. T., Van Cleve, K., and Foote, M. J. (1983). Vegetation, soils, and forest productivity in selected forest types in interior Alaska. *Can. J. For. Res.* **13,** 703–720.

Whittaker, R. H. (1960). Vegetation of the Siskiyou Mountains, Oregon and California. *Ecol. Monogr.* **30,** 279–338.

Zobel, D. B., McKee, A. W., Hawk, G. M., and Dyrness, C. T. (1976). Relationships of environment to composition, structure and diversity of forest communities of the central Western Cascades of Oregon. *Ecol. Monogr.* **46,** 135–156.

C H A P T E R 23

Global Warming and Human Impacts on Landscapes of Chile

EDUARDO R. FUENTES   MAURICIO R. MUÑOZ

I. Introduction

Global climate change will affect not only climate but also the whole geographic space of human lives, the abiotic component as well as the biotic one and their interrelationships. Some critical relationships between humans and their surroundings could change and affect their perception of the environment. In this case, humans can be expected to affect the landscapes where they live in patterns and with intensities that could be quite unlike the present ones. In this chapter we explore how these changes might develop and their landscape-level consequences. We concentrate on the Chilean population, but our analyses and conclusions could also be valid for other nonindustrialized countries where there is a very close coupling between landscape resources and human well-being.

We start by describing current and historical interactions between humans and landscapes and then propose some hypotheses on the dynamics that Chileans could have in association with global climate change. A cautionary note is necessary. Obviously, we cannot predict what will happen tomorrow, much less what will occur in the coming decades. At most we can extrapolate the ecological history of the past 500 years to the new climate scenarios.

In a way, historic extrapolation provides the worst possible scenarios because it assumes that nothing has been learned from past experience, that current environmental concerns will have no real impact on future activities, and that, just as in the past, there will be little or no effective environmental planning. This approach gives

329

a pessimistic, constrained forecast because it does not describe the almost infinite and truly unpredictable developments that may occur.

We can extrapolate from the past and show what could happen under the new circumstances of global climate change. Human actions have been so important and pervasive in structuring Chilean landscapes that current spatial patterns cannot be understood without acknowledging their influence. These actions are not likely to diminish in the near future, and it is of the highest priority to attempt to anticipate their future directions. Human impacts might well override any possible environmental benefits associated with climate change. Therefore, we believe our extrapolations could be useful, in spite of all their limitations, by uncovering some patterns and helping to design strategies to confront global change, thus avoiding the possibility that the worst possible scenario becomes real.

II. Current Situation

Three main attributes characterize the current scenario and the likely human impacts associates with global warming: (1) geographic assets and constraints, (2) human population trends, and (3) human attitudes toward the environment. We describe these attributes briefly and then state the hypotheses that relate them to global warming.

A. Geographic Constraints and Assets

Chile is a mountainous country (IGM, 1983) with more than 80% of its surface imposing topographic restrictions of some kind to agriculture, grazing, and even forestry (Fig. 1). Except for the central depression, coastal terraces, and larger valley, most of the country has topographic constraints for agricultural use (CORFO, 1966; Rodriquez, 1989).

Climate in Chile is largely constrained by two main factors: its latitudinal extension, from tropical (18°S) to temperate (55°S) latitudes, and its position in relation to the polar and subtropical (~ 25–30°S) high-pressure centers. The seasonal play between the large Antarctic cold high-pressure center and the subtropical high-pressure anticyclone, largely explains the current latitudinal rainfall belts in Chile (van Husen, 1967).

As a consequence of its latitudinal extension and location, Chile has two opposing continuous gradients: a north-to-south decreasing temperature gradient and a north-to-south increasing rainfall gradient (Fig. 2). Within these gradients at least three main geographic regions can be distinguished (Weischet, 1970; di Castri and Hajek, 1976):

1. The Norte Grande (Big North, 18 to 27°S). Rainfall-carrying fronts coming from the southwest rarely reach this region, making it a very dry desert. Temperatures tend to be high during the day and low during the night. Life is largely restricted to a few streams coming down from the high Andes, small patches of coastal highlands receiving fog from the ocean, and a small area of Altiplano, receiving summer rainfall

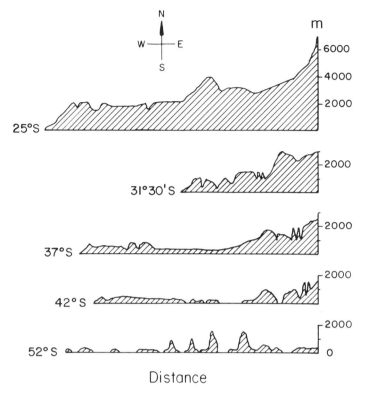

Fig. 1. Topographic profiles of Chile. From top to bottom, the Norte Grande, central Chile, and the Sur Grande. Notice the general absence of extended flatlands in the northern part of central Chile (second from top) and in the Sur Grande. The third graph shows the development of the central valley, between the Andes (east) and the coastal ranges (west). Some of the most productive landscapes are found in the valley. Desertification has been most severe in the coastal ranges and northern central Chile. See text for further discussion.

from the Atlantic climate system. People are scarcely settled in the Norte Grande (IGM, 1983), except for several coastal cities associated with copper mining and fishing.

2. Central Chile (~27 to 40°S). This area has a subtropical winter rainfall regime. With increasing latitude, fronts reach the region with increasing frequency (van Husen, 1967). At low latitudes (30 to 33°S) summers are dry, as are often the winters. Shrubby vegetation begins to occur on the interfluvial areas. Agriculture is mostly restricted to irrigated valleys. The interfluvial areas are being used today for dry farming and goat grazing (Gastó and Contreras, 1979).

Progressing southward, temperatures tend to decrease and rainfall tends to increase and show a less pronounced seasonal distribution. In the northern sector (~33°S) fronts reach the land systematically only during winters, and summers tend

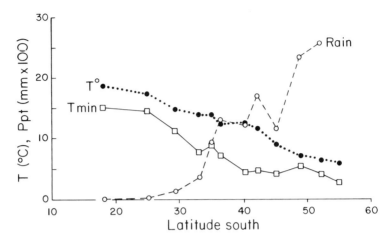

Fig. 2. Temperature and precipitation latitudinal gradients in Chile. Trends of annual mean (T) and minimum monthly temperatures (T min) and precipitation generate three distinctive climatic zones in the country: a warm zone where it almost never rains, the Norte Grande; a cool zone of constant rains, the Sur Grande; and an intermediate zone where productivity, species diversity, and human population are highest, central Chile.

to be dry. At latitudes above approximately 37°S conditions change enough for some summers to be wet also.

As latitude increases vegetation becomes lusher, initially taking the form of sclerophyllous shrublands and forests, then of winter deciduous *Nothofagus* forests, and eventually in the southern limit of the region of evergreen rain forests (Schmithusen, 1956). As a consequence of these two climate gradients, the highest productivity and species diversity are in the central part of the country and not at the extremes (Fig. 3).

Traditionally, the flat parts of central Chile have been rich lands for agriculture and forest plantations. This is also the area where most of the Chilean population lives (IGM, 1983) and where land degradation has been most severe (Peralta, 1978).

3. The Sur Grande (Big South, south of 40 to 42°S). Storm fronts reach this area throughout the year and temperatures tend to be low. Climate is of the oceanic type. Vegetation takes the forms of evergreen and winter deciduous forests, heaths, and grasslands (Schmithusen, 1956). Excessive precipitation, low temperatures, and high winds tend to restrict plant species diversity and agriculture. Human settlements are scant.

B. Human Population

A second component of the current scene concerns the human population. The Chilean human population is now about 12 million and is rapidly increasing (García-Vidal, 1982). The overall perception of Chileans seems to be that there is still room

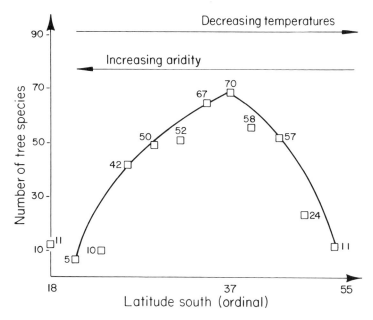

Fig. 3. Tree species diversity. Tree species diversity reaches a maximum where the opposing precipitation and temperature gradients are not extreme. Other groups show similar tendencies with their peaks somewhat displaced in one or the other direction. (Data from Rodriguez *et al.*, 1983.)

for many more people in the country, particularly in the Big South. The belief in ample space for population has strong and varied roots tied to traditional attitudes and customs, religion, and geopolitics. There are no estimates of how many people are enough and what the human carrying capacity of the land is.

Demands on landscapes are likely to increase in the near future, not only because of population density but also because of per capita increases in the standard of living. It is part of the Chilean belief that the standard of living of the whole population should increase and that socioeconomic development cannot be postponed. Chile, like other nonindustrialized countries, has paid and largely continues to pay for its imports and its foreign debt with the revenues from basic, non-value-added products. In the case of Chile these are mineral ore, agricultural products, fisheries, and lately wood chips and other nonprocessed forest products. Increasing international buying power, or maintaining it in an environmentally deteriorating situation, means increasing human impact on landscapes. This has happened in the past (Fuentes and Hajek, 1979) and it could also be the case in the immediate future. Another important element at present is that the Chilean population is politically very conscious and will not accept "solutions" that imply either staying at or returning to a subsistence economy or even reducing current economic growth rates because of some anticipated future global climate change.

C. Human Attitudes

A third component of the current scene is related to the perceptions that Chileans have of their environment. Several lines of evidence suggest that environmental awareness needs to be improved through education if the mistakes of the past are not to be repeated.

In Chile there is no tradition of naturalists, of people dedicated to the understanding of nature. The people that engage in these activities today are few. In general, the average person shows relatively little interest in the dynamics of landscapes that are not directly related to his or her immediate income. Natural history and ecology are fields of knowledge that much of Chilean society still does not fully appreciate. By contrast, Chileans tend to be more urban oriented and show high interest and high perception for the social dynamics of the country. More than 80% of the population of the country live in cities (Garcia-Vidal, 1982). Chile still has a high fraction of its population living under conditions that are not much above a subsistence economy (Tironi, 1989).

Surveys have shown that average citizens have very poor knowledge of wild species (Flip *et al.*, 1983). Even large plant configurations are not known to be native or introduced, part of the stable long-term vegetation, or recently perturbed (Fuentes *et al.*, 1984). In general, there is little concern about erosion, even in desertified areas.

It is not strange, then, that the so-called environmental problems have been discovered late in Chile and that many people complained until recently that environmental concerns were part of a strategy of leftist political groups.

In Chile there are still no environmental impact assessment laws for large engineering projects. Few people are informed about global warming and even fewer believe it is a real possibility. Most seem to think the suggestion of global warming is another pessimistic claim of environmental activists opposing development.

Things, however, seem to be changing. The first survey of environmental problems in the country has been completed (Hajek *et al.*, 1990). Environmental books and articles in newspapers and magazines are increasing in numbers and in quality. The newly elected government and the congress have committees that deal with environmental questions, and the general public is starting to become aware of the significance of environmental problems. To the extent that these tendencies begin to penetrate Chilean society, the "worst possible scenario" may prove to be false.

III. Landscape–Human Interactions

The environmental history of Chile clearly shows a close relationship between landscapes and human population. Human land occupancy is clearly dependent on geographic assets and constraints. Thus, the Norte Grande and the Sur Grande are scarcely settled, mostly for climatic and topographic reasons. The area with a subtropical winter rainfall type of climate, where most of the Chilean population lives, has been settled mostly around the flatter areas and near large rivers carrying snowmelt water from the nearly high Andes, even during the dry summer months.

Without the seasonal water reservoir of the high mountain areas, central Chile, with only 300 to 350 mm total annual rainfall, could not support large cities such as Santiago (about 4 million inhabitants) and the highly productive agriculture of the Central Valley. Water availability has been important in allowing settlements since pre-Columbian times, and since Spanish colonial times efforts have been made to increase the irrigated surfaces of the country by constructing a network of channels that distribute the water from snowmelt during the dry summer in the high Andes (Weischet and Schallhorn, 1974).

Climatic and topographic constraints have also been important in determining the rate and pattern of development of landscapes. The most agriculturally productive areas are those where slopes are gentle and water has been made available and also where soil erosion has been minimal (Peralta, 1978; Fuentes and Hajek, 1979). In contrast, the hilly interfluvial areas north of Santiago and coastal ranges south of 33°S have been used for woodcutting associated with mining operations first and then for rain-fed agriculture and grazing. All of this activity has contributed to classical desertification (UNCOD, 1977; Gastó and Contreras, 1979; Endlicher, 1988). In both cases, short-term endeavors such as mining with its associated ore smelting, cultivation of wheat for export, and subsistence activities of people in nearby areas have also contributed to the desertification process.

The first important aspect of this relationship between people and the land has been the lack of a conservation ethic on the part of the people. In spite of existing laws, since colonial times miners needing wood for their smelters, woodcutters, and farmers needing to clear land for agriculture have devastated the original forests and strongly modified all accessible landscapes (Cunill, 1971; Elizalde, 1970; Donoso, 1983). Land degradation has not been in the absence of forest protection laws, but in spite of them.

Another critical element of this process has been the existence of a large human population without access to flat, irrigated valleys. This population is relatively poor and without access to capital and know-how (UNCOD, 1977). They have few choices. In years of droughts they are forced to over exploit the land in spite of the climatically low productivity (Fuentes and Hajek, 1978). They partially compensate for these fluctuations by becoming small-scale miners, by cutting more wood to make and sell charcoal, and finally by migrating temporarily to other latitudes.

The use by this population of steep slopes for rain-fed agriculture in a land with high within- and between-year variability in rainfall is very critical. This practice, in conjunction with heavy woodcutting for domestic purposes and to make charcoal (Cunill, 1971), leaves the soil unprotected for at least part of the rainy season and has led to high soil erosion (Fuentes, 1990) and impoverishment of the land and of the people (Bahre, 1979).

Use of cleared lands for goat grazing without adjusting the stocking rate to the actual productivity is part of the degradation process. Further land degradation is produced because during dry years starving animals are forced to eat all available plant material and thus reduce the chances of vegetative recovery during the subsequent wet winters (Fuentes et al., 1989).

At the root of desertification is the absence of appropriate land use practices and effective environmental policies. These require, among other things, enough capital and technology to tune the time-varying local potentialities to the needs of the people. In the past too much land use has been of the *laissez-faire* type, in which either profit-making people have ignored conservation laws or people in a subsistence economy, without real options, have overexploited the land. It is not only a problem of not having appropriate laws and regulations. There is also a question of attitudes toward finite resources and of social responsibility for all of us. In many respects the impacts on the landscapes of subsistence economy people and profit-making entrepreneurs have been similarly destructive.

IV. Responses to Climate Change Times Human Actions

A. Assumptions

We start by stating the basic assumptions of our analysis. In a country like Chile, where human disturbance and landscape transformation have been pervasive in the past, global warming will in general have landscape effects that will be the product of an interaction of climate with human activities. That is, actual landscape changes will be the product of the interaction of climate change and human impacts. A single, simple response to climate change alone will probably be the exception in remote areas and not the rule. Global change must be put into the historical, geographic, and psychological context of the country before predictions of human impacts can be made.

Finally, we also assume that a country like Chile, with little capital resources, incipient scientific and technological capacities, attitudes toward nature that are not strongly developed, lack of strong regulatory enforcement, and insufficient higher education, will respond to climate rather conservatively, largely with current assets and with "strategies" that have been "successful" in the past. Both subsistence farmers and profit-making entrepreneurs are expected to respond with a limited short-term view. Moreover, current evidence suggests that the attitudes associated with "floating" international capital and local entrepreneurs do not differ much with regard to the use of short-term opportunities and the lack of long-term linkage with the natural resource productivity of the land.

B. Global Circulation Models

What do global circulation models (GCMs) tell us about future temperature and rainfall scenarios in Chile? Unfortunately, GCMs are still spatially crude and do not have the resolution needed to predict many important climatic parameters for a large part of the country. The climatic description from GCMs is particularly inaccurate for the Norte Grande (Fig. 4). In fact, it is not until south of La Serena, well into the area with a subtropical winter rain type of climate, that the descriptions of the current climate made by the GCMs is close enough to allow credible extrapolations to the doubling of CO_2 conditions (Fig. 5).

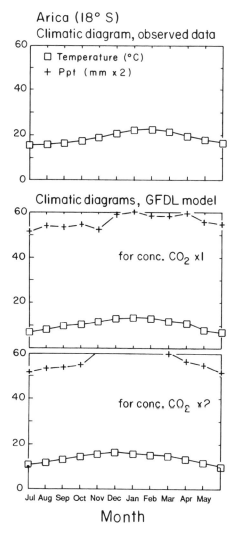

Fig. 4. Observed and modeled climate diagrams for Arica. The upper Walter–Gaussen diagram shows the observed data (Hajek and di Castri, 1975) and the two lower boxes show modeled (GFDL) data for Arica. Notice that for 1 CO_2 (current conditions) there is little agreement with the observed data. The climate diagram described by the model corresponds to a tropical forest, much like the one on the eastern side of the Andes. See the text for more explanations.

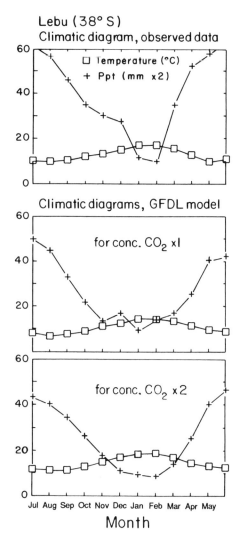

Fig. 5. Observed and modeled climate diagrams for Lebu. The upper Walter–Gaussen diagram shows the observed (Hajek and di Castri, 1975) and calculated (GFDL) data for Lebu Station on the coast of central Chile. Notice that for 1 CO_2 (current conditions) there is relatively good agreement with the observed data. See text for further explanation.

The Geophysical Fluid Dynamics Laboratory (GFDL) model (Princeton) provides better local descriptions of the current climate than the Goddard Institute for Space Studies (GISS) and UK Meterological Office (UKMO) models. Nevertheless, the spatial resolution constraints mentioned earlier still hold; the GFDL model has a resolution of only 4 × 5 degrees, somewhat better than the 8 × 10 degrees provided by the other two models. Consequently, climatic characterizations from this model are still crude. However, they are the best predictive tools available and we will use them as a basis for our discussion of the areas with winter rainfall and oceanic climates.

In general, for central Chile the GFDL model anticipates shifts toward somewhat lower rainfall, higher temperatures, and greater aridity for the areas with winter rainfall and oceanic climates (Fig. 6). We will now address what we think could be human responses and impacts on the three types of landscapes mentioned earlier, that is, the Norte Grande, central Chile with its subtropical winter rainfall climate, and the Sur Grande with its oceanic type of climate.

C. Predictions of Human × Climate Change Modifications

1. The Norte Grande. In the Norte Grande, the lack of spatial resolution, which does not allow a distinction to be made between the dry western side of the Andes and the wet eastern side, is critical in not providing an accurate picture of present-day

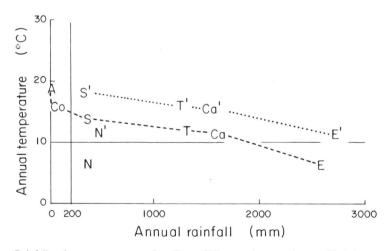

Fig. 6. Rainfall and temperature constraints. Seven Chilean stations are shown with their current total annual rainfall (abscissa) and mean annual temperature (ordinate): Arica (A), Copiapo (Co), Santiago (S), Temuco (T), Castro (Ca), Evangelistas (E), and Navarino (N). Solid line indicates the minimum rainfall required for rain-fed agriculture. Horizontal line indicates the 10°C temperature considered a threshold for agriculture. The locations S′, T′, Ca′, E′, and N′ indicate the new climate positions after doubling of CO_2 (GCM model). Notice that rainfall will not change the position of the localities in relation to the two thresholds, but temperature will produce significant changes, especially in the Big South.

conditions. Stations such as Arica (18°S), reputed for its aridity (Fig. 4), appear with a current hot and wet climate similar to what can be seen in Jujuy on the Argentinean side.

If the slightly wetter and hotter conditions that the GCMs predict for this area occur, actual conditions could improve by increasing the water availability for the irrigated low lands originating in the Altiplano. This assumption leads us to hypothesize that actual productivity in the irrigated lands could increase and water for settlements could become more plentiful.

In addition, we postulate that the altitude at which vegetation and grazing starts could decrease and agriculture could even be possible in some areas where temperatures are now too low. The landscape ecological consequences of these changes in the Andean realm can be postulated to be similar to those we will discuss in relation to the expected dynamics in other mountainous areas farther south.

2. Central Chile. The innate relationship between low rainfall and high variability and the attitudes of Chileans lead us to the following two hypotheses for areas showing high rainfall variability, mostly in the northern sector of central Chile.

If aridity increases, year-to-year variability is also likely to increase. Hence, even for informed people, the climate change signal will initially be difficult to separate from noises such as usual climate variability, usual population fluctuations and changes in plant and animal distributions and densities, and habitat degradation that imposes still little-understood dynamics on the landscapes. As an example, human disturbance (Armesto and Gutierrez, 1978; Fuentes, 1990) and grazing in particular (Milchunas *et al.*, 1988) are known to favor drought-tolerating vegetation, just as would be expected for a climate shift toward greater aridity. This noise could confound the climate change signal and allow debates of its reality and, hence, no action to mitigate climate change effects.

Most likely, entrepreneurs will not have a special interest in the degraded shrublands and sclerophyllous forests of the future, just as they currently have none. The better-quality, wetter sites are currently used for *Pinus radiata* plantations, and this use is likely to move farther south after the dry period becomes protracted in the more northerly sectors and southern climates become more suitable. The real threat to the landscapes of central Chile is from subsistence farmers, who do not have the mobility of firms working for profit.

How will subsistence farmers react to a dry year during this early period? Will governments implement large-scale and costly mitigation strategies in a situation that is ambiguous?

We postulate that responses of humans to global climate change will not be specific for some unknown amount of time. That is, farmers will respond to global change as if it were part of the normal variability in their region and therefore very much in the manner in which they have responded to hot spells, drought, and higher than normal rainfall events in the past. Even areas that currently exhibit moderate rainfall, like those near Santiago (350 mm/year), still have high interannual variability and should have confounding climatic behavior in the coming future. Only after a critical time will the new signal be strong enough and public awareness be raised high enough to produce appropriate changes in policy (Fig. 7). This hypothesis is based on the conservative attitudes seen in the past and the reluctance expected in a country with other pressing social demands.

The past history of responses to climate variability in mountainous areas, the observed human population trends, and the beliefs and current attitudes of Chileans led us to a second, complementary hypothesis. Human impact on the landscapes with high current climate variability will lead to more environmental degradation. The main reason is that as rainfall decreases and the frequency and severity of episodes of particularly dry years start to increase, overexploitation of the dry interfluves will increase.

Human use of the slope sites is likely to increase. People already subsisting on the interfluvial areas will see their crops and meager incomes further reduced. In an attempt to compensate, just as they do now, they will probably use the few relatively wet years to seed on even steeper slopes. They are also likely to cut more wood to make charcoal and intensify goat grazing. It also is not unlikely that the current trend toward migration out of the interfluvial areas to cities on the irrigated valleys (S. Miethke and E. R. Fuentes, unpublished) will be stopped or even partially reversed. All this is largely a consequence of the expected lower employment opportunities in the irrigated valleys once the winter snowline on the Andes is raised and water for irrigation and city consumption decreases.

On the other hand, the vegetation is likely to become progressively more stressed as warming increases and rainfall decreases, and this will make the vegetation more susceptible to human actions and less likely to recolonize cleared areas. This latter

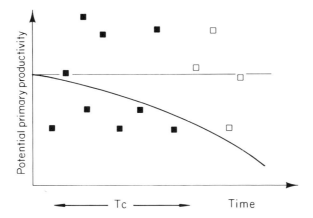

Fig. 7. Annual variability of potential primary productivity. At sites with a high between-year variation in potential primary productivity (PPP, filled squares) a critical time (T_c) will elapse before it is possible to discriminate between "normal" variability and the changes induced by global warming (open squares). During T_c people will largely respond to variation in PPP as if it were normal and, hence, try to survive with the old strategies during bad years and somehow compensate during the better ones. See text for more discussion.

tendency might be somewhat compensated for by the increased water use efficiency expected in some species, but this will occur only until some other element, such as nitrogen, becomes limiting (Graham *et al.*, 1990).

All of these changes will lead toward a net reduction of the vegetative ground cover in the arid and semiarid lands of central Chile and to soil erosion (Espinoza and Fuentes, 1984; Valdés, 1983; Fuentes, 1990). Degradation could be so severe that predictions of changes in vegetative cover based on Walter–Gaussen climate diagrams would be inaccurate (Fig. 8). That is, a more mesic vegetation would be predicted than what would actually be observed due to desertification.

3. The Sur Grande. Human population trends and a likely deteriorating environment in central Chile, in conjunction with the disproportionate expected warming in the Sur Grande region, leads us to predict that there will be increased human migration toward the latter region. In addition, capital-intensive firms are likely to be attracted by the higher timber growth rates and grazing opportunities available.

With the recent opening of a road that crosses the Sur Grande, even without global warming, an increase in the population and the commercial activities in those areas is expected. With global warming, the gradient that favored that migration and commercial use of the land will increase.

A second prediction for the Big South follows from this first one. The second prediction is based on past modes of human colonization and some current practices seen in the area. Because of the increase in temperatures above the threshold needed to plant wheat and a likely reduction in the circumpolar winds associated with a retreat of the Antarctic ice sheet, the Big South will become more equable, and part of the population will engage in agricultural and ranching operations previously precluded by climate. In addition, large forestry and ranching operations will look more attractive under the less restrictive temperature conditions (Figs. 9 and 10).

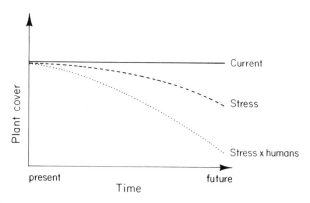

Fig. 8. Anticipated changes in plant cover. Expected cover if the current situation (current) is not modified, under stress induced by aridity and warming, and under the combined actions of humans and stress. Degradation could be most severe with human-combined effects (stress × humans), since there would be overexploitation of weakened plants.

Temperature and aridity constraints

Temperature
(mean monthly)
$$\begin{cases} T > 10\,C & \text{— warm} \\ 5 < T < 10\,C & \text{— cool} \\ T < 5\,C & \text{— cold} \end{cases}$$

Monthly aridity
(de Martonne's index)
$$\begin{cases} IM > 20 & \text{— wet} \\ 10 < IM < 20 & \text{— semi arid} \\ IM < 10 & \text{— arid} \end{cases}$$

Combined index
$$\begin{cases} \text{If warm and wet} & \text{— Favorable} \\ \text{If cold and dry} & \text{— Unfavorable} \\ \text{All other} & \text{— Semi favorable} \end{cases}$$

Fig. 9. Temperature and aridity constraints on productive systems. Commonly accepted thresholds for temperature (5°C) and aridity (10 and 20) constraints (de Martonnes index: ppt/T°C + 10) (see Hajek and di Castri, 1975) are shown. A combined index of mean monthly temperature and aridity defines favorable, unfavorable, and semifavorable months (Hajek and di Castri, 1975). For example, one area with a mean monthly temperature of 12°C and an index of 15 would be semifavorable, whereas if, for the same temperature constraint, IM > 20 it would be favorable, that is, without climatic constraints.

This leads us to a third prediction. Given the rather rugged topography of the Sur Grande and the general poor development of its soils, it is likely that these interactions of climate and humans could lead toward land degradation patterns similar to those that have already occurred in the regions with winter rainfall.

The Sur Grande is a mountainous area with many topographic restrictions for agriculture (Fig. 1). Winds can be very strong and there is a high potential for soil erosion. If people were left to colonize these areas and attempt to make a living or profit using the knowledge obtained at lower latitudes, they would be likely to repeat the modes of landscape use that they consider successful, the only ones they know. This could mean wholesale forest cutting for chips or logs for export, burning (the vegetation will be drier and more flammable!), planting on slopes, and grazing just as they are now occurring farther north, on Chiloé island.

In the past this pattern of land use was practiced in the coastal ranges of central Chile and in the area north of Santiago (33 to 29°S) (UNCOD, 1977; Gastó and Contreras, 1979; Endlicher, 1988). In all these cases desertification occurred, and it is not unlikely that an analogous process would also take place under topographically and climatically more rigorous conditions prevailing in the Sur Grande. In fact, some of this process has already been observed near Aysen (~45°S) (Donoso, 1983). In the

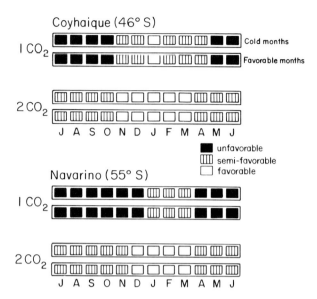

Fig. 10. Bioclimatic releases in the Big South. Stations of the Sur Grande are expected to have changed numbers of cold months and of favorable months under the doubling of CO_2 conditions. Cold months are defined as having a mean monthly temperature below 5°C; favorable months are defined as in Fig. 9. Under the new conditions, agriculture, including rain-fed agriculture, would be possible. Filled squares indicate cold or unfavorable months; dashed squares indicate mean monthly temperatures between 5 and 10°C, or semifavorable conditions; open squares indicate mean monthly temperature above 10°C and favorable months.

1940s the area between Aysen and Coihaique had its forests burned in an attempt to generate grazing grounds, with the consequence that massive erosion occurred and the sediments clogged the mouth of the Simpson River, which drains the watershed. Land degradation thus can occur even in a zone with an oceanic climate and fjords, and there is no reason to believe it could not extend to the rest of the Big South. Given the current high mobility of international capital, this is a serious risk.

With some modifications, this landscape degradation pattern that is likely to occur in the Sur Grande is also a possible scenario for the Altiplano of the Norte Grande. As we said, this area currently has temperatures that are too low for agriculture. With climate change, temperatures will rise, not as much as in the Sur Grande but enough to increase human demands. There is a real risk that these new demands, for example, for agriculture and overgrazing, will produce soil erosion and degradation just as in other mountainous and rainy areas farther south.

In all these cases, in the Big South as well as in the Big North, the use Walter–Gaussen diagrams would be of value in predicting climatic potential only and would thus underestimate the vegetation shift actually occurring when humans are part of the scenario.

D. What Use Are These Predictions?

It can be asked what value our predictions have. They are somewhat pessimistic; they do not say anything about the increases in agricultural and forestry production that might occur as temperature and rainfall increase. On the one hand, these changes could occur and they would be desirable. On the other, we think there is a real risk that we might not be able to make the best use of potential benefits of climate change if we do not learn from past experiences and do not design a strategy to adapt to or mitigate these changes.

We believe our predictions can serve as a starting point for research into their structure, assumptions, and so on and also serve as a basis for comparative research with other areas in which capital, know-how, and human attitudes are different. We believe that understanding human responses and new or modified human impacts on the landscape should be of prime importance in our understanding of the changes that will occur on the biosphere in the coming years.

A basic assumption behind all our predictions is that the response processes would again, just as in the historical past, be unplanned or largely unplanned and without appropriate know-how and education. These hypotheses, then, should serve at least as "neutral" models, neutral with regard to planning and to technological know-how. It will be in the hands of politicians to make one or the other anticipatory scenario real.

Some of the main areas in which research is badly needed are related to the attitudes, and potential changes of attitudes, of the social actors, that is, all of us. Education at various levels will be needed. Another area that is important is the development of ecological zonations emerging as a consequence of the assessment of the risks involved in various potential uses of the landscapes of the nondesertified areas of central Chile and especially of the Sur Grande.

We hope that realizing that our past tendency to "mine" renewable resources for profit or for subsistence could express itself again will trigger research on conservation-minded approaches. We also hope this realization, the new communication systems, and the modern approaches to the environment will prove that our extrapolations are false.

Acknowledgments

We would like to express our gratitude to P. Goss, P. Rodrigo, W. Jarrell, P. Alaback, A. E. Hoffmann, A. J. Hoffmann, and H. A. Mooney for kindly discussing earlier versions of the manuscript with us and helping us to improve it. They do not share all of our opinions, and therefore the responsibility for them is only ours. M. R. Munoz was financed by a Fundacion Andes doctoral fellowship. A grant from FONDECYT (Chile) covered the research reported.

References

Armesto, J. J., and Gutiérrez, J. R. (1978). El efecto del fuego en la estructura de la vegetación de Chile central. *An. Mus. Hist. Nat. (Valparaiso)* **11,** 43–48.
Bahre, C. (1979). "Destruction of the Natural Vegetation of North-Central Chile." Univ. of California Press, Berkeley.

Cunill, P. G. (1971). Factores en la destrucción del paisaje chileno: recolección, caza y tala coloniales. *Inf. Georgr.* (Número Especial), 235–264.

CORFO. (1966). "Geografia Economica de Chile." Editorial Universitaria, Santiago.

di Castri, F., and Hajek, E. R. (1976). "Bioclimatología de Chile." Vicerrectoria Académica, P. Univ. Católica de Chile, Santiago.

Donoso, C. (1983). "Historia del Paisaje Chileno." Facultad de Ciencias Forestales, Univ. Austral de Chile, Valdivia.

Elizalde, R. (1970). "La Sobrevivencia de Chile." Ministerio de Agricultura, Santiago, Chile.

Endlicher, W. (1988). "Geookologische Untersuchungen zur Landschaftsdegradation im Kustenbergland von Concepción (Chile)." Steiner-Verlag, Wiesbaden.

Espinoza, G., and Fuentes, E. R. (1984). Medidas de erosión en los Andes Centrales: efectos de pastos y arbustos. *Terra Austral.* **3,** 75–86.

Filp, J., Fuentes, E. R., Donoso, S., and Martinic, S. (1983). Environmental perception of mountain ecosystems in central Chile: an exploratory study. *Hum. Ecol.* **11,** 345–351.

Fuentes, E. R. (1990). *In* "Changing Landscapes: An Ecological Perspective" (I. S. Zonneveld and R. T. T. Forman, eds.), pp. 165–190. Springer-Verlag, Berlin.

Fuentes, E. R., and Hajek, E. R. (1978). Interacciones hombre–clima en la desertificación del norte chico chileno. *Cienc. Invest. Agraria* **5,** 137–142.

Fuentes, E. R., and Hajek, E. R. (1979). Patterns of landscapes modification in relation to agricultural practice in central Chile. *Environ. Conserv.* **6,** 265–271.

Fuentes, E. R., Espinoza, G., and Fuenzalida, I. (1984). Cambios vegetacionales recientes y percepción ambiental. El caso de Santiago de Chile. *Rev. Geogr. Norte Grande* **11,** 45–53.

Fuentes, E. R., Espinoza, G. A., and Hajek, E. R. (1989). *In* "Time Scales and Water Stress. Proceedings of 5th International Conference on Mediterranean Ecosystems" (F. di Castri, Ch. Cloret, S. Rambal, and J. Roy, eds.), pp. 347–360. International Union of Biological Sciences, Paris.

García-Vidal, H. (1982). "Chile. Esencia y Evolución." Instituto de Estudios Regionales, Univ. de Chile, Santiago.

Gastó, J., and Contreras, D. (1979). Un Caso de Desertificación en el Norte de Chile. El Ecosistema y su Fitocenosis. Boletín Técnico 42, Univ. de Chile, Santiago.

Graham, R. L., Turner, M. G., and Dale, V. H. (1990). How increasing CO_2 and climate change affect forests. *Bioscience* **40,** 575–587.

Hajek, E. R., and di Castri, F. (1975). "Bioclimatografía de Chile." Ediciones Univ. Católica de Chile. Santiago.

Hajek, E. R., Gross, P., and Espinoza, G. A. (1990). "Problemas Ambientales de Chile." Alfa-Beta Impresores, Santiago.

Instituto Geográfico Militar (IGM). (1983). "Atlas de la República de Chile," 2nd ed. IGM, Santiago.

Milchunas, D. G., Sala, O. E., and Lauenroth, W. K. (1988). A generalized model of the effects of grazing by large herbivores on grassland community structure. *Am. Nat.* **132,** 87–106.

Peralta, M. (1978). Procesos y áreas de desertificación en Chile continental. *Cienc. For.* **9,** 41–44.

Rodriguez, M. (1989). "Geografia Agricola de Chile." Editorial Universitaria, Santiago.

Rodriguez, R., Matthei, O., and Quezada, M. (1983). "Flora arbórea de Chile." Ediciones Univ. de Concepción, Concepción.

Schmithüsen, J. (1956). Die räumliche Ordnung der chilenischen vegetation. *Bonn. Geog. Abh.* **17,** 1–86.

Tironi, E. (1989). Es posible reducir la pobreza en Chile. Editora Zig, Zag, Santiago, Chile.

UNCOD. (1977). "Case-Study on Desertification, Region of Combarbalá, Chile." Nairobi, Kenya.

Valdés, J. (1983). Dinámica de la Desertificación en Tres Areas del Secano Interior de la IV Region. Tesis, Escuela de Ciencias Forestales, Univ. de Chile, Santiago.

van Husen, C. (1967). Klimagliederung in Chile auf der Basis von Haufigkeitsverteilungen der Niederschflagssumenn. *Freib. Geog. Hefte* **4,** 1–99.

Weischet, W. (1970). "Chile. Seine länderkundliche Individualitat und Struktur." Wissenschaftliche Landerkunde, Band 2/3, Wissenschaftliche Buchgesellschaft, Darmstadt.

Weischet, W., and Schallhorn, E. (1974). Altsiedlerkerne und frühkolonialer Ausbau in der Bewässerungskulturlandschaft Zentralchiles. *Erdkunde* **28,** 295–302.

C H A P T E R   24

North–South Comparisons: Managed Systems

PAUL B. ALABACK

The similarity of climate and ecosystem structure but the divergence in social and cultural influences and taxa cultivated along the west coast of North America and South America should make a valuable comparison for a long-term study of global change. Studies of managed systems along these analogous climatic gradients, where the same species or very similar species are often cultivated, should make comparisons much closer than would be possible with natural ecosystems. Since economic and cultural influences are critical to determining the nature and extent of managed systems, this comparison could be an important way to focus research on how these human factors influence resource management across this diverse climatic transect.

One of the principal contrasts between the west coasts of North and South America is the greater equitability of the climate of South America. Temperate South America has milder summers and winters than North America because of the greater oceanic influence and reduced land area. Summer temperatures and ecosystem distribution also vary between the two continents, with South America having equivalent ecosystems as much as 15° in latitude lower than in North America. As a consequence, the climate of South America would be expected to change much less rapidly than that of North America, both because of the buffering effect of the ocean and because of less warming at lower latitudes. (The lower tip of South America, for example, while having a climate similar to the northernmost rain forest environment of coastal Alaska, only occurs as high in latitude as the southern tip of Alaska.)

North America has a greater extent of cold- or snow-adapted ecosystems than South America. The boreal zones in Alaska and Canada have no equivalent in South

America. The alpine zone is also more extensive in North America than in South America and is closely related to avalanche activity. Global warming could have dramatic impacts on lowering the frequency of avalanches and decreasing the extent of permafrost in North America. These factors would be of less importance in South America.

Problems with pests and diseases would be expected to be less in Chile than in North America because of geography and history. Global warming would be expected to allow more virulent southern species to migrate northward in North America. In South America the lower diversity and greater isolation of the temperate zone would make for less opportunities for pest outbreaks or invasions (except those that might be imported from other temperate zones).

A major cultural contrast between the two regions is in the pattern of colonization and availability of capital for infrastructure. Subsistence agriculture and ranching are much more prevalent in Chile, whereas capital-intensive cultivation practices are more common in North America. Chile has a smaller population with a smaller density than in North America and has a potential for rapid growth. In North America the distribution of wealth is more equitable and consumption per capita is much higher than in Chile, implying greater potential for increasingly intensive land use. If social and political reforms continue at a rapid rate in Chile, major economic and social transformation could occur which will put increasing demands on the smaller land base, especially in the semiarid climatic zones. Large-scale clear-cut logging has been widespread throughout North America, whereas smaller-scale selective logging and burning have been more prevalent in Chile, except for the coastal *Pinus radiata* plantations. Assuming this pattern continues, we would expect significant contrasts in landscape pattern as both regions develop with profound implications for plant–animal relationships, spread of pests and diseases, and overall resilience to climatic change.

Chile will probably be a more conservative model of global climate change, whereas North America would provide a more reactive one. These differences would be driven primarily by subtle distinctions in climate but could be accentuated or even reversed by cultural influences. Over time we would hypothesize an increased divergence in functional characteristics of managed ecosystems on the two continents.

C H A P T E R   25

Concluding Remarks

EDUARDO R. FUENTES

There are evident similarities between the west coasts of North America and South America. Both continents are large continuous landmasses penetrating into high latitudes and have comparable climatic systems, driven by the relative positions of anticyclones and of cold polar systems. These similarities generate parallel climatic sequences and also antiparallel coastal currents with upwellings and El Niño–Southern Oscillation (ENSO) events along the coasts of both hemispheres.

In addition, both continents have mountain ranges running parallel to the coasts and thus have not only parallel latitudinal but also comparable longitudinal ecosystem types. At low latitudes highly fluctuating desert and semiarid ecosystems predominate, at midlatitudes climatically more even evergreen sclerophyllous shrublands are the rule, whereas at higher latitudes comparable oceanic climates produce evergreen rain forests that cover coasts constituted by fjords, channels, and small islands.

It has been shown (Mooney, 1977) that at least sclerophyllous shrublands in both continents exhibit a remarkable degree of ecological convergence in spite of their taxonomically rather distant floras and faunas. However, human use systems and their effects on landscape transformation in these areas with mediterranean-type of climates have not been as similar (Aschmann, 1991).

Although quite similar, and perhaps as similar as two such enormous landmasses on two hemispheres can be, the west coasts of South and North America are not mirror images of one another. In the Southern Hemisphere landmasses are smaller and polar and oceanic masses are higher than in the Northern Hemisphere. These differences suggest that the climatic changes anticipated for both areas are going to be different, that future changes might not be parallel, and that distortions in the time trajectories of currently comparable ecosystem types can be expected. In addition, human activities, which constitute such an important force shaping landscapes and ecosystems in both hemispheres, are different and are also likely to be play non-parallel roles in the future.

The west coasts of the Americas thus provide an unusual geographic framework to monitor changes and to generate and test hypotheses regarding the roles of the two main forces behind global change: direct and indirect (via climate) effects of human use of the biosphere. This book shows that similarities and differences between the two continents can be used to generate and test hypotheses about global change.

The best current estimates for climate changes suggest that changes between continents will not be parallel. In the Northern Hemisphere temperatures for the year 2030, when effective doubling of CO_2 is anticipated, could be of the order of 1 to 1.5°C higher than at present, whereas for the Southern Hemisphere only half as much increase can be expected. These rather modest amounts of temperature increase might be accompanied by significant changes in atmospheric circulation. In the Southern Hemisphere westerlies could become stronger, just the opposite of what is expected for the Northern Hemisphere. In addition, monsoonal effects could become stronger in the Northern Hemisphere. These atmospheric circulation differences could bring relatively drier conditions to the central part of Chile and, depending on the strength of the westerlies and on the superimposed monsoon effects, relatively more humid conditions for California. That is, future scenarios for the west coasts of North and South America are most likely not be as parallel as they are now, providing a unique opportunity to monitor and try to explain the differences produced in other components of the earth system.

The climate change scenarios given provided scientists from a number of other disciplines a target for exploring the consequences of change. In this book hydrologists, geomorphologists, biogeochemists, oceanographers, biologists, and scientists concerned with human use systems attempted to project the impact of changes within their study realms. These efforts gave rise to suggestions on how to approach the question of the different abiotic as well as biotic responses to climate change. The suitability of past ecological changes and of current ENSO and drought events as "windows" allowing the exploration of future events was suggested.

On the other hand, there were several specific suggestions regarding possible responses of the components of the earth system, such as speeding up of the hydrologic cycle, and the responses that certain plants and animals could show. These latter responses proved particularly difficult, given the short- and long-term responses that organisms show, the possible evolution of the attributes of short-lived organisms, the migration versus *in situ* responses of different taxa, and the indirect effects involved. These indirect effects include the different time responses of the various ecosystem components and the subsequent responses, not only to climate change but also to the presence of evolved plants and animals as well as of other organisms.

In addition, contrasts were made of potential responses between the two regions, further enriching the comparisons. In general, these comparisons tended to stress that in the future the two continents are likely to diverge because of the expected climate changes and also because of the role of humans in the two hemispheres.

The particular initial predictions made by the various scientists in the long run may not be as important as the demonstration of usefulness of a common geographic frame in which to pose and test truly interdisciplinary questions regarding global change.

Differences in the anticipated warming suggest that climate change could be slower and less dramatic than originally suspected. If these predictions are borne out, they suggest there is more time to investigate the change response and that we should concentrate on the effects of relatively small and slow changes rather than on very fast and dramatic effects.

On the other hand, direct human use of and human impact on the ecosystems might not be as slow in the future and the expected landscape transformations are most likely going to override climate change effects. This is more likely to be the case in South America, where climate change effects are like to be slower and less pronounced and direct human use is more likely to increase in the future. In the future, these differences in the main driving forces in the two hemispheres could be used as a key element providing tests of hypotheses on the role played by various factors in driving change.

A consideration that was stressed by various contributors and that should be used when planning future activities is that better use should be made of natural experiments such as ENSO or unusually heavy rainfall or drought events to study the consequences of climate change interacting with the human response to such changes. With the obvious caveats that such short-term "experiments" have in trying to anticipate what will be transient changes, they could nevertheless help us understand at least the kinds of responses that could be expected initially from complex and usually unmanageable systems from an experimental perspective.

In addition to using natural experiments in attempting to understand future changes, current global circulation model predictions can be used to design explicit experiments to unravel the causality of macrophenomena through careful designs and long-term monitoring of climatic, hydrologic, oceanographic, geomorphologic, biogeochemical, and biological variables. In these experiments parks and other protected areas, such as biosphere reserves, could serve as controls for direct human impact.

The meeting at La Serena generated such enthusiasm among the participants from Argentina, Canada, Chile, and the United States that it was decided to initiate a new interdisciplinary program aimed at increasing our understanding of global change and augmenting the training of scientists interested in this phenomenon. The Americas Interhemisphere Geo-Biosphere Association (AMIGO) was created with the specific aim of serving as a cradle for ideas and initiatives related to assessment, monitoring, testing, and training in relation to global change in the unique geographic frame provided by the west coast of the Americas.

References

Aschmann, H. (1991). Human impact on the biota of mediterranean-climate regions of Chile and Cali-
    fornia. *In* "Biogeography of Mediterranean Invasions" (R. H. Groves and F. di Castri, eds.). Cam-
    bridge Univ. Press, Cambridge, U.K.
Mooney, H. A. (1977). "Convergent Evolution in Chile and California. Mediterranean Climate Ecosys-
    tems." Dowden, Hutchinson, & Ross, Stroudsburg, Pennsylvania.

Index